College Lane, Hatfield, Herts. AL10 9AB
Information Hertfordshire
Services and Solutions for the University

For renewal of Standard and One Week Loans
please visit the web site http://www.voyager.herts.ac.uk

This item must be returned or the loan renewed by the due date.
A fine will be charged for the late return of items.

Networked Life

How does Google sell ad space and rank webpages? How does Netflix recommend movies, and Amazon rank products? How can you influence people on Facebook and Twitter, and can you really reach anyone in six steps? Why doesn't the Internet collapse under congestion, and does it have an Achilles' heel? Why are you charged per gigabyte for mobile data, and how can Skype and BitTorrent be free? How big is the "cloud" of cloud services, and why is WiFi slower at hotspots than at home?

Driven by 20 real-world questions about our networked lives, this book explores the technology behind the multi-trillion dollar Internet, wireless and online media industries. Providing easily understandable answers for the casually curious, alongside detailed explanations for those looking for in-depth discussion, this thought-provoking book is essential reading for students in engineering, science and economics, for network industry professionals, and for anyone curious about how technological and social networks really work.

Mung Chiang is a Professor of Electrical Engineering at Princeton University, and Director of the Princeton EDGE Lab. He has received the IEEE Kiyo Tomiyasu Award, and a US Presidential Early Career Award for Scientists and Engineers, for his research on networking. A co-founder and advisor to several startups, he also received a Technology Review TR35 Award for his contributions to network technology innovation. He is a fellow of the IEEE.

"We are entering a new Internet era – the era of the likes of Google, Amazon, Netflix, and Facebook – with entirely new types of problems. This book captures the new era, taking a fresh approach to both topic coverage and pedagogic style. Often at the end of a section it leaves the reader asking questions; then exactly those questions are answered in the subsequent section. Every university should offer a course based on this book. It could be taught out of both ECE or CS departments at the undergraduate or graduate levels."

Keith Ross, Polytechnic Institute of NYU

"How do the networks, which we increasingly rely upon in our everyday life, actually *work*? This book is an inspiring romp through the big ideas in networking, which is immediately rewarding and will motivate later courses."

Frank Kelly, University of Cambridge

Networked Life

20 Questions and Answers

MUNG CHIANG

Princeton University

CAMBRIDGE
UNIVERSITY PRESS

CAMBRIDGE UNIVERSITY PRESS
Cambridge, New York, Melbourne, Madrid, Cape Town
Singapore, São Paulo, Delhi, Mexico City

Cambridge University Press
The Edinburgh Building, Cambridge CB2 8RU, UK

Published in the United States of America by Cambridge University Press, New York

www.cambridge.org
Information on this title: www.cambridge.org/9781107024946

First published 2012

Printed in the United States by Edwards Brothers

A catalog record for this publication is available from the British Library

ISBN 978-1-107-02494-6 Hardback

Additional resources for this publication at www.cambridge.org/networkedlife

To my family

Contents

Preface

You pick up your iPhone while waiting in line at a coffee shop. You Google a not-so-famous actor and get linked to a Wikipedia entry listing his recent movies and popular YouTube clips. You check out user reviews on IMDb and pick one, download that movie on BitTorrent or stream it in Netflix. But for some reason the WiFi logo on your phone is gone and you're on 3G. Video quality starts to degrade a little, but you don't know whether it's the video server getting crowded in the cloud or the Internet is congested somewhere. In any case, it costs you $10 per gigabyte, and you decide to stop watching the movie, and instead multitask between sending tweets and calling your friend on Skype, while songs stream from iCloud to your phone. You're happy with the call quality, but get a little irritated when you see that you have no new followers on Twitter.

You've got a typical networked life, an online networked life.

And you might wonder how all these technologies "kind of" work, and why sometimes they don't. Just flip through the table of contents of this book. It's a mixture: some of these questions have well-defined formulations and clear answers while for others there is still a significant gap between the theoretical models and actual practice; a few don't even have widely accepted problem statements. This book is about formulating and answering these 20 questions.

This book is about the networking technologies we use each day as well as the fundamental ideas in the study of networks. Each question is selected not just for its relevance to our daily lives, but also for the core concepts and key methodologies in the field of networking that are illustrated by its answer. These concepts include aggregation and influence, distributed coordination, feedback control, and strategic equilibrium. And the analytic machineries are based on mathematical languages that people refer to as graph, optimization, game, and learning theories.

This is an undergraduate textbook for a new course created in 2011 at Princeton University: **Networks: Friends, Money, and Bytes**. The course targets primarily juniors and seniors in electrical engineering and computer science, but also beginning graduate students as well as students from mathematics, sciences, economics, and engineering in general. It can be viewed as a second course after the "signals and systems" course that anchors the undergraduate electrical and computer engineering curriculum today. Starting in September 2012, this course is also on free open access platforms, such as Stanford's coursera and the course's own open education website, as well as on YouTube and iTunes U.

This book weaves a diverse set of topics you would not normally see under the same cover into a coherent stream: from Arrow's impossibility and Rawls' fairness to Skype signaling and Clos networks, from collaborative filtering and firefly synchronization to MPEG/RTSP/TCP/IP and WiFi CSMA DCF. This begs a question: "So, what *is* the discipline of this book?" This is a question that most of the undergraduates do not care about. Neither does this book, which only wants to address these practical questions, using whatever modeling languages that have been observed to be the most relevant ones so far. It turns out that there is a small, common set of mathematics which we will need, but that's mostly because people have invented only a limited suite of modeling languages.

This is not a typical textbook for another reason. It does not start with general theories as do many books on these subjects, e.g., graph theory, game theory, and optimization theory, or with abstract concepts like feedback, coordination, and equilibrium. Instead it starts with concrete applications and practical answers, and sticks to them (almost) every step of the way. Theories and generalizations emerge, as if they were "accidental by-products," during the process of formulating and answering these questions.

This book can be supplemented with its website: `http://www.network20q.com`, including lecture slides, problem solutions, additional questions, examples of advanced material, further pointers to references, collections of news media coverage of the topics, "currency-earning" activities, course projects, blogs, tweets, surveys, and student-generated course material in wiki. We have created web features that turn this class into an online social network and a "networked economy."

This book can also be used by engineers, technology managers, and pretty much anyone with a keen interest in understanding how social and technological networks work. Often we sacrifice generality for accessibility, and supplement symbolic representation with numerical illustration.

- The first section of each chapter is a "short answer," and it is accessible by most people.

- Then there's a "long answer" section. If you remember differentiation and linear algebra (and occasionally a little bit of integration and basic probability), you can follow all the material there. We try to include only those symbols and equations that are really necessary to unambiguously express the ideas.

- The "examples" section contains detailed, numerical examples to reinforce the learning from the "long answer" section. On average, these first three sections of each chapter form the basis of one 80-minute lecture. Several of these lectures will go over 80 minutes while several others, including the last two, can be covered under 80 minutes.

- Each chapter concludes with a section on "advanced material," which requires the reader to be quite comfortable with symbolic operations and abstract reasoning, but can be skipped without losing the coherence and gist of the book. In the undergraduate course taught at Princeton, hardly any of the advanced material is covered. Covering all the "advanced material" sections would constitute an introductory graduate-level course. To keep the book thin, worked examples for these sections are pushed to the course website.
- At the end of each chapter, there are five homework questions, including easy drills, essential supplements, and some "out-of-syllabus" explorations about networks in biology, energy, magic, music, and transportation. The level of difficulty is indicated on a scale of one (easy) to three (hard) stars. On the course website, there is a much larger collection of additional homework problems, including many multiple-choice questions testing the basic understanding of the material.
- There are also five key references per chapter (yes, only five, in the hope that undergraduates may actually read some of these five, and my apologies to the authors of thousands of papers and books that could have been cited). These references open the door to further reading, including textbooks, research monographs, and survey articles.

This is a (relatively) thin book. It's a collage of snapshots, not an encyclopedia. It's an appetizer, not an entree. The majority of readers will not pursue a career specializing in the technical material in this book, so I take every opportunity to delete material that's very interesting to researchers but not essential to this undergraduate course. Each one of these 20 chapters deserves many books for a detailed treatment. I only highlight a few key ideas in the span of about 20 pages per chapter and 80 minutes per lecture. There are also many other mathematical languages in the study of networks, many other questions about a networked life, and many other types of networks that we do not have time to cover in one semester. But as the saying goes for a course: "It's more important to *uncover* than to cover a lot."

This is a book illustrating some pretty big ideas in networking, through 20 questions we can all relate to in our daily lives. Questions that tickle our imagination with surprises and incomplete answers. Questions that I wished I had known how to answer several years ago. Questions that are quickly becoming an essential part of modern education in electrical and computer engineering.

But above all, I hope this book is fun to read.

Mung Chiang
Princeton, NJ
July 2012

Acknowledgements

In so many ways I've been enjoying the process of writing this book and creating the new undergraduate course at Princeton University. The best part is that I got to, ironically in light of the content of this book, stay *offline* and focus on learning a few hours a day for several hundred days. I got to digest wonderful books and papers that I didn't have a chance to read before, to think about the essential points and simple structures behind the drowning sea of knowledge in my research fields, and to edit and re-edit each sentence I put down on paper. It reminded me of my own sophomore year at Stanford University one and a half decades ago. I often biked to the surreally beautiful Oval in the morning and dived into books of many kinds, most of which were not remotely related to my majors. As the saying goes, that was a pretty good approximation of paradise.

That paradise usually ends together with the college years. So I have many people to thank for granting me a precious opportunity to indulge myself again at this much later stage in life.

- The new course "Networks: Friends, Money, and Bytes" could not have been created in 2011 without the dedication of its three remarkable TAs: Jiasi Chen, Felix Wong, and Pei-yuan Wu. They did so much more for the course than a "normal" TA experience: creating examples and homework problems, initiating class activities, and running the online forum.

- Many students and postdocs in Princeton's EDGE Lab and EE Department worked with me in creating worked examples and proofreading the drafts: Chris Brinton, Amitabha Ghosh, Sangtae Ha, Joe Jiang, Carlee Joe-Wong, Yiannis Kamitsos, Haris Kremo, Chris Leberknight, Srinivas Narayana, Soumya Sen, Victoria Solomon, Arvid Wang, and Michael Wang.

- Princeton students in ELE/COS 381's first offering were brave enough to take a completely new course and contributed in many ways, not the least through the class website blogs and course projects. Students in the graduate course ELE539A also helped proofread the book draft and created multiple choice questions.

- Before I even got a chance to advertise the course, some colleagues already started planning to offer this course at their institutions in 2012: Jianwei Huang (Chinese University of Hong Kong), Hongseok Kim (Sogang University, Korea), Tian Lan (George Washington University), Walid Saad

(University of Miami), Chee Wei Tan (City University of Hong Kong), and Kevin Tang (Cornell University).

- The material in this book is inspired by discussions with colleagues in both academia and industry over the years. Since last summer, more than fifty colleagues has provided valuable suggestions directly for the course and the book. In particular, I received very detailed comments on earlier drafts of the book from Keith Cambron (AT&T Labs), Kaiser Fung (Sirius), Victor Glass (NECA), Jason Li (IAI), Jennifer Rexford (Princeton), Keith Ross (NYU Poly), Krishan Sabnani (Bell Labs), Walid Saad (University of Miami), Matthew Salganik (Princeton), Jacob Shapiro (Princeton), Kevin Tang (Cornell), and Walter Willinger (AT&T Labs), among others.

- Phil Meyler from Cambridge University Press encouraged me to turn the lecture notes into a textbook, and further connected me with a group of enthusiastic staff at CUP. I am grateful to the entire Cambridge editorial and marketing team across their UK and New York offices.

- This course was in part supported by a grant from the US National Science Foundation, in a program run by Darleen Fisher, for a team consisting of two engineers and two social scientists at Princeton. I'm glad to report that we achieved this educational goal in our proposal, and did that before the project's official end date.

My appreciation traces back to many of my teachers. For example, I've had the fortune to be co-advised in my Ph.D. study by Stephen Boyd and Tom Cover, two brilliant scholars who are also superb teachers. Their graduate-level textbooks, *Convex Optimization* by Boyd and Vandenberghe and *Elements of Information Theory* by Cover and Thomas, are two towering achievements in engineering education. Read these two books, and you'll "experience" the definition of "clarity," "accessibility," and "insight." When I was writing research papers with them, Tom would spend many iterations just to get one notation right, and Stephen would even pick out each and every LaTex inconsistency. It was a privilege to see first-hand how the masters established the benchmarks of technical writing.

Stephen and Tom were also the most effective lecturers in classroom, as was Paul Cohen, from whom I took a math course in my sophomore year. Pulling off the sweatshirt and writing with passion on the blackboard from the first moment he entered the classroom, Paul could put your breath on hold for 80 minutes. Even better, he forgot to give us a midterm and then gave a week-long, take-home final that the whole class couldn't solve. He made himself available for office hours on-demand to talk about pretty much anything related to math. The course was supposed to be on PDE. He spent just four lectures on that, and then introduced us to eighteen different topics that quarter. I've forgotten most of what I learned in that course, but I'll always remember that learning can be so much fun.

In the same winter quarter that I took Stephen's and Tom's courses, I also took from Richard Rorty a unique course at Stanford called "From Religion through Philosophy to Literature," which pulled me out of Platonism to which I had been increasingly attached as a teenager. Talking to Rorty drastically sharpened my appreciation of the pitfalls of mistaking representations for reality. A side-benefit of that awakening was a repositioning of my philosophy of science, which propagated to the undercurrents of this book.

Three more inspirations, from those I never met:

- Out of all the biographies I've read, the shortest one, by far, is by Paul Johnson on Churchill. And it's by far the most impactful one. Brevity is power.
- But even a short book feels infinitely long to the author until it goes to the press. What prevented me from getting paralyzed by procrastination is Frederick Terman's approach of writing textbooks while leading a much busier life (serving as a Dean and then the Provost at Stanford, and creating the whole Silicon Valley model): write one page each day.
- Almost exactly one century ago, my great grandfather, together with his brother, wrote some of the first modern textbooks in China on algebra and on astronomy. (And three decades ago, my grandfather wrote a textbook on econometrics at the age of seventy.) As I was writing this book, sometimes I couldn't help but picture the days and nights that they spent writing theirs.

For some reason, the many time commitments of a professor are often hard to compress. And I couldn't afford to cut back on sleep too much, for otherwise the number of mistakes and typos in this book would have been even larger. So it's probably fair to say that each hour I spent writing this book has been an hour of family time lost. Has that been a good tradeoff? Definitely not. So I'm glad that the book is done, and I'm grateful to my family for making that happen: my parents who helped take care of my toddler daughter when I was off to dwell in this book, my wife who supported me sitting there staring at my study's ceiling despite her more important job of curing the ill, and Novia who could have played with her Daddy a lot more in the past year. This book was written with my pen and their time.

Roadmap

This roadmap is written for course instructors, or perhaps as an *epilogue* for students who have already finished reading the book, especially Figure 0.3 and the list of 20 ideas below it. It starts with a taxonomy of the book and introduces its organization and notation. Then it discusses the similarities and differences between this book and some excellent books published over the last decade. Then it highlights three pedagogical principles guiding the book: Just In Time, Bridge Theory and Practice, and Book As a Network, and two contexts: the importance of domain-specific functionalities in network science and the need for undergraduate-curriculum evolution in electrical and computer engineering. It concludes with anecdotes of arranging this course as a social and economic network itself.

Taxonomy and Organization

The target audience of this book is both students and engineering professionals. For students, the primary audience are those from engineering, science, economics, operations research, and applied mathematics, but also those on the quantitative side of sociology and psychology.

There are three ways to use this book as a textbook.

- *An undergraduate general course at sophomore or junior level*: Go through all 20 chapters without reading the Advanced Material sections. This course serves as an introduction to networks before going further into senior-level courses in four possible directions: computer networking, wireless communication, social networks, or network economics.
- *An undergraduate specialized course at senior level*: Pick either the social and economic network chapters or the technology and economic network chapters, and go through the Advanced Material sections in those chapters.
- *A first-year graduate level course*: Go through all 20 chapters, including the Advanced Material sections.

While this book consists of 20 chapters, there are just four key recurring concepts underlying this array of topics. Table 0.1 summarizes the mapping from chapter number to the concept it illustrates.

Table 0.1 Key concepts: The chapters where each of the four key concepts show up for different types of networks.

Network Type	Aggregation & Influence	Distributed Coordination	Feedback Control	Strategic Equilibrium
Wireless		1	19	
Internet		10, 13, 16	14	
Content Distribution		15, 17	18	
Web	3, 4, 5			2
Online Social	6,8	9	7	
Internet Economics		20		11, 12

The modeling languages and analysis machineries originate from quite a few fields in applied mathematics, especially the four foundations summarized in Table 0.2.

Table 0.2 Main methodologies: The chapters where each of the four families of mathematical languages are used in different types of networks.

Network Type	Graph Theory	Optimization Theory	Game Theory	Learning Theory
Wireless		18, 19	1	
Internet	10	13, 14, 16		
Content Distribution		15, 17		
Web	3		2	4, 5
Online Social	7, 8, 9		6	
Internet Economics		11	20	12

The order of appearance of these 20 questions is arranged so that clusters of highly related topics appear next to each other. Therefore, we recommend going through the chapters in this sequence, unless you're OK with flipping back every now and then when key concepts from prior chapters are referenced. Figure 0.1 summarizes the "prerequisite" relationship among the chapters.

This book cuts across both networks among devices and networks among people. We examine networks among people that overlay on top of networks among devices, but also spend half of the book on wireless networks, content distribution networks, and the Internet itself. We'll illustrate important ideas and useful methodologies across both types of networks. We'll see striking parallels in the underlying analytic models, but also crucial differences due to domain-specific details.

We can also classify the 20 questions into three groups in terms of the stages of development in formulating and answering them:

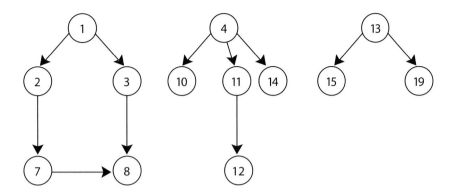

Figure 0.1 Dependence of mathematical background across some of the chapters is shown in these graphs. Each node is a chapter. Each directional link is a dependence relationship, e.g., Chapter 8's material requires that in Chapter 3 (which in turn requires that in Chapter 1) and that in Chapter 7 (which in turn requires that in Chapter 2, which in turn requires that in Chapter 1). Chapters 1, 4, and 13, the root nodes of these three trees, offer foundational material for ten other chapters. Some chapters aren't shown here because they don't form part of a dependence tree.

- Question well formulated, and theory-inspired answers adopted in practice: 1, 2, 3 4, 9, 10, 11, 13 14, 15, 16, 17, 18, 19.

- Question well formulated, but there's a gap between theory and practice (and we will discuss some possible bridges over the gaps): 12, 20.

- Question less well formulated (but certainly important to raise and explore): 5, 6, 7, 8.

It's comforting to see that most of our 20 chapters belong to the first group. Not surprisingly, questions about technological networks tend to belong to the first group, with those about social and economic networks gravitating more towards the second and third groups. It's often easier to model networked devices than networked human beings with predictive power.

Not all chapters explicitly study the impact of network topology, e.g., Chapter 7 studies influence models with decision externalities that are based on population sizes, while Chapter 8 looks at influence models with topology taken into account.

A quick word about the homework problems. There are five problems at the end of each chapter. These are a mixture of easy drills, simple extensions, challenging mini-research projects, and open-ended questions. Some important topics that we cannot readily fit into the main flow of the text are also postponed to the homework problem section. For those looking for more of the easy drills, the course website `www.network20q.com` offers additional questions.

Notation

We use **boldface** text for key terms when each is first defined. We use *italics* to highlight important, subtle, or potentially confusing points.

We use boldface math symbols to denote vectors or matrices, e.g., $\mathbf{x}$, $\mathbf{A}$. Vectors are column vectors by default. We do not use special fonts to represent sets when they are clear from the context. We use (t) to index iterations over continuous time, and $[t]$ or $[k]$ to index iterations over discrete time. We use $*$ to denote optimal or equilibrium quantities.

Some symbols have different meanings in different chapters, because they are the standard notation in different communities.

Related Books and Courses

There's no shortage of books on networks of many kinds. The popular ones that appeared in the past decade fall into two main groups:

- Popular science books, many of them filled with historical stories, empirical evidence, and sometimes a non-mathematical sketch of technical content. Some of the widely-read ones are *Bursts, Connected, Linked, Money Lab, Planet Google, Six Degrees, Sync, The Perfect Storm, The Tipping Point*, and *The Wisdom of Crowds*. Two other books, while not exactly on networks, provide important insights into many topics in networking: *Thinking, Fast and Slow* and *The Black Swan*. On the technology networks side, there are plenty of "for dummies" books, industry certification prep books, and entrepreneurship books. There are also several history-of-technology books, e.g., *Where the Geeks Stay Up Late* and *The Qualcomm Equation*.
- Popular undergraduate- or graduate-level textbooks. On the graph-theoretic and economic side of networking, three excellent textbooks appeared in 2010: *Networks, Crowds, and Markets* by Easley and Kleinberg, *Networks* by Newman, and *Social and Economic Networks* by Jackson. The last two are more on the graduate level. An earlier popular textbook is *Social Network Analysis: Methods and Applications* by Wasserman and Faust. On the computer networking side, there's a plethora of excellent textbooks. Two particularly popular ones today are *Computer Networking: A Top-Down Approach* by Kurose and Ross, and *Computer Networks: A Systems Approach* by Peterson and Davie. On wireless communications, several textbooks published in the last few years have become popular: *Wireless Communications* by Molisch, *Wireless Communications* by Goldsmith, and *Fundamentals of Wireless Communication* by Tse and Viswanath.

As illustrated in Figure 0.2, this book fills in the gap between existing groups of books.

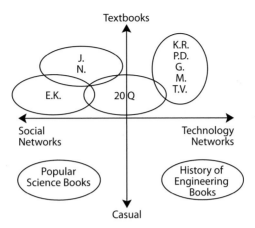

Figure 0.2 A cartoon illustrating roughly where some of the related books sit on two axes: one on the level of difficulty ranging from leisurely reading to graduate-level textbooks, and another on the mix of topics ranging from social and economic networks to technological networks. E.K. stands for Easley and Kleinberg, J. stands for Jackson, N. stands for Newman, K. R. stands for Kurose and Ross, P. D. stands for Peterson and Davie, G stands for Goldsmith, M stands for Molisch, and T.V. stands for Tse and Viswanath. 20Q stands for this book.

- Each chapter is driven by a practical question or observation, and the answers (or approximate answers) are explained using the rigorous language of mathematics. But mathematics never precedes practical problems.

- It also maintains a balance between social/economic networks and Internet/wireless networks, and between graph/economic theory and optimization/learning theory. For example, why WiFi works slower in hot spots is given as much attention as how Google auctions its ad spaces, and how IPTV networks operate is given as much detail as when information cascades initiate in a social group.

- A main goal of this book is to put social economic networks and technological networks side by side, and highlight their surprising similarities in spirit and subtle differences in detail. These examples range from the relation between Qualcomm's CDMA power control and Google's PageRank webpage ranking, to the connection between Galton's ox-weight estimation and 802.11n multiple-antenna WiFi.

These are also the differences between the Princeton undergraduate course and the seminal courses by Easley and Kleinberg at Cornell, and by Kearns at Penn. Those two courses have inspired a few similar courses at the interface between economics and computer science, such as those by by Acemoglu and Ozdaglar at MIT, by Chaintreau at Columbia, by Kempe at USC, by Parkes at Harvard, by Prabhakar at Stanford, by Spielman at Yale, by Wierman at Caltech...

These excellent courses have started structuring social and economic networking topics to undergraduates. On the other hand, both computer networking and wireless communications courses are standard, indeed often required, courses at many universities' Computer Science (CS) and Electrical Engineering (EE) departments. And across the EE, CS, Economics, Operations Research, and Applied Math departments, optimization theory, game theory, learning theory, and graph theory all have their separate courses. The new course at Princeton sits in-between the CS/Econ topics and the EE topics on networks. We hope there'll be more courses in EE and CS departments around the world that use unambigious languages to teach the concepts and methods common to social, economic, and technological networks.

Pedagogical Principles

This book and the associated course are also an experiment in three principles of teaching networks: JIT, BTP, and BAN.

Principle 1: Just In Time (JIT)

Models are often crippled by their own assumptions to start with, and frequently end up being largely irrelevant to what they set out to enable. Once in a while this is not true, but that's a low-probability event. That's why modeling is hard, especially for networks. So, before presenting any model, we first try to justify why the models are really necessary for the practical problems we face in each chapter. The material is arranged so that extensive mathematical machinery is introduced bit by bit, each bit presented just in time for the question raised. We enforce this "just-in-time" policy pretty strictly: no mathematical machinery is introduced unless it's used within the same section.

This might seem to be a rather unconventional way to write a textbook on the mathematical side of engineering. Usually a textbook asks the students to be patient with 50, 100, sometimes 200 pages of mathematics to lay the foundation first, and promises that motivating applications are coming after these pages. It's like asking a three-year-old to be patient for a long drive and promising ice-cream cones after many miles on the highway. In contrast, this book hands out an ice-cream cone every minute along the way, so that the three-year-old becomes very motivated to keep the journey going. It's more fun when gratification isn't delayed. "Fun right now" and "instant gratification" are what this book tries to achieve.

This book is an experiment motivated by this hypothesis: what professors call "fundamental knowledge" can be taught as "by-products" in the answers to practical questions that the undergraduates are interested in. A devoted sequence of lectures focusing exclusively (or predominantly) on the fundamental knowledge is not the *only* way to teach the material. Maybe we could also chop up the material and sprinkle it around. This does not "water-down" the material, it simply reorganizes it so that it shows up right next to the applications in each and every lecture. The downside is that the standard trains of thought running through the mathematical foundation of research communities are interrupted many times. This often leaves me feeling weird because I could not finish my normal teaching sequence, but the instructor feeling uncomfortable is probably a good sign. The upside is that undergraduates, who may not even be interested in a career in this field, view the course as completely driven by practical questions.

For example, the methodologies of optimization theory are introduced bit by bit in this book: linear programming and Perron-Frobenius theory in power control, convexity and least squares in Netflix recommendation, network utility maximization in Internet pricing, dynamic programming and multi-commodity flow in Internet routing, the gradient algorithm and dual decomposition in congestion control, and combinatorial optimization in peer-to-peer networks.

The methodologies of game theory are introduced bit by bit: the basic definitions in power control, auction theory in ad-space auctions, bargaining theory in Wikipedia consensus formation as well as in two-sided pricing of Internet access, and selfish maximization in tipping.

The methodologies of graph theory are introduced bit by bit: matching in ad-space auctions, consistency and PageRank in Google search, bipartite graph in Netflix recommendation, centrality, betweenness, and clustering measures in influence models, small worlds in social search, scale-free graphs in Internet topology, the Bellman–Ford algorithm and max flow min cut in Internet routing, and tree embedding in peer-to-peer networks.

The methodologies of learning theory are introduced bit by bit: collaborative filtering in Netflix recommendation, Bayesian estimation and adaptive boosting in ratings, and community detection in influence models.

Principle 2: BTP (Bridge Theory and Practice)

The size of the global industry touched upon by these 20 questions is many trillions of dollars. Just the market capitalizations of the 20 most relevant US companies to this book: Apple, Amazon, AT&T, Cisco, Comcast, Disney, eBay, EMC, Ericsson, Facebook, Google (including YouTube), Groupon, HP, Intel, LinkedIn, Microsoft (including Skype), Netflix, Qualcomm, Verizon, and Twitter added up to over $2.22 trillion as of July 4, 2012.

In theory, this book's theories are directly connected to the practice in this multi-trillion-dollar industry. In practice, that's not always true, especially in fields like networking where stable models, like the additive Gaussian noise channel for copper wire in communication theory, often do not exist.

Nonetheless, we try to strike a balance between

- presenting enough detail so that answers to these practical questions are grounded in actual practice rather than in "spherical cows" and "infinite planes," (although we couldn't help but keep "rational human beings" in several chapters), and

- avoiding too much detail that reduces the "signal-noise-ratio" in illustrating the fundamental principles.

This balance is demonstrated in the level of detail with which we treat network protocol descriptions, Wikipedia policies, Netflix recommendation algorithms, etc. And this tradeoff explains the (near) absence of random graph theory and of Internet protocols' header formats, two very popular sets of material in standard textbooks in math/CS-theory/sociology and in CS-systems/EE curricula, respectively.

Some of these 20 questions are currently trapped in particularly deep theory–practice gaps, especially those hard-to-formulate questions in Chapters 5 and 6, and those hard-to-falsify answers in Chapters 7 and 8. The network economics material in Chapters 11 and 12 also fits many of the jokes about economists, too many to quote here. (A good source of them is Taleb's *The Bed of Procrustes*.) Reverse engineering, shown across several chapters, has its own share of accurate jokes: "Normal people look at something that works in theory, and wonder if it'll also work in practice. Theoreticians look at something that works in practice, and wonder if it'll also work in (their) theory."

Time and time again, we skip the techniques of mathematical acrobats, and instead highlight the never-ending struggles between representations and realities during modeling: the process of "mathematical crystallization" where (most) parts of reality are thrown out of the window so that what remains becomes tractable using today's analytic machineries. What often remains unclear is whether the resulting answerable questions are still relevant and the resulting tractable models still have predictive powers. However, when modeling is done "right," engineering artifacts can be explained rather than just described, and better design can be carried out top-down rather than by "debug and tweak" It's often been quoted (mostly by theoreticians like me) that "there's nothing more practical than a good theory," and that "a good theory is the first-order exponent in the Taylor expansion of reality." Perhaps these can be interpreted as *definitions* of what constitutes a "good" theory. By such a definition, this book has traces of good theory, thanks to many researchers and practitioners who have been working hard on bridging the theory-practice gaps in networking.

Principle 3: BAN (Book As a Network)

Throughout the chapters, comparisons are constantly drawn with other chapters. This book itself is a network, a network of ideas living in nodes called chapters, and we grasp every opportunity to highlight each possible link between these nodes. The most interesting part of this book is perhaps this network effect among ideas: to see how curiously related, and yet crucially different they are.

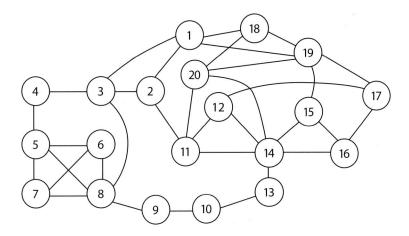

Figure 0.3 Intellectual connections across the chapters. Each node is a chapter, and each bidirectional link is an intellectual connection, via either similar concepts or common methodologies. Cliques of nodes and multiple paths from one node to another are particularly interesting to observe in this graph.

Figure 0.3 shows the main connections among the chapters. This is what the book is about: weave a network of ideas (about networks), and the positive network effect comes out of that.

We can extract the top 20 ideas across the chapters. The first 10 are features of networks, the next 5 design ideas, and the last 5 modeling approaches.

1. Resource sharing (such as statistical multiplexing and fairness): Chapters 1, 11, 13, 14, 15, 16, 17, 18, 20.
2. Opinion aggregation and consensus formation: Chapters 3, 4, 5, 6, 18.
3. Positive network effect (such as resource pooling and economy of scale): Chapters 9, 11, 13, 15, 16.
4. Negative network effect (such as tragedy of the commons): Chapters 11, 20.
5. The wisdom of crowds (diversity gain and efficiency gain): Chapters 7, 8, 18, 19.
6. The fallacy of crowds (cascade and contagion): Chapters 7, 8.
7. Functional hierarchy and layering: Chapters 13, 14, 15, 17, 19.
8. Spatial hierarchy and overlaying: Chapters 10, 13, 15, 16, 17.
9. From local actions to global property: Chapters 1, 6, 7, 8, 13, 14, 15, 18.
10. Overprovision capacity vs. overprovision connectivity: Chapters 14, 15, 16.
11. Feedback control: Chapters 1, 7, 13, 14.
12. Utility maximization: Chapters 1, 2, 11, 12, 14, 20.
13. Protocols: Chapters 14, 15, 17, 19.
14. Signaling: Chapters 6, 19.
15. Randomization: Chapters 3, 15, 18.
16. Graph consistency models: Chapters 3, 13.

17. Strategic equilibrium models: Chapters 1, 2, 15.
18. Generative model (and reverse engineering): Chapters 9, 10, 14.
19. Latent-factor models: Chapter 4.
20. Axiomatization models: Chapters 6, 20.

In the first offering of this course at Princeton, the undergrads voted (by Borda count) "resource sharing," "opinion aggregation," and "positive network effect" as the top three concepts they found most useful. They also voted the key equations in PageRank, distributed power control, and Bellman–Ford as the top three equations.

Almost every one of these 20 ideas cuts across social/economic networks and technological networks. Some examples are given below.

- The emergence of global coordination through local actions based on local views is a recurring theme, from influence models in social networks to routing and congestion control in the Internet, and from consumer reaction to pricing signals to power control in wireless networks.
- Resource sharing models, in the form of additive sharing $x + y \leq 1$, or multiplicative sharing $x/y \geq 1$, or binary sharing $x, y \in \{0, 1\}, x + y \leq 1$, are introduced for network pricing as well as the classic problems of congestion control, power control, and contention control.
- The (positive) network effect is often highlighted in social and economic networks. It also finds a concrete realization in how content is shared over the Internet through peer-to-peer protocols and how data centers are scaled up.
- "The wisdom of (independent and unbiased) crowds" is another common theme. There are two types of "wisdom" here. (1) Diversity gain in reducing the chance of some bad event (typically represented mathematically by $1 - (1 - p)^N$, where N is the size of the crowd and p the probability of some bad event). (2) Efficiency gain in smoothing out some average metric (typically represented mathematically as a factor-N in the metric). Both types are observed in social networks and in the latest generation of 4G and WiFi wireless networks.
- Consensus formation is used in computing webpage importance scores in PageRank as well as in discovering the right time to transmit in WiFi.
- Spatial hierarchy is used both in how search is done in a small world and in how the Internet is structured.
- The design methodology of feedback control is used in influence models in social networks and congestion control in the Internet.
- Utility maximization is used in auctioning advertising spots and setting Internet access pricing.
- The power method is used both in Google's PageRank and in Qualcomm's distributed power control.
- Randomization is used in PageRank and 802.11 CSMA.
- Strategic equilibrium models are used in auctioning and BitTorrent.

- Reverse engineering is used in studying scale-free networks and TCP.
- Axiomatization is used in voting procedure and fairness evaluation.

This list goes on. Yet equally important are the subtle differences between technological and socio-economic networks. Exhibit A for this alert is the (non-existence of) the Achilles' heel of the Internet and the debate between two generative models (preferential attachment vs. constrained optimization) of scale-free networks.

Two Bigger Pictures

There are two broader themes in the backdrop of this book:

- *Instill domain-specific functionalities to a generic network science.* A "network science" around these 20 questions must be based on domain-specific models and on the pursuit of falsification. For example, while a random graph is elegant, it's often neither a relevant approach to design nor the only generative model to explain what we see in this book. And as much as metrics of a static graph are important, engineering protocols governing the *functionalities* of feedback, coordination, and robustness are just as crucial as the *topological* properties of the graph like the degree distribution.
- *Revisit the Electrical and Computer Engineering (ECE) undergraduate curriculum.* In the standard curriculum in ECE since around the 1960s, a "signals and systems" course is one of the first foundational courses. As networks of various kinds play an increasingly important role both in engineering design and in the society, it's time to capture fundamental concepts in networking in a second systems course. Just as linear time-invariant systems, sampling, integral transforms, and filter design have laid the foundation of ECE curriculum since the 1960s, we think the following concepts have now become fundamental to teach to future ECE students (whether they are taught in the JIT way or not): patterns of connections among nodes, modularization and hierarchy in networked systems, consistency and consensus in graphs, distributed coordination by pricing feedback, strategic equilibrium in competition and cooperation, pros and cons of scaling up, etc.

  So this book is an experiment in both *what* to teach and *how* to teach in an ECE undergraduate course in systems: what constitutes core knowledge that needs to be taught and how to teach it in a context that enhances learning efficiency. As much as we appreciate FIR and IIR filter design, Laplace and Z transforms, etc., maybe it's about time to explore the possibility of reducing the coverage of these topics by just a tiny bit to make room for mathematical notions just as fundamental to engineering today. And we believe the best way to drive home these ideas is to tie in with applications that teenagers, and many of the older folks, use every day.

Class as a Social and Economic Network

The class "Networks: Friends, Money, and Bytes," created in parallel to this book in Fall 2011 and cross-listed in EE and CS at Princeton University, was a social and economic network itself. We tweeted, we blogged, and we created wikis. On the first day of the class, we drew a class social graph, where each node is a student, and a link represents a "know by first name before coming to the first lecture" relationship. After the last lecture, we drew the graph again.

We also created our own currency called "nuggets." The TAs and I "printed" our money as we saw fit. There were several standard ways to earn nuggets, including catching typos in lecture notes and writing popular blogs. There were ten class activities beyond homework problems that were rewarded by nuggets, including one activity in which the students were allowed to buy and sell their homework solutions using auctions. The matching of students and class project topics was also run through bidding with nuggets. Eventually the nugget balances translate into an upward adjustment of grades. To see some of the fun of ELE/COS 381 at Princeton University, visit www.network20q.com.

Starting in Fall 2012, this course is offered on Stanford's coursera and its own open education course website, and on YouTube and iTunes U too. The Princeton offering adopts the approach of flipped classroom advocated by the Khan Academy. With many parallel, on-going efforts from different universities and companies (CodeAcademy, coursera, EdX, udacity, etc.), it will take a few years before the landscape of open online education becomes stable. Higher education will be enhanced by this movement that has been gathering momentum since MIT's open courseware initiative in 2002. Yet many issues remain to settle at the time of writing this book: the modes and efficiency of students' learning, the boundary and reach of higher education, the prioritization and value propositions of a university's missions, the roles of faculty and the nature of classroom teaching, the differentiation between self-education and branded-certification, the authenticity and ownership of student-activity records, the tuition revenue to universities and business models of open access platforms, the drawbacks of monetization on for-profit platforms... What is already clear, however, is that this mode of education cannot be feasible without the widespread use of mobile data devices, the video-watching habit of the YouTube generation, or the crowd-sourcing of social-networked online study group. These are exactly some of the key topics studied in this course. From voting of popular questions to distributing video over 4G networks, this is a course *about* these networks and taught *through* these networks.

1 What makes CDMA work for my smartphone?

1.1 A Short Answer

Take a look at your iPhone, Android phone, or a smartphone running on some other operating system. It embodies a remarkable story of technology innovations. The rise of wireless networks, the Internet, and the web over the last five decades, coupled with advances in chip design, touchscreen material, battery packaging, software systems, business models... led to this amazing device you are holding in your hand. It symbolizes our age of networked life.

These phones have become the mobile, lightweight, smart centers of focus in our lives. They are used not just for voice calls, but also for **data applications**: texting, emailing, browsing the web, streaming videos, downloading books, uploading photos, playing games, or video-conferencing friends. The throughputs of these applications are measured in bits per second (bps). These data fly through a **cellular network** and the **Internet**. The cellular network in turn consists of the radio air-interface and the core network. We focus on the air-interface part in this chapter, and turn to the cellular core network in Chapter 19.

Terrestrial wireless communication started back in the 1940s, and cellular networks have gone through generations of evolution since the 1970s, moving into what we hear as 4G these days. Back in the 1980s, some estimated that there would be 1 million cellular users in the USA by 2000. That turned out to be one of those way-off under-estimates that did not even get close to the actual impact of networking technologies.

Over more than three decades of evolution, a fundamental concept of cellular architecture has remained essentially the same. The entire space of deployment is divided into smaller regions called **cells**, which are often represented by hexagons as in Figure 1.1, thus the name cellular networks and cell phones. There is one **base station** (BS) in each cell, connected on the one side to switches in the core network, and on the other side to the **mobile stations** (MSs) assigned to this cell. An MS could be a smart phone, a tablet, a laptop with a dongle, or any device with antennas that can transmit and receive in the right frequencies following a cellular network standard. There are a few other names for them, for example, sometimes an MS is also called a User Equipment (UE) and a BS called a Node B (NB) or an evolved Node B (eNB).

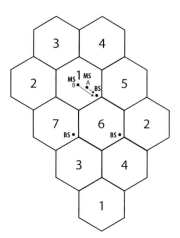

Figure 1.1 Part of a typical cellular network with a frequency reuse of 7. Each cell is a hexagon with a base station (BS) and multiple mobile stations (MSs). Only a few of them are drawn in the figure. Each BS has three directional antennas, each of which covers a 120-degree sector. Some mobile stations, like MS A', are close to the base station with strong channels to the BS. Others, like MS B, are on the cell edge with weak channels. Attenuation enables frequency reuse, but the variability of and the inability to control attenuation pose challenges to wireless cellular network design.

We see a clear hierarchy, a fixed infrastructure, and one-hop radio links in cellular networks. This is in contrast to other types of wireless networks. Moreover, the deployment of base stations is based on careful radio engineering and tightly controlled by a wireless provider, in contrast to WiFi networks in Chapter 18.

Why do we divide the space into smaller regions? Because the wireless spectrum is scarce and radio signals weaken over space.

Transmitting signals over the air means emitting energy over different parts of the electromagnetic **spectrum**. Certain regions of the spectrum are allocated by different countries to cellular communications, just like other parts of the spectrum are allocated to AM and FM radio. For example, the 900 MHz range is allocated for the most popular 2G standard called GSM, and in Europe the 1.95 GHz range and 2.15 GHz range are allocated for UMTS, a version of the 3G standard. Some part of the spectrum is unlicensed, like in WiFi, as we will see in Chapter 18. Other parts are licensed, like those for cellular networks, and a wireless service provider needs to purchase these limited resources with hefty prices. The spectrum for cellular networks is further divided into chunks, since it is often easier for transmitters and receivers to work with narrower frequency bands, e.g., on the order of 10 MHz in 3G.

The signals sent in the air become weaker as they travel over longer distances. The amount of this attenuation is often proportional to the square, or even the fourth power, of the distance traversed. So, in a typical cellular network, the signals become too weak to be accurately detected after a couple of miles. At

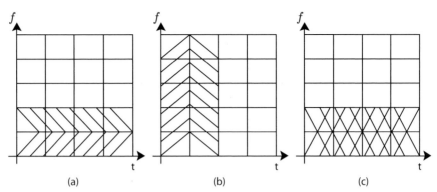

Figure 1.2 Part of a time–frequency grid is shown in each graph. For visual clarity, we only show two slices of resources being used. (a) FDMA and (b) TDMA are dedicated resource allocation: each frequency band or timeslot is given to a user. In contrast, (c) CDMA is shared resource allocation: each time–frequency bin is shared by multiple users, all transmitting and receiving over the same frequency band and at the same time. These users are differentiated by signal processing. Power control also helps differentiate their signals.

first glance, this may sound like bad news. But it also means that the frequency band used by base station A can be reused by another base station B sufficiently far away from A. All we need to do is tessellate the frequency bands, as illustrated in Figure 1.1, so that no two cells share the same frequency band if they are too close. In Figure 1.1, we say that there is a **frequency reuse factor** of 7, since we need that many frequency bands in order to avoid having two close cells sharing the same frequency band. **Cellular architecture** enables the network to scale up over space. We will visit several other ways to scale up a network later.

Now, how can the users in the *same* cell share the same frequency band? There are two main approaches: orthogonal and non-orthogonal allocation of resources.

Frequency is clearly one type of resource, and time is another. In **orthogonal allocation**, each user is given a small band of frequency in Frequency-Division Multiple Access (**FDMA**), or a timeslot in Time-Division Multiple Access (**TDMA**). Each user's allocation is distinct from others, as shown in Figure 1.2(a) and (b). This often leads to an inefficient use of resources. We will see in later chapters a recurring theme: a dedicated assignment of resources to users becomes inefficient when users come and go frequently.

The alternative, **non-orthogonal allocation**, allows all users to transmit at the same time over the same frequency band, as in Code-Division Multiple Access. **CDMA** went through many ups and downs with technology adoption from 1989 to 1995, but is now found in all the 3G cellular standards as part of the design. In CDMA's first standard, IS-95 in the 2G family, the same frequency band is reused in all the cells, as illustrated in Figure 1.2(c). But how can we distinguish the users if their signals overlap with each other?

Think of a cocktail party with many pairs of people trying to carry out individual conversations. If each pair takes turns in communicating, and only one person gets to talk during each timeslot, we have a TDMA system. If all pairs can communicate at the same time, and each uses a different language to avoid confusion, we have a CDMA system. But there are not enough languages whose pronunciations do not cause confusion, and human ears are not that good at decoding, so interference is still an issue.

How about controlling each person's volume? Each transmitter adjusts the volume of its voice according to the relative distances among the speakers and listeners. In a real cocktail party, unless there is some politeness protocol or it hurts people's vocal chord to raise their voice, we end up in a situation where everyone is shouting and yet most people cannot hear well. Transmit power control should mitigate this problem.

The core idea behind the CDMA standards follows our intuition about the cocktail party. First, the transmitter multiplies the digital signals by a sequence of 1s and minus 1s, a sequence we call the **spreading code**. The receiver multiplies the received bits by the same spreading code to recover the original signals. This is straightforward to see: 1×1 is 1, and -1×-1 is also 1. What is non-trivial is that a family of spreading codes can be designed such that only *one* spreading code, the original one used by the transmitter, can recover the signals. If you use any other spreading code in this family, you will get noise-like, meaningless bits. We call this a family of **orthogonal codes**. Users are still separated by orthogonalization, just along the "code dimension" as opposed to the more intuitive "time dimension" and "frequency dimension." This procedure is called direct sequence **spread spectrum**, one of the standard ways to enable CDMA.

However, there may not be enough orthogonal spreading codes for all the mobile stations. Families of orthogonal codes are limited in their sizes. Furthermore, a slight shift on the time axis can scramble the recovered bits at the receiver. We need the clocks on all the devices to be synchronized. But this is infeasible for the **uplink**, where mobiles talk to the base station: MSs cannot easily coordinate their clocks. It is difficult even in the **downlink**, where the base station talks to the mobiles: the BS has a single clock but the wireless channel distorts the bits. Either way, we do not have perfectly orthogonal spreading codes, even though these imperfect codes still provide significant "coding gain" in differentiating the signals.

We need an alternative mechanism to differentiate the users and to tackle the **interference** problem. Wireless signals are just energy propagating in the air, and one user's signal is every other user's interference. Interference, together with the attenuation of signals over distance and the fading of signals along multiple paths, are the the top three issues we have to address in wireless channels. Interference is an example of **negative externality** that we will encounter many times in this book, together with ways to "internalize" it by designing the right mechanism.

Here is an example of significant interference. As shown in Figure 1.1, a user standing right next to the BS can easily overwhelm another user far away at the edge of the cell. This is the classic **near-far problem** in CDMA networks. It was solved in the IS-95 standard by Qualcomm in 1989. This solution has been one of the cornerstones in realizing the potential of CDMA since then.

Qualcomm's solution to the near-far problem is simple and effective. The receiver infers the channel quality and sends that back to the transmitter as feedback. Consider an uplink transmission: multiple MSs trying to send signals to the BS in a particular cell. The BS can estimate the channel quality from each MS to itself, e.g., by looking at the ratio of the received signal power to the transmitted power, the latter being pre-configured to some value during the channel-estimation timeslot. Then, the BS inverts the channel quality and sends that value, on some feedback control channel, back to the MSs, telling them that these are the gain parameters they should use in setting their transmit powers. In this way, all the received signal strengths will be made equal. This is the basic MS **transmit power control** algorithm in CDMA.

But what if equalization of the received signal powers is *not* the right goal? For voice calls, the typical application on cell phones in 2G networks in the 1990s, there is often a target value of the received signal quality that each call needs to achieve. This signal quality factor is called the Signal to Interference Ratio (**SIR**). It is the ratio between the received signal strength and the sum strength of all the interference (plus the receiver noise strength). Of course, it is easy to raise the SIR for just one user: just increase its transmitter's power. But that translates into higher interference for everyone else, which further leads to higher transmit powers being used by them if they also want to maintain or improve their SIRs. This positive feedback escalates into a transmit power "arms race" until each user is transmitting at the maximum power. That would not be a desirable state to operate in.

If each user fixes a reasonable target SIR, can we do better than this "arms race" through a more intelligent power control? Here, "being reasonable" means that the SIRs targeted by all the users in a cell are mutually compatible; they *can* be simultaneously achieved by some configuration of transmit powers at the MSs.

The answer is yes. In 1992-1993, a sequence of research results developed the basic version of **Distributed Power Control** (DPC), a fully **distributed algorithm**. We will discuss later what we mean by "distributed" and "fully distributed." For now, it suffices to say that, in DPC, each pair of transmitter (e.g., an MS) and receiver (e.g., the BS) does not need to know the transmit power or channel quality of any other pair. At each timeslot, all it needs to know is the actual SIR it currently achieves at the receiver. Then, by taking the ratio between the fixed, target SIR and the variable, actual SIR value measured for this timeslot, and multiplying the current transmit power by that ratio, we get the transmit power for the next timeslot. This update happens simultaneously at each pair of transmitter and receiver.

This simple method is an **iterative algorithm**; the updates continue from one timeslot to the next, unlike with the one-shot, received-power-equalization algorithm. But it is still simple, and when the target SIRs can be simultaneously achieved, it has been proven to **converge**: the iterative procedure will stop over time. When it stops, it stops at the right solution: a power-minimal configuration of transmit powers that achieves the target SIRs for all. DPC converges quite fast, approaching the right power levels with an error that decays as a geometric series. DPC can even be carried out asynchronously: each radio has a different clock and therefore different definitions of what timeslot it is now.

Of course, in real systems the timeslots are indeed *asynchronous* and power levels are *discrete*. Asynchronous and quantized versions of DPC have been implemented in all the CDMA standards in 3G networks. Some standards run power control 1500 times every second, while others run 800 times a second. Some discretize power levels to 0.1 dB, while others between 0.2 and 0.5 dB. Without CDMA, our cellular networks today could not work as efficiently. Without power control algorithms (and the associated handoff method to support user mobility), CDMA could not function properly. In Chapter 19, we will discuss a 4G standard called LTE. It uses a technology called OFDM instead of CDMA, but power control is still employed for interference reduction and for energy management.

Later, in Chapter 18, we will discuss some of the latest ideas that help further push the data rates in new wireless network standards, ranging from splitting, shrinking, and adjusting the cells to overlaying small cells on top of large ones for traffic offloading, and from leveraging multiple antennas and tilting their positions to "chopping up" the frequency bands for more efficient signal processing.

1.2   A Long Answer

1.2.1   Distributed power control

Before we proceed to a general discussion of the Distributed Power Control (DPC) algorithm, we must first define some symbols.

Consider N pairs of transmitters and receivers. Each pair forms a (logical) link, indexed by i. The transmit power of the transmitter of link i is p_i, some positive number, usually capped at a maximum value: $p_i \leq p_{max}$ (although we will not consider the effect of this cap in the analysis of the algorithm). The transmitted power impacts both the received power at the intended receiver and the received interference at the receivers of all other pairs.

Now, consider the channel from the transmitter of link (i.e., transmitter–receiver pair) j to the receiver of link i, and denote the **channel gain** by G_{ij}. So G_{ii} is the direct channel gain; the bigger the better, since it is the channel for the intended transmission for the transmitter–receiver pair of link i. All the other $\{G_{ij}\}$, for j not equal to i, are gains for interference channels, so the smaller the better. We call these channel "gains", but actually they are less than 1, so maybe a better term is channel "loss."

This notation is visualized in Figure 1.3 for a simple case of two MSs talking to a BS, which can be thought of as two different (logically separated) receivers physically located together.

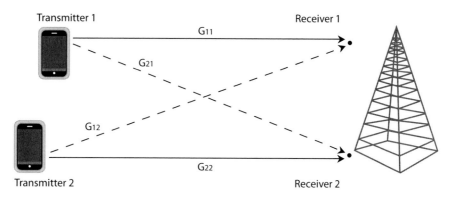

Figure 1.3 Uplink interference between two mobile stations at the base station. We can think of the base station as two (logically separated) receivers collocated. G_{11} and G_{22} are direct channel gains, the bigger the better. G_{12} and G_{21} are interference channel gains, the smaller the better.

Each G_{ij} is determined by two main factors: (1) location of the transmitter and receiver and (2) the quality of the channel in between. G_{ii} is also enhanced by the CDMA spreading codes that help the intended receivers decode more accurately.

The received power of the intended transmission at the receiver is therefore $G_{ii}p_i$. What about the interference? It is the sum of $G_{ij}p_j$ over all transmitters j (other than the intended one i): $\sum_{j \neq i} G_{ij}p_j$. There is also noise n_i in the receiver electronics for each receiver i. So we can write the SIR, a unit-less ratio, at the receiver of logical link i as

$$\text{SIR}_i = \frac{G_{ii}p_i}{\sum_{j \neq i} G_{ij}p_j + n_i}. \tag{1.1}$$

For proper decoding of the packets, the receiver needs to maintain a target level of SIR. We will denote that as γ_i for link i, and we want $\text{SIR}_i \geq \gamma_i$ for all i. Clearly, increasing p_1 raises the SIR for receiver 1 but lowers the SIR for all other receivers.

As in a typical algorithm we will encounter throughout this book, we assume that time is divided into discrete slots, each indexed by $[t]$. At each timeslot t, the receiver on link i can measure the received SIR readily, and feeds back that number, $\text{SIR}_i[t]$, to the transmitter.

The DPC algorithm can be described through a simple equation: each transmitter simply multiplies the current power level $p_i[t]$ by the ratio between the target SIR, γ_i, and the current measured $\text{SIR}_i[t]$, to obtain the power level to use in the next timeslot:

$$p_i[t+1] = \frac{\gamma_i}{\text{SIR}_i[t]} p_i[t], \quad \text{for each } i. \tag{1.2}$$

We will use the symbols $\forall i$ later to say "for each i."

We see that each receiver i needs to measure only its own SIR at each iteration, and each transmitter only needs to remember its own target SIR. There is no need for passing any control message around, like telling other users what power level it is using. Simple in *communication*, it is a *very* distributed algorithm, and we will later encounter many types of distributed algorithms in various kinds of networks.

This algorithm is also simple in its *computation*: just one division and one multiplication. And it is simple in its parameter *configuration*: there are actually no parameters in the algorithm that need to be tuned, unlike in quite a few other algorithms in later chapters. Simplicity in communication, computation, and configuration is a key reason why certain algorithms are widely adopted in practice.

Intuitively, this algorithm makes sense. First, when the iterations stop because no one's power is changing any more, i.e., when we have convergence to an **equilibrium**, we can see that $SIR_i = \gamma_i$ for all i.

Second, there is hope that the algorithm will actually converge, given the direction in which the power levels are moving. The transmit power moves up when the received SIR is below the target, and moves down when it is above the target. *Proving* that convergence will happen is not as easy. As one transmitter changes its power, the other transmitters are doing the same, and it is unclear what the next timeslot's SIR values will be. In fact, this algorithm does *not* converge if too many γ_i are too large, i.e., when too many users request large SIRs as their targets.

Third, if satisfying the target SIRs is the only criterion, there are many transmit power configurations that can do that. If $p_1 = p_2 = 1$ mW achieves these two users' target SIRs, $p_1 = p_2 = 10$ mW will do so too. We would like to pick the configuration that uses the least amount of power; we want a power-minimal solution. And the algorithm above seems to be lowering the power when a high power is unnecessary.

DPC illustrates a recurring theme in this book. We will see in almost every chapter that individual's behaviors driven by self-interest can often aggregate into a fair and efficient state globally across all users, especially when there are proper feedback signals. In contrast, either a centralized control or purely random individual actions would have imposed significant downsides.

1.2.2   DPC as an optimization solution

In general, "will it converge?" and "will it converge to the right solution?" are the top two questions that we would like to address in the design of all iterative algorithms. Of course, what "the right solution" means will depend on the definition of optimality. In this case, power-minimal transmit powers that achieve the target SIRs for all users are the "right solution." Power minimization is the **objective** and achieving target SIRs for all users is the **constraint**.

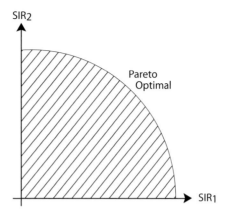

Figure 1.4 An illustration of the SIR feasibility region. It is a constraint set for power control optimization, and visualizes the competition among users. Every point strictly inside the shaded region is a feasible vector of target SIRs. Every point outside is infeasible. And every point on the boundary of the curve is Pareto optimal: you cannot increase one user's SIR without reducing another user's SIR.

In this case, there are many ways to address these questions, for example, using machineries from optimization theory or game theory. Either way, we can show that, under the condition that the target SIR values are indeed **achievable** at all (i.e., there are some values of the transmit powers that can achieve the target SIRs for all users), DPC will converge, and converge to the right solution.

We can illustrate a typical set of feasible SIRs in the SIR feasibility region shown in Figure 1.4. Clearly, we want to operate on the boundary of this region, and each point on the boundary is called **Pareto optimal**. Along this boundary, one user's higher SIR can be achieved, but at the expense of a lower SIR for another user. This highlights another recurrent theme in this book: the need to tackle tradeoffs among competing users and across different design objectives, and the importance of providing incentives for people to react to. There is no free lunch; and we have to balance the benefits with the costs.

It turns out that DPC also solves a global optimization problem for the network. Here, "global" means that the interests of *all* the users are incorporated. In this case, it is the sum of transmit powers that is minimized, and every user's target SIR must be met.

Once we have an objective function and a set of constraints, and we have defined which quantities are variables and which are constants, an **optimization** problem is formulated. In this case, the transmit powers are the variables. The achieved SIRs are also variables, but are derived from the powers. All the other quantities are the constants: they are not degrees of freedom under your control.

If the variable vector $\mathbf{x}_0$ satisfies all the constraints, it is called a **feasible solution**. If an optimization problem's constraints are not mutually compatible, it is called **infeasible**. If an $\mathbf{x}^*$ is both feasible and better than any other feasible solution, i.e., gives the smallest objective function value for a minimization problem (or the largest objective function value for a maximization problem), it is called an **optimal** solution. An optimal solution might not exist, e.g., minimize $1/x$ for $x \in \mathcal{R}$. And optimal solutions may not be unique either.

Here is the optimization problem of varying transmit power to satisfy fixed target SIR constraints and then minimize the total power:

$$
\begin{array}{ll}
\text{minimize} & \sum_i p_i \\
\text{subject to} & \text{SIR}_i(\mathbf{p}) \geq \gamma_i, \quad \forall i \\
\text{variables} & \mathbf{p}.
\end{array}
\tag{1.3}
$$

Problem (1.3) looks complicated if we substitute the definition (1.1) of the SIR as a function of the whole vector $\mathbf{p}$:

$$
\begin{array}{ll}
\text{minimize} & \sum_i p_i \\
\text{subject to} & \dfrac{G_{ii}p_i}{\sum_{j \neq i} G_{ij}p_j + n_i} \geq \gamma_i, \quad \forall i \\
\text{variables} & \mathbf{p}.
\end{array}
$$

But it can be simplified through a different representation. We can easily rewrite it as a **linear programming** problem: minimizing a linear function of the variables subject to linear constraints of the variables:

$$
\begin{array}{ll}
\text{minimize} & \sum_i p_i \\
\text{subject to} & G_{ii}p_i - \gamma_i(\sum_{j \neq i} G_{ij}p_j + n_i) \geq 0, \quad \forall i \\
\text{variables} & \mathbf{p}.
\end{array}
$$

Linear programming problems are easy optimization problems. More generally, convex optimization (to be introduced in Chapter 4) is easy; easy in theory in terms of complexity and easy in practice with fast solution software. In the Advanced Material, we will derive DPC as the solution to problem (1.3).

We will be formulating and solving optimization problems many times in future chapters. Generally, solving a global optimization via local actions by each user requires explicit signaling. But in this case, it turns out that the selfish behavior of users in their own power minimization also solves the global, constrained optimization problem; their interests are correctly aligned already. This is more of an exception than the norm.

1.2.3   DPC as a game

Power control is a competition. One user's received power is another's interference. Each player searches for the right "move" (or, in this case, the right transmit power) so that its "payoff" is optimized (in this case, the transmit power is the smallest possible while providing the user with its target SIR γ_i). We also hope that the whole network reaches some desirable equilibrium as each player strategizes. The concepts of "players," "move," "payoff," and "equilibrium" can be defined in a precise and useful way.

We can model *competition* as a **game**. The word "game" here carries a technical meaning. The study of games is a branch of mathematics called game theory. If the competition is among human beings, a game might actually correspond to people's strategies. If it is among devices, as in this case among radios, a game is more like an angle of interpretation and a tool for analysis. It turns out that *cooperation* can also be modeled in the language of game theory, as we will show in Chapter 6.

In the formal definition, a game is specified by three elements:

1. a set of **players** $\{1, 2, \ldots, N\}$,
2. a **strategy space** A_i for each player i, and
3. a **payoff function**, or utility function, U_i for each player to maximize (or a **cost function** to minimize). Function U_i maps each combination of all players' strategies to a real number, the payoff (or cost), to player i.

	Not Confess	Confess
Not Confess	$(-1, -1)$	$(-5, 0)$
Confess	$(0, -5)$	$(-3, -3)$

Table 1.1 Prisoner's dilemma. This is a famous game in which there is a unique and undesirable Nash equilibrium. Player A's two strategies are the two rows. Player B's two strategies are the two columns. The values in the table represent the payoffs to the two players in each scenario.

Now consider the two-player game in Table 1.1. This is the famous **prisoner's dilemma** game, which we will also encounter later in voting paradoxes, tragedy of the commons, and P2P file sharing. The two players are two prisoners. Player A's strategies are shown in rows and player B's in columns. Each entry in the 2×2 table has two numbers, (x, y), where x is the payoff to A and y that to B if the two players pick the corresponding strategies. As you would expect from the coupling between the players, each payoff value is determined jointly by the strategies of both players. For example, the payoff function maps (Not Confess, Not Confess) to -1 for both players A and B. These payoffs are negative because they are the numbers of years the two prisoners are going to serve in prison. If one confesses but the other does not, the one who confesses gets a deal to walk away free and the other one is heavily penalized. If both confess, both serve three years. If neither confesses, only a lesser conviction can be pursued and both serve only one year. Both players know this table, but they cannot communicate with each other.

If player A chooses the strategy Not Confess, player B should choose the strategy Confess, since $0 > -1$. This is called the **best response strategy** by player B, in response to player A choosing the strategy Not Confess.

If player A chooses the strategy Confess, player B's best response strategy is still Confess, since $-3 > -5$. When the best response strategy of a player is the *same* no matter what strategy the other player chooses, we call that a **dominant strategy**. It might not exist. But, when it does, a player will obviously pick a dominant strategy.

In this case, `Confess` is the dominant strategy for player B. By symmetry, it is also the dominant strategy for player A. So both players will pick `Confess`, and (`Confess`, `Confess`) is an **equilibrium** for the game. This is a slightly different definition of "equilibrium" from what we saw before, where equilibrium means an update equation reaches a fixed point.

Clearly, this equilibrium is undesirable: (`Not Confess`, `Not Confess`) gives a higher payoff value to both players: -1 instead of -3. But the two prisoners could not have coordinated to achieve (`Not Confess`, `Not Confess`). An equilibrium might not be **socially optimal**, i.e., a set of strategies maximizing the sum of payoffs $\sum_i U_i$ of all the players. It might not even be Pareto optimal, i.e., a set of strategies such that no player's payoff can be increased without hurting another player's payoff.

	Action Movie	Romance Movie
Action Movie	$(2, 1)$	$(0, 0)$
Romance Movie	$(0, 0)$	$(1, 2)$

Table 1.2 Coordination Game. In this game, there are two Nash equilibria. Neither player has an incentive to unilaterally change strategy in either equilibrium. Rows are your strategies and columns your friend's.

Consider a different game in Table 1.2. This is a typical game model for **coordination**, a task we will see many times in future chapters. You and your friend are trying to coordinate which movie to watch together. If you disagree, you will not go to any movie and the payoff is zero for both of you. If you agree, each will get some positive payoff but the values are different, as you prefer the action movie and your friend prefers the romance movie. (By the way, we will try to understand how to predict a person's preference for different types of movies in Chapter 4.)

In this game, there is no dominant strategy, but it so happens that there are pairs of best response strategies that match each other. If my best response to my friend picking strategy a is strategy b, and my friend's best response to my picking strategy b is strategy a, then the (a, b) pair "matches."

In the coordination game, (`Action`, `Action`) is such a pair, and (`Romance`, `Romance`) is another pair. For both pairs, neither player has an incentive to *unilaterally* move away from its choice in this pair of strategies. If both move at the same time, they could both benefit, but neither wants to do that *alone*. This creates an equilibrium in strategic thinking: I will not move unless you move, and you think the same way too. This is called a **Nash equilibrium**. In the prisoner's dilemma, (`Confess`, `Confess`) is a Nash equilibrium. In the coordination game, both (`Action`, `Action`) and (`Romance`, `Romance`) are Nash equilibria.

Symbolically, for a two-user game, suppose the two payoff functions are (U_1, U_2) and the two strategy spaces are $(\mathcal{A}, \mathcal{B})$ for the two players, respectively. We say $(a^* \in \mathcal{A}, b^* \in \mathcal{B})$ is a Nash equilibrium if

$$U_1(a^*, b^*) \geq U_1(a, b^*), \text{ for any } a \in \mathcal{A},$$

and

$$U_2(a^*, b^*) \geq U_2(a^*, b), \text{ for any } b \in \mathcal{B}.$$

A Nash equilibrium might not exist in a game. And when it exists, it need not be unique (like in the coordination game above), or socially optimal, or Pareto optimal. But if the players are allowed to throw a coin and decide *probabilistically* which strategy to play, i.e., a **mixed strategy**, it is guaranteed, by Nash's famous result, that a Nash equilibrium always exists.

We will expand and use our game-theoretic language in several future chapters. In our current case of power control as a game, the set of players is the set of transmitters. The strategy for each player is its transmit power level, and the strategy space is the set of transmit powers so that the target SIR is achieved. The cost function to minimize is the power level itself.

In this power control game, while the cost function is independent across the players, each player's strategy space A_i depends on the transmit powers of all the other players. This coupling across the players is introduced by the very nature of interference, and leads to strategic moves by the players.

Here is a simple fact, which is important to realize and easy to verify: iterating by the best response strategy of each player in the power control game is given precisely by (1.2). Given that all the other transmitters transmit at a certain power level, my best response strategy is to pick the power level that is my current power level times the ratio between my target and the current SIRs.

Now, consider player 1. If the transmit powers of all the other players become smaller, player 1's strategy space A_1 will be larger. (This is just another way to say there is negative externality.) There are more transmit powers to pick while still being able to maintain the target SIR, since other players' transmit powers are smaller and the denominator in the SIR is smaller. Given this sense of monotonicity, we suspect that maybe best responses converge. Indeed it can be shown that, for games with this property (and under some technicalities), the sequence of best response strategies converge.

This game-theoretic approach is one of the ways to prove the convergence of DPC. Another way is through the angle of global optimization and with the help of linear algebra, as we will present in the Advanced Material. But first, a small and detailed example.

1.3 Examples

A word about all the examples in this book. Most of them are completely numerical and presented in great detail, so as to alleviate any "symbol-phobia" a reader may have. This means that we have to strike a tradeoff between a realistic size for illustration and a small size to fit in a few pages. In some cases, these examples do not represent the actual scale of the problems, while scaling-up can actually be a core difficulty.

Suppose we have four (transmitter, receiver) pairs. Let the channel gains $\{G_{ij}\}$ be given in Table 1.3. As the table suggests, we can also represent these gains in a matrix. You can see that, in general, $G_{ij} \neq G_{ji}$ because the interference channels do not have to be symmetric.

| Receiver | Transmitter of Link | | | |
of Link	1	2	3	4
1	1	0.1	0.2	0.3
2	0.2	1	0.1	0.1
3	0.2	0.1	1	0.1
4	0.1	0.1	0.1	1

Table 1.3 Channel gains in an example of DPC. The entries are for illustrating the algorithm. They do not represent actual numerical values typically observed in real cellular networks.

Suppose that the initial power level is 1.0 mW on each link, and that the noise on each link is 0.1 mW. Then, the initial signal-to-interference ratios are given by

$$\text{SIR}_1[0] = \frac{1 \times 1.0}{0.1 \times 1.0 + 0.2 \times 1.0 + 0.3 \times 1.0 + 0.1} = 1.43,$$

$$\text{SIR}_2[0] = \frac{1 \times 1.0}{0.2 \times 1.0 + 0.1 \times 1.0 + 0.1 \times 1.0 + 0.1} = 2.00,$$

$$\text{SIR}_3[0] = \frac{1 \times 1.0}{0.2 \times 1.0 + 0.1 \times 1.0 + 0.1 \times 1.0 + 0.1} = 2.00,$$

$$\text{SIR}_4[0] = \frac{1 \times 1.0}{0.1 \times 1.0 + 0.1 \times 1.0 + 0.1 \times 1.0 + 0.1} = 2.50,$$

where we use the formula

$$\text{SIR}_i = \frac{G_{ii} p_i}{\sum_{j \neq i} G_{ij} p_j + n_i},$$

with p_i representing the power level of link i and n_i the noise on link i. These SIR values are shown on a linear scale rather than the log (dB) scale in this example.

We will use DPC to adjust the power levels. Suppose that the target SIRs are

$$\gamma_1 = 2.0,$$
$$\gamma_2 = 2.5,$$
$$\gamma_3 = 1.5,$$
$$\gamma_4 = 2.0.$$

Then the new power levels are, in mW,

$$p_1[1] = \frac{\gamma_1}{\text{SIR}_1[0]} p_1[0] = \frac{2.0}{1.43} \times 1.0 = 1.40,$$

$$p_2[1] = \frac{\gamma_2}{\text{SIR}_2[0]} p_2[0] = \frac{2.5}{2.00} \times 1.0 = 1.25,$$

$$p_3[1] = \frac{\gamma_3}{\text{SIR}_3[0]} p_3[0] = \frac{1.5}{2.00} \times 1.0 = 0.75,$$

$$p_4[1] = \frac{\gamma_4}{\text{SIR}_4[0]} p_4[0] = \frac{2.0}{2.5} \times 1.0 = 0.80.$$

Now each receiver calculates the new SIR and feeds it back to its transmitter:

$$\text{SIR}_1[1] = \frac{1 \times 1.40}{0.1 \times 1.25 + 0.2 \times 0.75 + 0.3 \times 0.8 + 0.1} = 2.28,$$

$$\text{SIR}_2[1] = \frac{1 \times 1.25}{0.2 \times 1.40 + 0.1 \times 0.75 + 0.1 \times 0.8 + 0.1} = 2.34,$$

$$\text{SIR}_3[1] = \frac{1 \times 0.75}{0.2 \times 1.40 + 0.1 \times 1.25 + 0.1 \times 0.8 + 0.1} = 1.28,$$

$$\text{SIR}_4[1] = \frac{1 \times 0.80}{0.1 \times 1.40 + 0.1 \times 1.25 + 0.1 \times 0.75 + 0.1} = 1.82.$$

The new power levels in the next timeslot become, in mW,

$$p_1[2] = \frac{\gamma_1}{\text{SIR}_1[1]} p_1[1] = \frac{2.0}{2.28} \times 1.40 = 1.23,$$

$$p_2[2] = \frac{\gamma_2}{\text{SIR}_2[1]} p_2[1] = \frac{2.5}{2.33} \times 1.25 = 1.34,$$

$$p_3[2] = \frac{\gamma_3}{\text{SIR}_3[1]} p_3[1] = \frac{1.5}{1.28} \times 0.75 = 0.88,$$

$$p_4[2] = \frac{\gamma_4}{\text{SIR}_4[1]} p_4[1] = \frac{2.0}{1.82} \times 0.80 = 0.88,$$

with the corresponding SIRs as follows:

$$\text{SIR}_1[2] = \frac{1 \times 1.23}{0.1 \times 1.34 + 0.2 \times 0.88 + 0.3 \times 0.88 + 0.1} = 1.83,$$

$$\text{SIR}_2[2] = \frac{1 \times 1.34}{0.2 \times 1.23 + 0.1 \times 0.88 + 0.1 \times 0.88 + 0.1} = 2.56,$$

$$\text{SIR}_3[2] = \frac{1 \times 0.88}{0.2 \times 1.23 + 0.1 \times 1.34 + 0.1 \times 0.88 + 0.1} = 1.55,$$

$$\text{SIR}_4[2] = \frac{1 \times 0.88}{0.1 \times 1.23 + 0.1 \times 1.34 + 0.1 \times 0.88 + 0.1} = 1.98.$$

Calculating the new power levels again, we have, in mW,

$$p_1[3] = \frac{\gamma_1}{\text{SIR}_1[2]} p_1[2] = \frac{2.0}{1.83} \times 1.23 = 1.35,$$

$$p_2[3] = \frac{\gamma_2}{\text{SIR}_2[2]} p_2[2] = \frac{2.5}{2.56} \times 1.34 = 1.30,$$

$$p_3[3] = \frac{\gamma_3}{\text{SIR}_3[2]} p_3[2] = \frac{1.5}{1.55} \times 0.88 = 0.85,$$

$$p_4[3] = \frac{\gamma_4}{\text{SIR}_4[2]} p_4[2] = \frac{2.0}{1.98} \times 0.88 = 0.89.$$

Then the new SIRs are

$$\text{SIR}_1[3] = \frac{1 \times 1.35}{0.1 \times 1.30 + 0.2 \times 0.85 + 0.3 \times 0.89 + 0.1} = 2.02,$$

$$\text{SIR}_2[3] = \frac{1 \times 1.30}{0.2 \times 1.35 + 0.1 \times 0.85 + 0.1 \times 0.89 + 0.1} = 2.40,$$

$$\text{SIR}_3[3] = \frac{1 \times 0.85}{0.2 \times 1.35 + 0.1 \times 1.30 + 0.1 \times 0.89 + 0.1} = 1.45,$$

$$\text{SIR}_4[3] = \frac{1 \times 0.89}{0.1 \times 1.35 + 0.1 \times 1.30 + 0.1 \times 0.85 + 0.1} = 1.97,$$

and the new power levels, in mW, are

$$p_1[4] = \frac{\gamma_1}{\text{SIR}_1[3]}p_1[3] = \frac{2.0}{2.02} \times 1.35 = 1.33,$$

$$p_2[4] = \frac{\gamma_2}{\text{SIR}_2[3]}p_2[3] = \frac{2.5}{2.40} \times 1.30 = 1.36,$$

$$p_3[4] = \frac{\gamma_3}{\text{SIR}_3[3]}p_3[3] = \frac{1.5}{1.45} \times 0.85 = 0.88,$$

$$p_4[4] = \frac{\gamma_4}{\text{SIR}_4[3]}p_4[3] = \frac{2.0}{1.97} \times 0.89 = 0.90.$$

We see that the power levels are beginning to converge: p_1, p_2, p_3 and p_4 all change by less than 0.1 mW now. The new SIRs are

$$\text{SIR}_1[4] = \frac{1 \times 1.33}{0.1 \times 1.36 + 0.2 \times 0.88 + 0.3 \times 0.90 + 0.1} = 1.96,$$

$$\text{SIR}_2[4] = \frac{1 \times 1.36}{0.2 \times 1.33 + 0.1 \times 0.88 + 0.1 \times 0.90 + 0.1} = 2.49,$$

$$\text{SIR}_3[4] = \frac{1 \times 0.88}{0.2 \times 1.33 + 0.1 \times 1.36 + 0.1 \times 0.90 + 0.1} = 1.49,$$

$$\text{SIR}_4[4] = \frac{1 \times 0.90}{0.1 \times 1.33 + 0.1 \times 1.36 + 0.1 \times 0.88 + 0.1} = 1.97.$$

Iterating one more time, the new power levels, in mW, are

$$p_1[5] = \frac{\gamma_1}{\text{SIR}_1[4]}p_1[4] = \frac{2.0}{1.96} \times 1.33 = 1.37,$$

$$p_2[5] = \frac{\gamma_2}{\text{SIR}_2[4]}p_2[4] = \frac{2.5}{2.49} \times 1.36 = 1.36,$$

$$p_3[5] = \frac{\gamma_3}{\text{SIR}_3[4]}p_3[4] = \frac{1.5}{1.49} \times 0.88 = 0.89,$$

$$p_4[5] = \frac{\gamma_4}{\text{SIR}_4[4]}p_4[4] = \frac{2.0}{1.97} \times 0.90 = 0.92,$$

with corresponding SIRs:

$$\text{SIR}_1[5] = \frac{1 \times 1.37}{0.1 \times 1.36 + 0.2 \times 0.89 + 0.3 \times 0.92 + 0.1} = 1.98,$$

$$\text{SIR}_2[5] = \frac{1 \times 1.36}{0.2 \times 1.37 + 0.1 \times 0.89 + 0.1 \times 0.92 + 0.1} = 2.45,$$

$$\text{SIR}_3[5] = \frac{1 \times 0.89}{0.2 \times 1.37 + 0.1 \times 1.36 + 0.1 \times 0.92 + 0.1} = 1.48,$$

$$\text{SIR}_4[5] = \frac{1 \times 0.92}{0.1 \times 1.37 + 0.1 \times 1.36 + 0.1 \times 0.89 + 0.1} = 1.98.$$

All the SIRs are now within 0.05 of the target. The power levels keep iterating, taking the SIRs closer to the target. Figure 1.5 shows the graph of power level versus the number of iterations. After about 20 iterations, the change is too small to be seen on the graph; the power levels at that time are

$$p_1 = 1.46 \text{ mW},$$
$$p_2 = 1.46 \text{ mW},$$
$$p_3 = 0.95 \text{ mW},$$
$$p_4 = 0.97 \text{ mW}.$$

The resulting SIRs are shown in Figure 1.6. We get very close to the target SIRs, by visual inspection, after about 10 iterations.

While we are at this example, let us also walk through a compact, matrix representation of the target SIR constraints. This will be useful in the next section.

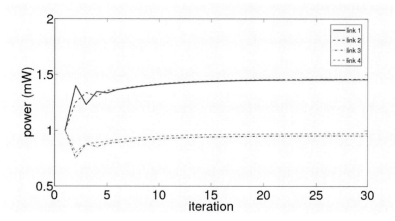

Figure 1.5 The convergence of power levels in an example of DPC.

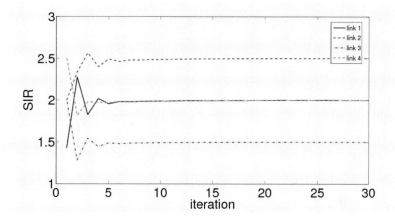

Figure 1.6 The convergence of SIRs in an example of DPC.

If the target SIR γ_i is achieved or exceeded by $\{p_i\}$, we have

$$\frac{G_{ii}p_i}{\sum_{j \neq i} G_{ij}p_j + n_i} \geq \gamma_i,$$

for $i = 1, 2, 3, 4$. On multiplying both sides by $\sum_{j \neq i} G_{ij}p_j + n_i$ and dividing by G_{ii}, we have

$$p_i \geq \frac{\gamma_i}{G_{ii}} \left(\sum_{j \neq i} G_{ij}p_j + n_i \right),$$

which can be written as

$$p_i \geq \gamma_i \sum_{j \neq i} \frac{G_{ij}}{G_{ii}} p_j + \frac{\gamma_i}{G_{ii}} n_i. \tag{1.4}$$

Now we define the variable vector

$$\mathbf{p} = \begin{bmatrix} p_1 \\ p_2 \\ p_3 \\ p_4 \end{bmatrix},$$

and a constant vector

$$\mathbf{v} = \begin{bmatrix} \frac{\gamma_1 n_1}{G_{11}} \\ \frac{\gamma_2 n_2}{G_{22}} \\ \frac{\gamma_3 n_3}{G_{33}} \\ \frac{\gamma_4 n_4}{G_{44}} \end{bmatrix} = \begin{bmatrix} \frac{2.0 \times 0.1}{1.0} \\ \frac{2.5 \times 0.1}{1.0} \\ \frac{1.5 \times 0.1}{1.0} \\ \frac{2.0 \times 0.1}{1.0} \end{bmatrix} = \begin{bmatrix} 0.20 \\ 0.25 \\ 0.15 \\ 0.20 \end{bmatrix}.$$

Define also a 4×4 diagonal matrix $\mathbf{D}$ with γ_i on the diagonal, and another 4×4 matrix $\mathbf{F}$ where $F_{ij} = G_{ij}/G_{ii}$ for $i \neq j$, and the diagonal entries of F are zero. Plugging in the numbers, we have

$$\mathbf{D} = \begin{bmatrix} 2.0 & 0 & 0 & 0 \\ 0 & 2.5 & 0 & 0 \\ 0 & 0 & 1.5 & 0 \\ 0 & 0 & 0 & 2.0 \end{bmatrix},$$

$$\mathbf{F} = \begin{bmatrix} 0 & 0.1 & 0.2 & 0.3 \\ 0.2 & 0 & 0.1 & 0.1 \\ 0.2 & 0.1 & 0 & 0.1 \\ 0.1 & 0.1 & 0.1 & 0 \end{bmatrix}.$$

We can now rewrite (1.4) as

$$\mathbf{p} \geq \mathbf{DFp} + \mathbf{v} = \begin{bmatrix} 0 & 0.20 & 0.40 & 0.60 \\ 0.50 & 0 & 0.25 & 0.25 \\ 0.30 & 0.15 & 0 & 0.15 \\ 0.20 & 0.20 & 0.20 & 0 \end{bmatrix} \mathbf{p} + \begin{bmatrix} 0.20 \\ 0.25 \\ 0.15 \\ 0.20 \end{bmatrix},$$

where $\geq$ between two vectors (of equal length) simply represents component-wise inequality between the corresponding entries of the two vectors.

We can check that the power levels in the last iteration shown above satisfy this inequality tightly:

$$\begin{bmatrix} 1.46 \\ 1.46 \\ 0.95 \\ 0.97 \end{bmatrix} \geq \begin{bmatrix} 0 & 0.20 & 0.40 & 0.60 \\ 0.50 & 0 & 0.25 & 0.25 \\ 0.30 & 0.15 & 0 & 0.15 \\ 0.20 & 0.20 & 0.20 & 0 \end{bmatrix} \begin{bmatrix} 1.46 \\ 1.46 \\ 0.95 \\ 0.97 \end{bmatrix} + \begin{bmatrix} 0.20 \\ 0.25 \\ 0.15 \\ 0.20 \end{bmatrix} = \begin{bmatrix} 1.46 \\ 1.46 \\ 0.95 \\ 0.97 \end{bmatrix}.$$

We will use this matrix representation as we go from problem representation (1.3) to problem representation (1.5) in the next section. We will also see in later chapters many different matrices that summarize the *topology* of a network, and operations involving these matrices that model *functionalities* running on the network.

1.4   Advanced Material

1.4.1   Iterative power method

We can generalize the vector notation in the last example. Let $\mathbf{1}$ represent a vector of 1s, so the objective function is simply $\mathbf{1}^T\mathbf{p} = \sum_i p_i$. Let $\mathbf{I}$ be the identity matrix, $\mathbf{D}$ a diagonal matrix with the diagonal entries being the target SIR values $\{\gamma_i\}$, and $\mathbf{F}$ be a matrix capturing the given channel conditions: $F_{ij} = G_{ij}/G_{ii}$ if $i \neq j$, and $F_{ii} = 0$. The constant vector $\mathbf{v}$ captures the normalized noise levels, with $v_i = \gamma_i n_i/G_{ii}$. We will soon see why this shorthand notation is useful.

Equipped with the notation above, we can represent the target SIR constraints in problem (1.3) as

$$\mathbf{p} \geq \mathbf{DFp} + \mathbf{v},$$

and further group all the terms involving the variables $\mathbf{p}$. The linear programming problem we have now becomes

$$\begin{array}{ll} \text{minimize} & \mathbf{1}^T\mathbf{p} \\ \text{subject to} & (\mathbf{I} - \mathbf{DF})\mathbf{p} \geq \mathbf{v} \\ \text{variables} & \mathbf{p}. \end{array} \qquad (1.5)$$

You should verify that problem (1.3) and problem (1.5) are indeed equivalent, by using the definitions of the SIR, of matrices $(\mathbf{D}, \mathbf{F})$, and of vector $\mathbf{v}$.

Linear programming problems are conceptually and computationally easy to solve in general. Our special case here has even more structure. This $\mathbf{DF}$ matrix is a **non-negative matrix**, since all entries of the matrix are non-negative numbers. Non-negative matrices are a powerful modeling tool in linear algebra, with applications from economics to ecology. They have been well-studied in matrix analysis through the **Perron–Frobenius theory**.

If the largest eigenvalue of the $\mathbf{DF}$ matrix, denoted as $\rho(\mathbf{DF})$, is less than 1, then the following three statements are true.

- (a) We can guarantee that the set of target SIRs can indeed be achieved simultaneously, which makes sense since $\rho(\mathbf{DF}) < 1$ means that the $\{\gamma_i\}$ in $\mathbf{D}$ are not "too big," relative to the given channel conditions captured in $\mathbf{F}$.

- (b) We can invert the matrix defining the linear constraints in our optimization problem (1.5): solve the problem by computing $(\mathbf{I}-\mathbf{DF})^{-1}\mathbf{v}$. But, of course, there is no easy way to directly run this matrix inversion distributively across the MSs.

- (c) The inversion of the matrix can be expressed as a sum of terms, each term a multiplication of matrix $\mathbf{DF}$ by itself. More precisely: $(\mathbf{I} - \mathbf{DF})^{-1} = \sum_{k=0}^{\infty}(\mathbf{DF})^k$. This is an infinite sum, so we say that the partial sum of K

terms, $\sum_{k=0}^{K}(\mathbf{DF})^k$, will converge as K becomes very large. Furthermore, the tail term in this sum, $(\mathbf{DF})^k$, approaches 0 as k becomes large:

$$\lim_{k \to \infty} (\mathbf{DF})^k = 0.$$

The key insight is that we want to invert a matrix $(\mathbf{I} - \mathbf{DF})$, because that will lead us to a power-minimal solution to achieving all the target SIRs:

$$\mathbf{p}^* = (\mathbf{I} - \mathbf{DF})^{-1} \mathbf{v}$$

is a solution of problem (1.3), i.e., for any solution $\hat{\mathbf{p}}$ satisfying the constraints in problem (1.3), $\mathbf{p}^*$ is better:

$$\mathbf{p}^* \le \hat{\mathbf{p}}.$$

Now, the matrix-inversion step is not readily implementable in a distributed fashion, as we need in cellular network power control. Fortunately, it can be achieved by applying the following update.

(1) First, $\mathbf{p}^* = (\mathbf{I} - \mathbf{DF})^{-1} \mathbf{v}$ can be represented as a power series, as stated in Statement (c) above:

$$\mathbf{p}^* = \sum_{k=0}^{\infty} (\mathbf{DF})^k \mathbf{v}. \tag{1.6}$$

(2) Then, you can readily check that the following iteration over time gives exactly the above power series (1.6), as time t goes on:

$$\mathbf{p}[t+1] = \mathbf{DF}\mathbf{p}[t] + \mathbf{v}. \tag{1.7}$$

One way to check this is to substitute the above recursive formula of $\mathbf{p}$ (1.7) all the way to $\mathbf{p}[0]$, the initialization of the iteration. Then, at any time t, we have

$$\mathbf{p}[t] = (\mathbf{DF})^t \mathbf{p}[0] + \sum_{k=0}^{t-1} (\mathbf{DF})^k \mathbf{v},$$

which converges, as $t \to \infty$, to

$$\lim_{t \to \infty} \mathbf{p}[t] = 0 + \sum_{k=0}^{\infty} (\mathbf{DF})^k \mathbf{v} = \mathbf{p}^*,$$

since $(\mathbf{DF})^t \mathbf{p}[0]$ approaches 0 as $t \to \infty$. So, we know (1.7) is right. This also shows that it does not matter what the initialization vector $\mathbf{p}[0]$ is. Its effect will be washed away as time goes on.

(3) Finally, rewrite the vector form of update equation (1.7) in scalar form for each transmitter i, and you will see that it is exactly the DPC algorithm (1.2):

$$p_i[t+1] = \frac{\gamma_i}{\mathrm{SIR}_i[t]} p_i[t], \quad \text{for each } i.$$

We just completed a development and convergence analysis of the DPC as the solution of a global optimization problem through the language of linear algebra. In general, for any square matrix $\mathbf{A}$, the following statements are equivalent.

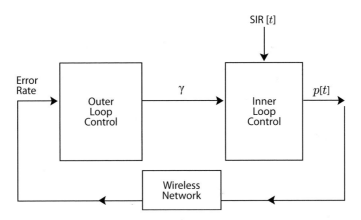

Figure 1.7 Inner and outer loops of power control in cellular networks. The inner loop takes in a fixed target SIR, compares it with the current SIR, and updates the transmit power. The outer loop adjusts the target SIR on the basis of the performance measured over a longer timescale. We have focused on the inner-loop power control in this chapter.

1. The largest eigenvalue is less than 1.
2. The limit of this matrix multiplying itself is 0: $\lim_{k \to \infty} \mathbf{A}^k = 0$.
3. The infinite sum of powers $\sum_{k=0}^{\infty} \mathbf{A}^k$ exists and equals $(\mathbf{I} - \mathbf{A})^{-1}$.

What we saw in DPC is an iterative **power method** (the word "power" here has nothing to do with transmit powers, but instead refers to raising a matrix to some power, i.e., multiplying a matrix by itself many times). It is a common method used to develop iterative algorithms arising from linear systems models. We formulate the problem as a linear program, then we define the solution to the problem as the solution to a linear equation, implement the matrix inversion through a sequence of matrix powers, turn each of those steps into an easy computation at each timeslot, and, finally, achieve the matrix inversion through an iteration in time. This will be used again when we talk about Google's PageRank algorithm in Chapter 3.

1.4.2 Outer loop power control

As shown in the block diagram in Figure 1.7, in cellular networks there are two timescales of power control. What we have been discussing so far is the **inner-loop power control**. Nested outside of that is the **outer-loop power control**, where the target SIRs $\{\gamma_i\}$ are determined.

One standard way to determine target SIRs is to measure the received signal quality, in terms of decoding error probabilities, at the receiver. If the error rate is too high, the target SIR needs to be increased. And if the error rate is lower than necessary, the target SIR can be reduced.

Alternatively, we can also consider optimizing target SIRs as optimization *variables*. This is particularly useful for 3G and 4G networks where data traffic dominates voice traffic in cellular networks. A higher SIR can provide a higher rate at the same signal quality. But every user wants to achieve a higher SIR. So we need to model this either as a network-wide optimization, maximizing the sum of all users' payoff functions, or as a game, with each user maximizing its own payoff function of SIR.

Summary

Box 1 Distributed power control

Different users' signals interfere with each other in the air, leading to a feasible SIR region with a Pareto-optimal boundary. Interference coordination in CDMA networks can be achieved through distributed power control with implicit feedback. It solves an optimization problem for the network in the form of linear programming, and can also be modeled as a non-cooperative game.

Further Reading

This chapter introduces several foundational methodologies: optimization, games, and algorithms, as well as the basics of cellular wireless networks. As a result, there are many interesting texts to read.

1. An early paper quantifying the benefits of CDMA was written by a Qualcomm team, including Andrew Viterbi:
I. M. Jacobs, R. Padovani, A. J. Viterbi, L. A. Weaver, C. E. Wheatley, "On the capacity of a cellular CDMA system," *IEEE Transactions on Vehicular Technology*, vol. 40, no. 2, pp. 303–312, May 1991.

2. The DPC algorithm in the core of this chapter appeared in the following seminal paper:
G. J. Foschini and Z. Miljanic, "A simple distributed autonomous power control algorithm and its convergence," *IEEE Transactions on Vehicular Technology*, vol. 42, no. 3, pp. 641–646, November 1993.

3. Much more discussion on power control algorithms in cellular networks can be found in the following monograph:
M. Chiang, P. Hande, T. Lan, and C. W. Tan, "Power control for cellular wireless networks," *Foundation and Trends in Networking*, vol. 2, no. 4, pp. 381-533, July 2008.

4. A standard reference on linear algebra is the following mathematics textbook, which includes a chapter on non-negative matrices and Perron–Frobenius theory:

R. A. Horn and C. R. Johnson, *Matrix Analysis*, Cambridge University Press, 1990.

5. There are many textbooks on game theory, in all types of styles. A comprehensive introduction is

R. B. Myerson, *Game Theory: Analysis of Conflict*, Harvard University Press, 1997.

Problems

1.1 *Distributed power control* ⋆

(a) Consider three pairs of transmitters and receivers in a cell, with the following channel gain matrix **G** and noise of 0.1 mW for all the receivers. The target SIRs are also shown below.

$$
\mathbf{G} = \begin{bmatrix} 1 & 0.1 & 0.3 \\ 0.2 & 1 & 0.3 \\ 0.2 & 0.2 & 1 \end{bmatrix}, \qquad \gamma = \begin{bmatrix} 1 \\ 1.5 \\ 1 \end{bmatrix} .
$$

With an initialization of all transmit powers at 1 mW, run DPC for ten iterations and plot the evolution of transmit powers and received SIRs. You can use any programming language, or even write the steps out by hand.

(b) Now suppose the power levels for logical links 1, 2, and 3 have converged to the equilibrium in (a). A new pair of transmitter and receiver, labeled as logical link 4, shows up in the same cell, with an initial transmit power of 1 mW and demands a target SIR of 1. The new channel gain matrix is shown below.

$$
\mathbf{G} = \begin{bmatrix} 1 & 0.1 & 0.3 & 0.1 \\ 0.2 & 1 & 0.3 & 0.1 \\ 0.2 & 0.2 & 1 & 0.1 \\ 0.1 & 0.1 & 0.1 & 1 \end{bmatrix} .
$$

Similarly to what you did in (a), show what happens in the next ten timeslots. What happens at the new equilibrium?

1.2 *Power control infeasibility* ⋆⋆

Consider a three-link cell with the link gains G_{ij} shown below. The receivers request $\gamma_1 = 1, \gamma_2 = 2$, and $\gamma_3 = 1$. The noise $n_i = 0.1$ for all i.

$$G = \begin{bmatrix} 1 & 0.5 & 0.5 \\ 0.5 & 1 & 0.5 \\ 0.5 & 0.5 & 1 \end{bmatrix}.$$

Prove this set of target SIRs is infeasible.

1.3 *A zero-sum game* ⋆

In the following two-user game, the payoffs of users Alice and Bob are exactly negative of each other in all the combinations of strategies (a,a), (a,b), (b,a), (b,b). This models an extreme case of competition, and is called a **zero-sum game**. Is there any pure strategy equilibrium? How many are there?

	a	b
a	$(2, -2)$	$(3, -3)$
b	$(3, -3)$	$(4, -4)$

1.4 *Mechanism design* ⋆⋆

Consider the game below. There are two players, Alice and Bob, each with two strategies, and the payoffs are shown below. Consider only pure strategy equilibria.

	a	b
a	$(0, 2)$	$(2, 0)$
b	$(6, 0)$	$(3, 2)$

(a) Is there a Nash equilibrium, and, if so, what is it?

(b) We want to make this game a "better" one. What entries in the table would you change to make the resulting Nash equilibrium unique and socially optimal?

This is an example of **mechanism design**: change the game so as to induce movement of players to a desirable equilibrium. We will see a lot more of mechanism design in future chapters.

1.5 *Repeating prisoner's dilemma* ⋆ ⋆ ⋆

(a) Suppose the two prisoners know that they will somehow be caught in the same situation five more times in future years. What will each prisoner's strategy be in choosing between confession and no confession?

(b) Suppose the two prisoners have infinite lifetimes, and there is always a 90% chance that they will be caught in the same situation after each round of this game. What will each prisoner's strategy be now?

2   How does Google sell ad spaces?

2.1   A Short Answer

Much of the web services and online information is "free" today because of the advertisements shown on the websites. It is estimated that the online ad industry worldwide reached $94.2 billion in 2012. Compared with traditional media, online advertisements' revenue ranked right below TV and above newspapers.

In the early days of the web, i.e., 1994-1995, online advertisements were sold as banners on a per-thousand-impression basis. But seeing an ad does not mean clicking on it or buying the advertised product or service afterwards. In 1997, GoTo (which later became Overture) started selling advertisement spaces on a per-click basis. This middle ground between ad revenue (what the website cares about) and effectiveness of ad (what the advertisers care about) became a commonly accepted foundation for online advertising.

With the rise of Google came one of the most stable online ad market segments: **search ads**, also called *sponsored search*. In 2002, Google started the AdWords service where you can create your ad, attach keywords to it, and send it to Google's database. When someone searches for a keyword, Google will return a list of search results, as well as a list of ads on the right panel, or even the main panel, if that keyword matches any of the keywords of ads in its database. This process takes place continuously and each advertiser can adjust her bids frequently. There are often many ad auctions happening at the same time too. We will skip these important factors in the basic models in this chapter, focusing just on a single auction.

Now we face three key questions. First, where will your ad appear on the list? We all know that the order of appearance makes a big difference. You will have to pay more to have your ad placed higher in the list. For example, when I did a search for "Banff National Park" on Google in September 2011, I saw an ad for `www.banfflakelouise.com`, a vacation-planning company. This ad was right on top of the main panel, above all the search results on websites and images of Banff National Park. (By the way, how those "real" search results are ordered is the subject of the next chapter.) You also see a list of ads on the right panel, starting with the top one for `www.rockymountaineer.com`, a tourist train company. These two companies probably get most of the clicks, and pay more than the other advertisers for each click. The rest of this chapter delves into the auction methods that allocate these ad spaces according to how much each advertiser is willing to pay.

Second, when will these advertisers pay Google? Only when someone clicks on the link and visits their websites (like I just did, thus contributing to Google's revenue). The average number of times that a viewer of the search result page clicks an ad link, over say one hour, is called the **clickthrough rate**. In a general webpage layout, it may be difficult to rank ad spaces by their positions along a line, but we can always rank them by their clickthrough rates. Let us say the payment by advertisers to Google is *proportional* to the clickthrough rates.

Third, what is in it for the advertisers then? Their revenue derived from placing this particular ad is the product of two things: C, the number of clicks (per unit time, say, one hour), and R, the average revenue (in dollars) generated from each click.

- Let us assume that the number of clicks per hour actually observed is indeed the estimated clickthrough rate, both denoted as C. This is of course not true in general, but it is a reasonable assumption to make for our purpose. We also assume that C is independent of the content in the actual advertisement placed, again a shaky assumption to make the model more tractable.
- As for the average revenue R generated from each click (averaged over all clicks), that highly depends on the nature of the goods or services being advertised and sold. R for each ad-space buyer is assumed to be independent of which ad space she ends up buying, a more reasonable assumption than the one on the independence of C of the advertisement content.

This product, $C \times R$, is the buyer's expected revenue from a particular ad space. It is the **valuation** of the ad space to the buyer. For example, if C is 20 clicks per hour for an ad space and R is \$8 generated per click for an ad-space buyer, the valuation of that space to this buyer is \$160 (per hour). For multiple ad spaces, the valuation of each buyer is a *vector*, with one entry per ad space.

In this discussion, there is one seller, which is Google, many buyers/bidders (the advertisers), and many "goods" (the ad spaces). Each bidder can bid for the ad spaces, and Google will then allocate the ad spaces among the bidders according to some rule, and charge the bidders accordingly.

This process is an **auction**. In general, you can have S sellers, N bidders, and K items in an auction. We will consider only the case with $S = 1$. An ad space auction is a special case of general auctions. Auctions can be analyzed as games, i.e., with a set of players, a strategy set per player, and a payoff function per player.

Depending on the rules of an auction, each bidder may choose to bid in different ways, maybe bidding her true valuation of the ad spaces. It would be nice to design the rules so that such a **truthful bidding** behavior is encouraged. But Google has other considerations too, such as maximizing its total revenue: the sum of the revenue from each bidder, which is in turn the product of the number

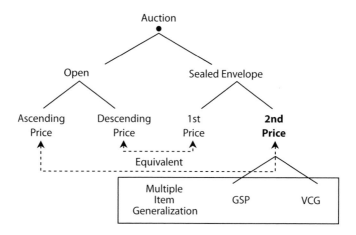

Figure 2.1 A taxonomy of major types of auction in this chapter. The second price, sealed envelope auction is equivalent to (a simpler version of) ascending price open auction, and can be generalized in two different ways to multiple-item auctions: (1) a simple extension to the Generalized Second Price (GSP) auction, and (2) a more sophisticated extension to the Vickrey–Clarke–Groves (VCG) auction that preserves the truthful bidding property.

of clicks (actually observed in real time) and the per-click charge (determined during the auction).

Before we analyze a K-item auction, we will first study auctions with only $K = 1$ item. There are two main types of such one-item auctions: ascending-price and descending-price. These intuitive types of auction, in use since the Roman Empire, require a public venue for announcing the bids.

- In an **ascending price auction**, an auctioneer announces a base price, and then each bidder can raise her hand to bid a higher price. This price war keeps escalating until one bidder submits a price, no other bidder raises a hand, and the auctioneer calls out "gone." The last bidder is the winning bidder, and she pays the price she bid in the last round.

- In a **descending price auction**, an auctioneer announces a high price first, so high that no bidder is willing to accept it. The auctioneer then starts to lower the price, until there is one bidder who shouts out "OK." That bidder is allocated the item, and pays the price announced when she said "OK."

The alternative to a public venue is private disclosure of bids, called **sealed envelope** auctions. This is much more practical in many settings, including selling ad spaces by Google and auctioning goods on eBay. There are two types of such auctions, but it turns out that their results are essentially equivalent to the two types of open auctions we just discussed.

Each bid b_i is submitted by bidder i in a sealed envelope. All bids are then revealed simultaneously to the auctioneer, who will then decide

- the allocation, and
- how much to charge for each item.

The allocation part is easy: the highest bidder gets the item; but the amount charged can vary.

- In a **first price auction**, the winner pays the highest bid, i.e., her own bid.
- In a **second price auction**, the winner pays the second highest bid, i.e., the bid from the next highest bidder.

Second price auction sounds "wrong." If I know I will be paying the next highest bid, why not bid extremely high so that I can win the item, and then pay a much lower price for it? As it turns out, this intuition *itself* is wrong. The assumption of "much lower prices being bid by other bidders" does not hold when everyone engages in the same strategic thinking.

Instead, a second price auction is effectively equivalent to the highly intuitive ascending price auction, and can induce truthful-bidding behavior from the bidders. That is why second price auction is used so often, from auctioning major municipal projects to auctioning wireless spectrum.

Finally, we come back to auctions of K items (still with 1 seller and N bidders). If we follow the basic mechanism of second price auction, we obtain what is called the **Generalized Second Price** (GSP) for ad space auction: the ith ad space goes to the bidder that puts in the ith highest bid, and the charge, per clickthrough rate, is the $(i+1)$th bid. If the webpage layout shows the ads vertically, the advertiser in a given ad space is paying a price that is the same as the bid from the advertiser in the ad space *right below* hers. This simple method is used by Google in selling its ad spaces.

But it turns out GSP is *not* an auction that induces truthful bidding, and there can be many Nash equilibria if we analyze it as a game. An alternative is the **Vickrey–Clarke–Groves** (VCG) auction, which actually extends the second price auction's property of truthful bidding to multiple-item auctions. A VCG auction charges on the basis of negative externality, a principle that we will see many times throughout this book. The relationships between these types of auction are summarized in Figure 2.1.

Throughout the chapter, we focus on the simplest case, where there is a single round of bidding. In reality, there are multiple related bids going on at the same time, e.g., www.banfflakelouise.com may be bidding for multiple related keywords, such as "Banff," "Lake Louise," and "Canadian vacation," simultaneously. In a homework problem, we will go into a little more detail on one aspect of simultaneous auction in the context of spectrum auctioning.

2.2   A Long Answer

2.2.1   When do we need auctions?

No matter which format, an auction runs a resource-allocation process. It allocates items among bidders and sets the prices. We assume each bidder has a valuation of the item, and that the valuation is private and independent. **Private valuation** means that the value is unknown to others (all the other bidders and the auctioneer). **Independent valuation** means that one bidder's valuation does not depend on other bidders' valuations.

You will see that all assumptions are false, which is *why* we call them "assumptions" rather than "facts." But some assumptions are so false that the theory built on them loses predictive power. In this chapter, the private- and independent-valuation assumptions for auction will be used quite fruitfully, but it is still worthwhile to point out that they may be false in reality.

- In many instances, valuations often *depend* on each other, especially when there is a secondary market for you to resell the items. If you are bidding on a foreclosure house, you are probably dealing with dependent valuation.
- Valuations are sometimes *public*. In fact, some eBay bidding strategies attempt to reveal others' valuations, a particularly helpful strategy when you do not know how to value an item and you think other bidders might be experts who know more.
- Valuations in some cases are actually *not precisely known* even to the bidder herself.

There are several criteria used to compare the different outcomes of an auction.

- Seller's revenue, which clearly the seller would like to maximize. It often turns out that the revenue-maximizing strategy is very hard to characterize.
- The sum of payoffs received by all the bidders. This metric will become clearer as we model auctions as games. There is a tradeoff between seller revenue and bidders' payoff.
- Truthful bidding, a property of an auction where each bidder bids her true valuation.

2.2.2   Auctions as games

We can view an auction as a game.

1. The set of players is the set of N bidders, indexed by i.
2. The strategy is the bid b_i of each bidder, and each has a strategy space being the range of bids that she might put forward.
3. Each bidder's payoff function, U_i, is the difference between her valuation v_i of the item and the price p_i she has to pay, if she wins the auction:

$$U_i(\mathbf{b}) = v_i - p_i(\mathbf{b}).$$

On the other hand, the payoff is

$$U_i(\mathbf{b}) = 0$$

if bidder i loses the auction, since she is neither paying anything nor getting the item. Obviously, winning or losing the auction is also an outcome determined by $\mathbf{b}$.

Here, the coupling of the bidder's bidding choices is shown through the dependence of U_i on the entire *vector* $\mathbf{b}$, even though the valuation is independent: v_i depends only on i. Different auction rules lead to different $p_i(\mathbf{b})$; that is the mechanism-design part. By changing the rules of an auction, we can induce different bidding behaviors. For a given mechanism, each bidder will pick her b_i to maximize U_i:

$$b_i^* = \text{argmax}_{b_i} U_i(b_1, b_2, \ldots, b_N).$$

In the case of a first price (sealed envelope) auction, the price p_i which the bidder has to pay is simply her own bid value b_i, since that is the highest bid. So the payoff for bidder i, when winning, is $v_i - b_i$. From this bidder's perspective, she wants to pick the "right" b_i: not so big that the payoff is small (possibly zero) when winning, and not so small that she does not win the auction at all and receives 0 payoff. It is not easy to solve this optimization since it involves the strategies of all other bidders. But one thing is clear: she should bid less than her true valuation, for otherwise the best payoff she can receive is zero.

In the case of a second price (sealed envelope) auction, the price p_i one has to pay (when winning the auction) is the second highest price, i.e., the bid from say bidder j. The payoff is $v_i - b_j$ if bidder i wins, and 0 otherwise. In this game, it turns out that, *no matter* what other bidders might bid, each bidder can maximize her payoff by bidding her true valuation: $b_i = v_i$. Truthful bidding is a dominant strategy for the game. In other words, if the auction rule says that $p_i(\mathbf{b})$ is the second-highest entry in $\mathbf{b}$, then setting $b_i = v_i$ maximizes U_i for each bidder, no matter what the values of the rest of the entries in $\mathbf{b}$ are.

2.2.3   Single-item auction: Second price

There are several ways to view the somewhat counter-intuitive result of bidding behavior in a second price auction. Fundamentally, it is precisely the *decoupling* of the payoff amount (how much you gain if you win) from the auction result (whether you win at all) that induces each bidder to bid her true valuation. This is also what happens in the familiar format of an open auction with ascending price. Your bid determines how long you will stay in the price war. But when it stops, you pay the bid of the next-highest bidder plus a small amount capped by the minimum increment per new bid (unless you overbid much more than the minimum increment), for that is the auctioneer's announcement when the price war stops as a result of the next-highest bidder dropping out.

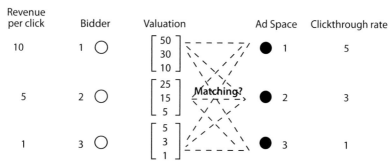

Figure 2.2 An example of three bidders and three ad spaces. Each bidder's valuation is a vector now. Each valuation vector is proportional to the vector of clickthrough rates $[C_1, C_2, C_3]$, where the proportionality constant is the average revenue per click, R, for that bidder of ad space. We assume that the clickthrough rates are independent of the content of the ad placed at each position. If the bidders and ad spaces are listed in descending order of their R and C, and each player bids true valuations, GSP simply matches "horizontally."

We can readily convince ourselves that truthful bidding is a dominant strategy. Say you want to bid your true valuation: $b = v$. Suppose that, as your advisor, I suggest lowering that bid to be something less than v. Call this new bid $\tilde{b}$. You should realize that such an action will change the outcome (auction result or payment amount) only if the next highest bid, say user 2's bid, b_2, is between b and $\tilde{b}$: $b > b_2 > \tilde{b}$. And in this case, you lose the auction, which you could have won and received a positive payoff of $v - p = v - b_2 = b - b_2 > 0$. So you would rather not take my advice to lower your bid.

Suppose instead I suggest raising your bid to be something more than v. Call this new bid $\hat{b}$. Again, such an action will change the outcome only if the highest bid, let us say user 2's bid, b_2, is in between b and $\hat{b}$: $\hat{b} > b_2 > b$. And in this case, you now win the auction, but receive a negative payoff, as you end up paying more than you value the item: $v - p = v - b_2 = b - b_2 < 0$. Had you bid b, you would have lost and received a payoff of 0, a better outcome than a negative payoff. So you would rather not take my advice to raise your bid.

Therefore, no matter what other bidders do, you would rather neither lower nor raise your bid. You would rather simply bid v.

This argument seals the math, but what is the intuition behind the math? For example, while it is clear why a first price auction is not truthful-bid-inducing, what about a third price auction? We will see in the Advanced Material that the intuition is that second price here can capture the negative externality, the "damage," caused by the winner to the other bidders.

2.2.4   Multiple-item auction: Generalized second price (GSP)

So far in this section, we have been examining auctions with only one item to sell. Search ad markets have multiple ad spaces to sell in each auction: multiple items facing multiple bidders. For simplicity, we assume that there is only one auction going on. Let us say there are three ad spaces in this auction, with an average clickthrough rates of 5, 3, and 1, as shown in Figure 2.2. And there are three advertisers (bidders for ad spaces), with different expected revenues per click, say 10, 5, and 1 (all in dollars). Then the valuation of each advertiser is as follows: the first advertiser has valuations of [50, 30, 10] for the three spaces, the second advertiser has valuations of [25, 15, 5], and the third advertiser has valuations of [5, 3, 1]. The job of a multiple-item auction is to assign one item to each advertiser.

If the number of advertisers and the number of ad spaces are different, some advertisers may get no ad space or some ad space will be left unsold. We will consider these cases as simple extensions in a homework problem.

We will later encounter other types of **bipartite graphs**, like the one we saw for the relation between the auctioned items and bidders in Figure 2.2. In a bipartite graph, all the nodes belong to either the left column or the right column, and all the links are only between the two columns but not within each column.

Now, back to multiple-ad-space auctions. A bidder, i.e., an advertiser, i sends in a bid b_i, indicating how much she is willing to pay per click. This is actually a significant simplification in an ad auction. Since there are multiple ad spaces, why not ask each advertiser to submit many bids, one bid per ad space, effectively presenting a scale of their preferences? We will encounter such vector preferences later, but for ad auction, the industry thought that a scalar representation of this vector preference suffices. Each entry in the valuation vector is just a multiple of this scalar.

So, advertiser i sends in a vector of bids $[b_iC_1, b_iC_2, \ldots, b_iC_K]$ for the K ad spaces with clickthrough rates $C_j, j = 1, 2, \ldots, K$. She may *not* pick b_i to be the same as the expected revenue per click.

In GSP, the ith ad space is allocated to the ith highest bidder. As we will see in the Advanced Material, this rule of ad space allocation is the same for VCG. But the amount each bidder is charged is different depending on whether GSP or VCG is used, and the strategic thinking in bidders' minds is consequently different too. In GSP, the charges are easy to determine: the ith ad space winner pays the per click price of the $(i+1)$th highest bidder.

In summary, for GSP the following criteria apply.

- The "allocation part" is easy to see: advertiser i's bidding vector is proportional to the vector of clickthrough rates $[C_1, C_2, \ldots, C_K]$, where the proportionality constant is bid b_i. So the advertiser who sends in the ith highest bid for each click gets the ith most-valuable ad space. Advertiser i then drops out of future bidding after winning an ad space.
- The "charge part" is simply a straight-forward extension of the second price approach. (We could have also generalized the first price approach, as Overture did in the late 1990s, but soon realized that it was an unstable mechanism.)

Often, a minimum bid is also mandated, say, $0.5. So in the above example in Figure 2.2, if all the advertisers bid their true valuations, advertiser 1 is matched to ad space 1, with a price of $5 per click, advertiser 2 to ad space 2, with a price of $1 per click, and advertiser 3 to ad space 3 with a price of $0.5 per click, the minimum bid allowed.

We can also consider GSP as a game too. Despite the apparent similarities between GSP auctions (for multiple items) and second price auctions (for a single item), there are substantial differences. There can be multiple Nash equilibria in the GSP game, and it is possible that none of them involves truthful bidding. The example in the next section illustrates these properties of GSP. This is in sharp contrast to second price auction's desirable property: truthful bidding as a dominant strategy. For a multiple-item auction that preserves this property, we will have to wait until the Advanced Material, where we will describe the VCG auction. It is not "second price" that matters, but charging on the basis of the damage caused to other bidders.

2.3   Examples

2.3.1   Single-item auction on eBay

Let us start with an examle of single-item auctions. Founded in 1995 and with over 40 million users now, eBay runs online auctions for all kinds of goods. The eBay auction style largely follows the second price auction, but is not exactly the same. There are four main differences:

- A seller announces a start price, but can also choose to specify a secret *reserve price*: the minimum price below which she will not sell the good. Bidders are aware of the existence of the reserve price but not the actual value. This allows a seller to set a start price low enough to attract interest but still maintain a minimum revenue upon the actual sale of the item. In the rest of this section, for simplicity, we will assume the reserve price is the same as the start price. There is also a *minimal increment* of δ dollars: the winner pays the second-highest bid plus this δ (unless that exceeds her own bid, in which she just pays her own bid, the highest bid).

- An eBay auction is *not* sealed envelope. Some information about the current bids is continuously released to the public. This is particularly helpful when some bidders are not sure about the valuation of a good. It also generates more "fun" in the auction process. So what information is displayed? eBay looks at the price the winner would need to pay, had the auction been concluded right now, i.e., the smaller of (1) the current highest bid, b_1, and (2) the second highest bid, b_2, plus δ. This price is announced in public. The next bid has to exceed it by another minimum increment δ, which becomes the new *ask price* displayed. So, the announced price is $\min\{b_1, b_2 + \delta\} + \delta$.

- It has a fixed and publicly announced time horizon, e.g., 3 days. This *hard closing* rule changes bidding behavior. For example, many bidders choose to wait until the last 10 minutes to enter their bids, knowing that it is the time period that really matters. There are even third-party tools for automated "sniping," where bids are sent on your behalf in the last seconds before closing. It is advantageous to wait if you would like to surprise your competitors or to learn something about their valuations. It is advantageous not to wait if you would like to scare the competitors away with a very high bid to start with and avoid a dragged-out bidding war.

- It allows automated "proxy agent" bidding to simplify the user interface. A bidder can enter the *maximum bid* she is willing to put in, and then let the proxy run the course. When the bidder is no longer the highest bidder, yet the displayed ask price is less than or equal to this indicated maximum bid, a bid is automatically entered to take the ask price.

Here is an example of a timeline illustrating an eBay auction with three bidders: Alice, Bob, and Chris.

- *Start*

  Sam, the seller, lists a lamp for sale on eBay, with a start price of $5.00 and a duration of 5 days. The minimum increment is $1.00. The reserve price is set to be the same as the start price.
  Highest bid: n/a; Ask price: $5.00.

- *Day 1*

  The first bid is from Alice, who uses a proxy agent with the maximum bid set to $12.00. eBay therefore bids $5.00 on Alice's behalf. The ask price becomes $5.00 + $1.00 = 6.00.
  Highest bid: Alice, $5.00; Ask price: $6.00.

- *Day 2*

  The second bid is from Bob, who bids $8.00 even though the ask price is $6.00. eBay immediately raises bid to $8.00 + $1.00 = 9.00 on Alice's behalf, since she is using proxy-agent bidding and the ask price is not greater than her maximum bid. The ask price becomes $\min\{\$9.00, \$8.00 + \$1.00\} + \$1.00 = \$10.00$.
  Highest bid: Alice, $9.00; Ask price: $10.00.

- *Day 3*

  Bob tries again, bidding $10.50 this time. eBay immediately raises the bid to $10.50 + $1.00 = 11.50 on Alice's behalf. The ask price becomes $\min\{\$11.50, \$10.50 + \$1.00\} + \$1.00 = \$12.50$.
  Highest bid: Alice, $11.50; Ask price: $12.50.

- *Day 4*

    Bob gets frustrated and raises his bid to $17.50. The highest bidder now changes to Bob and the ask price becomes min$\{\$17.50, \$11.50 + \$1.00\} + \$1.00 = \$13.50$.

    Highest bid: Bob, $17.50; Ask price: $13.50.

- *Day 5*

    It so happens that Chris enters the auction at the last moment and bids $18.00, even though he does not know $17.50 is the current highest bid. The ask price becomes min$\{\$18.00, \$17.50 + \$1.00\} + \$1.00 = \$19.00$.

    Highest bid: Chris, $18.00; Ask price: $19.00.

- *End*

    Nobody takes the ask price before the auction terminates at the hard closing time announced before. The auction ends and Chris wins the lamp with a price of min$\{\$18.00, \$17.50 + \$1.00\} = \$18.00$.

The bidding history is summarized in Table 2.1:

	Day 1	Day 2	Day 3	Day 4	Day 5
Alice	$5.00	$9.00	$11.50	–	–
Bob	–	$8.00	$10.50	$17.50	–
Chris	–	–	–	–	$18.00
Ask price	$6.00	$10.00	$12.50	$13.50	$19.00

Table 2.1 The bidding history and ask price evolution in an example of an eBay auction. There are three bidders, each using a different bidding strategy, to compete for a single item. The auction lasts 5 days.

2.3.2 Multiple-item GSP auction in Google

Now we move on to multiple-item auctions in the GSP style. Let us consider the bipartite graph in Figure 2.2 again.

- *Bidding*: Assume truthful bidding, i.e., **b** is the same **v** for each of the three bidders as shown in the graph. For example, bidder 1 bids $50 (per hour) = $10 per click × 5 clicks per hour, for ad space 1, $30 for ad space 2, etc.
- *Auction outcome (matching, or allocation)*: By the GSP mechanism, the most valuable ad space is allocated to the highest bidder, which clearly is bidder 1 (since bidder 1 bids 50, bidder 2 bids 25, and bidder 3 bids 5 for this ad space). Similarly, ad space 2 is allocated to bidder 2, and ad space 3 to bidder 3. This matching is not surprising, since we have already ordered the bidders and the ad spaces in descending order.
- *Auction outcome (Charging, or pricing, or payment)*: Assume the actual number of clicks per hour is the same as the estimated clickthrough rate: bidder

1 pays the second-highest bid, which is $5 (per click). So she pays $5 × 5 (the second 5 here refers to the clickthrough rate of 5 per hour for ad space 1)= $25 per hour for ad space 1. Bidder 2 payers $1 per click. So she pays $1 × 3 = $3 per hour for ad space 2. Bidder 3 pays just the minimum bid, e.g., $0.5 per click as set by Google. So she pays $0.5 × 1 = $0.5 per hour.

- *Revenue to the seller*: Google has collected a revenue of $25 + 3 + 0.5 = $28.5 per hour.
- *Payoff for each bidder*: Payoff is valuation v minus price p. So bidder 1's payoff is $10 − $5 = $5 per click, or equivalently, $5 × 5 = $25 per hour. Bidder 2's payoff is $5 − $1 = $4 per click, or $4 × 3 = $12 per hour. Bidder 3's payoff is $1 − 0.5 = $0.5 per click, or $0.5 × 1 = $0.5 per hour.

In summary, the total payoff is $25 + 12 + 0.5 = $37.5 per hour.

2.3.3   Another example of GSP

Suppose there are two ad spaces on a webpage and three bidders. An ad in the first space receives 400 clicks per hour, while the second space gets 200. Bidders 1, 2, and 3 have values per click of $12, $8, and $4, respectively.

If all advertisers bid truthfully, then the bids are (in dollars) $[4800, 2400]$ from bidder 1, $[3200, 1600]$ from bidder 2, and $[1600, 800]$ from bidder 3. Bidder 1 wins the first ad space, paying $8 per click, while bidder 2 wins the second space, paying $4 per click. Bidder 3 does not get any ad space. If the actual clickthrough rates are the same as the estimated ones, the payments of bidders 1 and 2 are $3200 and $800, respectively. And the payoffs are $1600 and $800, respectively.

Truth-telling is indeed an equilibrium in this example, as you can verify that no bidder can benefit by changing her bids.

But, in general, truth-telling is not a dominant strategy under GSP. For example, consider a slight modification: the first ad space receives 400 clicks a day, and the second one 300. If all players bid truthfully, then bidder 1's payoff, as before, is ($12 − $8) ∗ 400 = $1600. If, instead, she shades her bid and bids only $7 per click to get the second ad space, her payoff will be equal to ($12 − $4) × 300 = $2400, which is bigger than $1600. Therefore, she would bid below her valuation and receive a higher payoff. The difference between the first and second ad spaces' clickthrough rates is simply not large enough relative to the difference in per-click payment for her to bid truthfully.

2.4   Advanced Material

2.4.1   VCG auction

As we just saw, the GSP auction does not guarantee truthful bidding for multiple-item auctions. If that property is desired, the proper generalization of second price auction is the VCG auction.

To search for the correct intuition, we revisit single-item second price auctions. Had the highest bidder not been there in the auction, the second-highest bidder would have obtained the item, while the other bidders face the same outcome of losing the auction. So the "damage" done by the highest bidder to the entire system (other than the highest bidder herself) is the valuation of the second highest bidder, which is exactly how much the highest bidder is charged. For example, suppose the three valuations are 10, 5, and 1, respectively from Alice, Bob, and Chris. If Alice were not there in the system, Bob would have won the item, and Chris would have lost the auction anyway. Now that Alice is in the system, Bob loses the chance to receive a valuation of 5, and Chris does not suffer any lost valuation. So the total valuation lost to the system (other than Alice) is 5, which is the price Alice pays according to the second price auction's rule.

The key idea to realize is that, in a second price auction, the winner is charged for how much her winning reduces the payoffs of the other bidders. This is another instance of internalizing negative externality. This time the negative externality is imposed by the winner on all the other bidders. It is this property that enables truthful bidding.

In Chapter 1, power control in wireless networks compensates for negative externality due to signal interference. We will later see negative externalities characterized and compensated for in voting systems, in tragedy of the commons, and in TCP congestion control. In multiple-item auctions, we compare two scenarios and take the difference as the quantification of the "damage" done by bidder i:

- The revenue generated by Google if bidder i were not in the system.
- The revenue generated by Google from all other bidders when bidder i is in the system.

In VCG auctions, there are three main steps, as summarized in Figure 2.3.

1. The first step is to invite each bidder to send in her bids, one for each of the items. In Google's case, they are proportional to a common scalar. Let v_{ij} be the value of the bid submitted to the seller by bidder i for item j. Now, these bids might not be the true valuations. But, as we will soon discover, by properly matching the bidders with the items and then charging the right prices, we can induce the bidders to submit the true valuations as $\{v_{ij}\}$. Again, auction design can be viewed as a mechanism design problem.

2. Second, an optimization is carried out by the seller. A **matching** is computed through this optimization, in which each item is matched to a bidder.

   For example, if we draw three horizontal lines from bidders to items in Figure 2.2, that would be a matching, denoted as $\{(1,1),(2,2),(3,3)\}$. This matching returns a total valuation of $50 + 15 + 1 = 66$ to the bidders.

   In VCG, we want the matching to maximize the total valuation of the *whole system*: we maximize $\sum_{(ij)} v_{ij}$ (where the sum is over all the matched

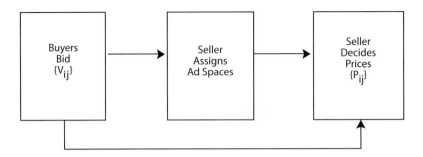

Figure 2.3 Three steps in VCG auctions. First, each bidder sends in a vector of bids for the K items. Then the seller solves an optimization problem of finding the matching between bidders and items (and computes the corresponding maximized value V for the system). Finally, given the matching, the seller determines the price p_{ij} that bidder j pays for item i that she is matched to, on the basis of the amount of negative externality caused by this matching.

pairs) over all the possible matchings. For example, a different matching of $\{(1,2),(2,3),(3,1)\}$ would return a total valuation of $\sum_{(ij)} v_{ij} = 30+5+5 = 40$, an inferior matching to $\{(1,1),(2,2),(3,3)\}$, as illustrated in Figure 2.4.

Denote by V the resulting maximized total value. Suppose item j is allocated to bidder i in this matching. Obviously, we can write V as the sum of the value v_{ij} and $\hat{V}_{i\leftarrow j}$, the sum of the values of all the *other* (item, bidder) pairs in the matching given that item j goes to bidder i:

$$V = v_{ij} + \hat{V}_{i\leftarrow j}.$$

This notation will become useful soon.

3. Matching decides the allocation, and pricing is determined only after the matching. For a given matching, each pair of (item, bidder) is charged a price of p_{ij}, which is the amount of damage caused by this matching. So, how do we quantify the damage done by a bidder i getting item j?

Imagine two alternative systems. (a) A system *without* bidder i, and there is a corresponding $V_{no\ i}$ as the maximum total valuation for everyone else. (b) A system *with* bidder i, who will get item j, and the corresponding $\hat{V}_{i\leftarrow j}$ as the total maximum valuation for everyone else. The difference between these two valuations, $V_{no\ i} - \hat{V}_{i\leftarrow j}$, is the "damage" caused to everyone else, and that is the price p_{ij} the seller charges user i for being matched to item j:

$$p_{ij} = V_{no\ i} - \hat{V}_{i\leftarrow j}. \tag{2.1}$$

This completes the description of a VCG auction.

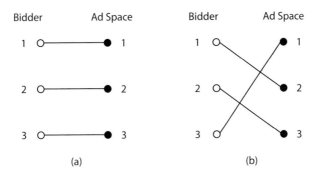

Figure 2.4 Two of the many possible matchings between bidders (on the left) and ad spaces (on the right) of a multi-item auction's bipartite graph. The weight of link (ij) is $v_{ij} = R_i C_j$. A maximum weight matching selects a subset of those links so that each bidder is matched to an item, and the sum of these links' weights is the largest possible among all matchings. With the parameter values in Figure 2.2, matching (a) turns out to maximize $\sum_{(ij)} v_{ij}$ (and is thus chosen in step 2 of a VCG auction) while matching (b) does not.

2.4.2   An example

As in the GSP example we just saw in the last section, suppose there are two ad spaces on a webpage and three bidders. An ad in the first space receives 400 clicks per hour, while the second space gets 200. Bidders 1, 2, and 3 have valuations per click of $12, $8, and $4, respectively. Suppose the bids are the true valuations for now; we will later *prove* that it is indeed a dominant strategy in the auction game.

The matching by VCG happens to be the same as by GSP in this case: bidder 1 gets the first ad space, and bidder 2 the second, with bidder 3 not matched to any ad space. The second bidder's payment is still $800 because the damage it causes to the system is that bidder 3 lost the second ad space, thus $4 × 200 = $800 value. However, the payment of bidder 1 is now $2400: $800 for the damage to bidder 3 and $1600 for the damage to bidder 2: bidder 2 moves from position 1 to position 2, thus causing her $(400-200) = 200$ clicks per day at the valuation of $8 a click.

In this example, the seller's revenue under VCG is $2400+$800 = $3200, lower than that under GSP ($3200 + $800 = $4000), if the advertisers bid truthfully in both cases.

2.4.3   Truthful bidding

So back to this question: what would each (rational) bidder i bid for item j in a VCG auction? We claim it must be the true valuation, i.e., the expected revenue per click times the estimated clickthrough rate.

Suppose that was not true, and bidder i bids some other number. Again, it is important to realize that this would make a difference in a VCG auction's matching only if it resulted in bidder i getting a different item, say item h, e.g., as in Figure 2.4(b). Obviously,

$$v_{ij} + \hat{V}_{i \leftarrow j} \geq v_{ih} + \hat{V}_{i \leftarrow h},$$

since the left side is V, the *maximized* total valuation over *all* possible matchings. Subtracting $V_{no\ i}$ from both sides,

$$v_{ij} + \hat{V}_{i \leftarrow j} - V_{no\ i} \geq v_{ih} + \hat{V}_{i \leftarrow h} - V_{no\ i},$$

and using the definition of p_{ij} in (2.1), we have

$$v_{ij} - p_{ij} \geq v_{ih} - p_{ih},$$

i.e., the payoff of bidding true valuation is higher. This argument shows that VCG induces truthful bidding as a dominant strategy for each bidder.

2.4.4   Other considerations

What about the seller's perspective: to maximize (over all the possible matchings) $\sum_{(ij)} p_{ij} C(j)$, the sum of per-click price times the estimated clickthrough rate across the (item, bidder) pairs matched. As we just saw in the example, GSP may generate more revenue than VCG, or less, depending on the problem parameters' values. If the actual clickthrough rate, as a function of the ad space location j, also depends on who is the winning bidder i, i.e., $C(j)$ becomes $C_i(j)$ and depends on i, then the seller's revenue maximization problem becomes even more challenging.

Which one to use: GSP or VCG? Google uses GSP, while many web 2.0 companies use VCG. Companies like AppNexus run variants of VCG auctions on large-scale (billions of auctions each day), real-time (millisecond matching), and user-profile-based targeted-ad placement. The comparison between GSP and VCG needs to include various considerations:

- GSP is simpler for Google to explain to advertisers than VCG.
- While GSP is simple when $C(j)$ is independent of winner i, it becomes more complicated when the clickthrough rate of an ad space depends on the actual ad placed there.
- While VCG guarantees truthful bidding, that might not be the case if there are multiple auctions held in parallel across many websites.
- If Google switches from GSP to VCG, the advertisers may act irrationally. They may ignore this change, continue to shade their bids, and use the same bids as before. For the same set of bids, VCG revenue is lower than GSP revenue. Relative to the potential benefit, the cost of transitioning from GSP to VCG may be too high for Google.

What is clear from this chapter is that different transaction mechanisms can induce very different behaviors from people. People react to different prices with different demands, react to different allocation methods with different strategies, and react to different expectations of tomorrow's events with different actions today. More generally, our design of a network and its functions may induce different optimization problems for the operators or the individual players. This is a recurring theme throughout this book.

Summary

Box 2 Second price auctions

Auctions allocate items among competing bidders. Different mechanisms of allocation and charging leads to different bidding behaviors and strategic equilibira. Charging based on externality, like in a second price auction of a single item or in a VCG auction of multiple items, induces truthful bidding as a dominant strategy in bidders' strategic thinking.

Further Reading

Auction theory is a well-studied discipline in economics and in computer science. There are many variations of the basic auctions we covered here: reverse auction with one bidder and many sellers, double auctions with many bidders and many sellers, and multiple-winner auctions. Researchers have also studied topics like revenue maximization strategies and collusion among bidders.

1. The following is a standard textbook on the subject:
V. Krishna, *Auction Theory*, 2nd edn., Academic Press, 2009.

2. Another insightful survey with special emphasis on applications is
P. Milgrom, *Putting Auction Theory to Work*, Cambridge University Press, 2004.

3. The following book provides an in-depth discussion of eBay auctions and the surprising behaviors found there:
K. Steiglitz, *Snipers, Shills, and Sharks*, Princeton University Press, 2007.

4. The following short paper provides a concise survey of the key ideas in Google advertising and sponsored search:
H. R. Varian, "The economics of Internet search," *Rivista di Politica Economica*, vol. 96, no. 6, pp. 9–23, 2006.

5. A more technical discussion focusing on GSP can be found here, especially the viewpoint of auctions as a dynamic game:

B. Edelman, M. Ostrovsky, and M. Schwarz, "Internet advertising and the generalized second-price auction: Selling billions of dollars worth of keywords," *The American Economic Review*, vol. 97, no. 1, pp. 242–259, 2007.

Problems

2.1 *A simple ad space auction* ⋆

Three advertisers (1, 2, 3) bid for two ad spaces (A, B). The average revenues per click are $6, $4, $3 for the bidders, respectively, and the clickthrough rate of the ad spaces are 500, 300 clicks per hour respectively.

(a) Draw the bipartite graph with nodes indicating advertisers/ad spaces and edges indicating values per hour. Indicate the maximum matching with bold lines.

(b) Assume a GSP auction with truthful bidding, what is the result of the auction in terms of the allocation, the prices charged, and the payoffs received?

2.2 *eBay Auction* ⋆⋆

Alice lists a lamp for sale on eBay via auction with both the start price and reserve price set to $7.00 and a duration of 5 days. The minimal increment is $0.25 and the following events happen during the auction:

- *Day 1* Bidder 1 uses a proxy agent, setting the maximum bid up to $11.00.
- *Day 2* Bidder 2 bids $9.25.
- *Day 3* Bidder 3 uses a proxy agent, setting the maximum bid up to $17.25.
- *Day 4* Bidder 2 bids $13.65.
- *Day 5* Bidder 1 bids $27.45.

List the bidding history of all three bidders over each day of the auction. Who is the winner and what price does she pay?

2.3 *More items than bidders* ⋆

Alice and Bob are bidding for three ad slots on a webpage, and one bidder can win at most one slot. Suppose the clickthrough rates are 500, 300, and 200 per hour, respectively. Assume that Alice receives $r per click.

(a) Denote by b_1 and b_2 the bids by Alice and Bob respectively. In GSP auction, discuss Alice's payoff in terms of b_1 and b_2.

(b) Does Alice have a dominant strategy? If so, what is it?

2.4   *Reverse auction* ⋆

Reverse auction is a type of auction where there are multiple sellers and only one bidder. The roles of bidders and sellers are reversed, that is, sellers lower their bids during auction and the one with the lowest bid sells her item.

Suppose there are three sellers in a reverse auction with one bidder. Denote b_i as the price seller i bids, and v_i as the value seller i attaches to the item.

(a) In the case of second price auction, what is the payoff function for seller i, as a function of b_1, b_2, and b_3?

(b) Is truthful bidding a dominant strategy?

2.5   *Spectrum auction and package bidding* ⋆⋆

Wireless cellular technologies rely on spectrum assets. Around the world, auctions have emerged as the primary means of assigning spectrum licenses to companies wishing to provide wireless communication services. For example, from July 1994 to July 2011, the US Federal Communications Commission (FCC) conducted 92 spectrum auctions, raising over $60 billion for the US Treasury, and assigned thousands of licenses to hundreds of firms to different parts of the spectrum and different geographic regions of the country.

The US FCC uses **simultaneous ascending auction**, in which groups of related licenses are auctioned simultaneously and the winner pays the highest bid. The British OfCom, in contrast, runs **package bidding**, where each potential spectrum bidder can bid on a joint set of frequency bands.

Among the many issues involved in spectrum auctioning is the debate between simultaneous ascending auction and package bidding auction. We will illustrate the inefficiency resulting from disallowing package bidding in a toy example. The root cause for this inefficiency is "bidder-specific complementarity" and the lack of competition.

Suppose that there are two bidders for two adjacent seats in a movie theater. Bidder 1 is planning to watch the movie together with her spouse as part of a date. She values the two spots *jointly* at $15, and a single spot is worth nothing. Bidder 2 plans to watch the movie by himself, and values each seat at $10, and the two seats together at $12 (since it is a little nicer to have no one sitting next to him on one side of his seat).

(a) Assume a simultaneous ascending auction is used for the seats, and Bidder 1 correctly guesses that Bidder 2 values $10 for one seat and $12 for two seats together. What strategy will Bidder 1 take? What is the result of the auction, in terms of the allocation, the price charged, and the payoffs received?

(b) Repeat part (a) but now assume package bidding is used. In particular, Bidder 1 can bid on a package consisting of both seats. Explain the differences with (a).

3 How does Google rank webpages?

3.1 A Short Answer

Now we turn to the other links you see on a search-result webpage; not the ads or sponsored search results, but the actual ranking of webpages by search engines such as Google. We will see that, each time you search on www.google.com, Google solves a very big system of linear equation to rank the webpages.

The idea of embedding links in text dates back to the middle of the last century. As the Internet scaled up, and with the introduction of the web in 1989, the browser in 1990, and the web portal in 1994, this vision was realized on an unprecedented scale. The network of webpages is huge: somewhere between 40 billion and 60 billion according to various estimates. And most of them are connected to each other in a giant component of this network. It is also sparse: most webpages have only a few hyperlinks pointing inward from other webpages or pointing outward to other webpages. Google search organizes this huge and sparse network by ranking the webpages.

More important webpages should be ranked higher. But how do you quantify *how* important a webpage is? Well, if there are many other important webpages pointing towards webpage A, A is probably important. This argument implicitly assumes two ideas:

- Webpages form a network, where a webpage is a node, and a hyperlink is a *directed* link in the network: webpage A may point to webpage B without B pointing back to A.
- We can turn the seemingly circular logic of "important webpages pointing to you means you are important" into a set of equations that characterize the *equilibrium* (a fixed-point equilibrium, not a game-theoretic Nash equilibrium) in terms of a *recursive definition* of "importance." This importance score will then act as an approximation of the ultimate test of search engines: how useful a user finds the search results.

As mentioned in Chapter 1, a network consists of both a *topology* and *functionalities*. Topology is often represented by a graph and various matrices, several of which will be introduced in this chapter and a few more in later chapters. And we will assume some models of the "search and navigation" functionality in this chapter.

Suppose there are N webpages. Each webpage i has O_i **outgoing links** and I_i **incoming links**. We cannot just count the number of webpages pointing to a given webpage A, because that number, the **in-degree** of the node in the hyperlinked graph, is often not the right measure of importance.

Let us denote the "importance score" of each webpage by π_i. If important webpages point to webpage A, maybe webpage A should be important too, i.e., $\pi_A = \sum_{i \to A} \pi_i$, where the sum is taken over all the webpages pointing to A. However, this is not quite right either, since node i may be pointing to many other nodes in this graph, and that means each of these nodes receives only a small portion of node i's importance score.

Let us assume that each node's importance score is *evenly* distributed across all the outgoing links from that node, i.e., each of the outgoing neighbors of node i receives π_i/O_i importance score. Now each node's importance score can also be written as the sum of the importance scores received *from* all of the incoming neighbors, indexed by j, e.g., for node 1,

$$\sum_{j \to 1} \frac{\pi_j}{O_j}.$$

If this sum is indeed also π_1, we have *consistency* of the scores. But it is not clear whether we can readily compute these scores, or even whether there is a consistent set of scores at all.

It turns out that, with a couple of modifications to the basic idea above, there is always a unique set of consistent scores, denoted as $\{\pi_i^*\}$. These scores determine the ranking of the webpages: the higher the score, the higher the webpage is ranked.

For example, consider a very small graph with just four webpages and six hyperlinks, shown in Figure 3.1. This is a directed graph where each node is a webpage and each link a hyperlink. A consistent set of importance scores turns out to be (0.125, 0.125, 0.375, 0.375): webpages 3 and 4 are more important than webpages 1 and 2. In this small example, it so happens that webpages 3 and 4, linking each other, push both webpages' rankings higher.

Intuitively, the scores make sense. First, by symmetry of the graph, webpages 1 and 2 should have the same importance score. We can view webpages 3 and 4 as if they form one webpage first, a supernode 3+4. Since node 3+4 has two incoming links, and each of nodes 1 and 2 has only one incoming link, node 3+4 should have a higher importance score. Since node 3 points to node 4 and vice versa, these two nodes' importance scores mix into an equal division at equilibrium. This line of reasoning qualitatively explains the actual scores we see.

But how do we calculate the exact scores? In this small example, it boils down to two simple linear equations. Let the score of node 1 (and 2) be x, and that of node 3 (and 4) be y. Looking at node 1's incoming links, we see that there is only one such link, coming from node 4 that points to three nodes. So we know

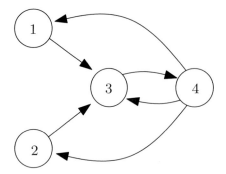

Figure 3.1 A simple example of importance score with four webpages and six hyperlinks. It is a small graph with much symmetry, leading to a simple calculation of the importance scores of the nodes.

$x = y/3$. By normalization, all scores must add up to $2x + 2y = 1$. So we have $x = 0.125$ and $y = 0.375$.

Now, how do we compute this set of consistent scores in a large, sparse, general graph of hyperlink connectivity?

3.2 A Long Answer

In any search engine, there are two main activities constantly occurring behind the scenes: (1) crawling the hyperlinked web space to get the webpage information, and (2) indexing this information into concise representations and storing the indices.

When you search in Google, it triggers a ranking procedure that takes into account two main factors:

- a **relevance score**: how relevant to the search the content is on each webpage, and
- an **importance score**: how important the webpage is.

It is the composite score of these two factors that determines the ranking. We focus on the importance score, since that usually determines the order of the top few webpages in any reasonably popular search, and has a tremendous impact on how people obtain information and how online businesses generate traffic.

We will be constructing several related matrices: $\mathbf{H}, \hat{\mathbf{H}}$, and $\mathbf{G}$, step by step (this matrix $\mathbf{G}$ is not the channel gain matrix of Chapter 1; it denotes the Google matrix in this chapter). Eventually, we will be computing an eigenvector of $\mathbf{G}$ as the importance-score vector. Each matrix is $N \times N$, where N is the number of the relevant webpages. These are extremely large matrices, and we will discuss the computational challenge of scaling-up in the Advanced Material.

3.2.1 Constructing H

The first matrix we define is $\mathbf{H}$: its (i, j)th entry is $1/O_i$ if there is a hyperlink from webpage i to webpage j, and 0 otherwise. This matrix describes the network topology: which webpages point to which. It also evenly spreads the importance of each webpage among its outgoing neighbors, or the webpages that it points to.

Let π be an $N \times 1$ column vector denoting the importance scores of the N webpages. We start by guessing that the consistent score vector is $\mathbf{1}$, simply a vector of 1s because each webpage is equally important. So we have an initial vector $\pi[0] = \mathbf{1}$, where 0 denotes the 0th iteration, i.e., the initial condition.

Then, we multiply π^T on the right by the matrix $\mathbf{H}$. (By convention, a vector is a column vector. So when we multiply a vector on the *right* by a matrix, we put the transpose symbol T on top of the vector.) You can write out this matrix multiplication, and see that this is spreading the importance score (from the last iteration) evenly among the outgoing links, and re-calculating the importance score of each webpage in this iteration by summing up the importance scores from the incoming links. For example, $\pi_1[2]$ (for webpage 1 in the second iteration) can be expressed as the following weighted sum of importance scores from the first iteration:

$$\pi_1[2] = \sum_{j \to 1} \frac{\pi_j[1]}{O_j},$$

i.e., the inner-product of π vector from the previous iteration and the first column of $\mathbf{H}$:

$$\pi_1[2] = (\pi[1])^T (\text{column 1 of } \mathbf{H}).$$

Similarly,

$$\pi_i[2] = (\pi[1])^T (\text{column } i \text{ of } \mathbf{H}), \quad \forall i.$$

If we index the iterations by k, the update at each iteration is simply

$$\pi^T[k] = \pi^T[k-1]\mathbf{H}. \tag{3.1}$$

Again, we followed the (visually rather clumsy) convention in this research field that defined $\mathbf{H}$ such that the update is a multiplication of row vector π^T by $\mathbf{H}$ from the right.

Since the absolute values of the entries in π do not matter, only the ranked order, we can also normalize the resulting π vector so that its entries add up to 1.

Now the quesetion is: Do the iterations in (3.1) converge? Is there a K sufficiently large that, for all $k \geq K$, the $\pi[k]$ vector is arbitrarily close to $\pi[k-1]$ (no matter what the initial guess $\pi[0]$ is)? If so, we have a way to compute a consistent score vector as accurately as we want.

But the answer is "not quite yet." We need two adjustments to $\mathbf{H}$.

3.2.2 Constructing $\hat{\mathbf{H}}$

First, some webpages do not point to any other webpages. These are "dangling nodes" in the hyperlink graph. For example, in Figure 3.2, node 4 is a dangling

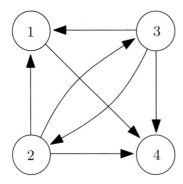

Figure 3.2 A network of hyperlinked webpages with a dangling node 4.

node, and its row is all 0s in the **H** matrix:

$$\mathbf{H} = \begin{bmatrix} 0 & 0 & 0 & 1 \\ 1/3 & 0 & 1/3 & 1/3 \\ 1/3 & 1/3 & 0 & 1/3 \\ 0 & 0 & 0 & 0 \end{bmatrix}.$$

There are no consistent scores. To see this, we write out the system of linear equations $\pi^T = \pi^T \mathbf{H}$:

$$\begin{cases} \frac{1}{3}(\pi_2 + \pi_3) = \pi_1 \\ \frac{1}{3}\pi_3 = \pi_2 \\ \frac{1}{3}\pi_2 = \pi_3 \\ \pi_1 + \frac{1}{3}(\pi_2 + \pi_3) = \pi_4. \end{cases}$$

Solving these equations gives $\pi_1 = \pi_2 = \pi_3 = \pi_4 = 0$, which violates the normalization requirement $\sum_i \pi_i = 1$.

One solution is to replace each row of 0s, like the last row in **H** above, with a row of $1/N$. Intuitively, this is saying that even if a webpage does not point to any other webpage, we will force it to spread its importance score evenly among all the webpages out there.

Mathematically, this amounts to adding the matrix $\frac{1}{N}(\mathbf{w1}^T)$ to **H**, where **1** is simply a vector of 1s, and **w** is an indicator vector with the ith entry being 1 if webpage i points to no other webpages (a dangling node) and 0 otherwise (not a dangling node). This is an *outer product* between two N-dimensional vectors, which leads to an $N \times N$ matrix. For example, if $N = 2$ and $\mathbf{w} = [1\ 0]^T$, we have

$$\frac{1}{2}\begin{pmatrix} 1 \\ 0 \end{pmatrix}(1\ 1) = \begin{pmatrix} 1/2 & 1/2 \\ 0 & 0 \end{pmatrix}.$$

This new matrix we add to **H** is clearly simple. Even though it is big, $N \times N$, it is actually the same vector **w** repeated N times. We call it a **rank-1 matrix**.

The resulting matrix,

$$\hat{\mathbf{H}} = \mathbf{H} + \frac{1}{N}(\mathbf{w1}^T),$$

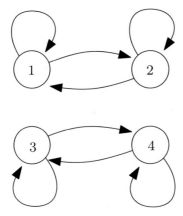

Figure 3.3 A network of hyperlinked webpages with multiple consistent score vectors.

has non-negative entries and each row adds up to 1. So we can think of each row as a probability vector, with the (i, j)th entry of $\hat{\mathbf{H}}$ indicating the probability that, if you are currently on webpage i, you will click on a link and go to webpage j.

Well, the structure of the matrix says that you are equally likely to click on any link shown on a webpage, and, if there is no link at all, you will be equally likely to visit any other webpage. Such behavior is called a **random walk on graphs** and can be studied as **Markov chains** in probability theory. Clearly this does not model web browsing behavior exactly, but it does strike a quite effective balance between the *simplicity* of the model and the *usefulness* of the resulting webpage ranking. We will see a similar model for influence in social networks in Chapter 8.

3.2.3   Constructing **G**

We mentioned that there were two issues with **H**. The second is that there might be *many* consistent score vectors, all compatible with a given $\hat{\mathbf{H}}$. For example, for the graph in Figure 3.3, we have

$$\mathbf{H} = \begin{bmatrix} 1/2 & 1/2 & 0 & 0 \\ 1/2 & 1/2 & 0 & 0 \\ 0 & 0 & 1/2 & 1/2 \\ 0 & 0 & 1/2 & 1/2 \end{bmatrix}.$$

Different choices of $\pi[0]$ result in different π^*, which are all consistent. For example, if $\pi[0] = [1\ 0\ 0\ 0]^T$, then $\pi^* = [0.5\ 0.5\ 0\ 0]^T$. If $\pi[0] = [0\ 0.3\ 0.7\ 0]^T$, then $\pi^* = [0.15\ 0.15\ 0.35\ 0.35]^T$.

One solution to this problem is to add a little *randomization* to the iterative procedure and the recursive definition of importance. Intuitively, we say there is a chance of $(1 - \theta)$ that you will jump to some other random webpage, without clicking on any of the links on the current webpage.

Mathematically, we add yet another matrix $\frac{1}{N}\mathbf{11}^T$, a matrix of 1s scaled by $1/N$ (clearly a simple, rank-1 matrix), to $\hat{\mathbf{H}}$. But this time it is a *weighted* sum, with a weight $\theta \in [0, 1]$. $(1 - \theta)$ quantifies how likely it is that you will randomly jump to some other webpage. The resulting matrix is called the **Google matrix**:

$$\mathbf{G} = \theta\hat{\mathbf{H}} + (1 - \theta)\frac{1}{N}\mathbf{11}^T. \tag{3.2}$$

Now we can show that, independently of the initialization vector $\pi[0]$, the iterative procedure:

$$\pi^T[k] = \pi^T[k - 1]\mathbf{G} \tag{3.3}$$

will converge as $k \to \infty$, and converge to the unique vector π^* representing the consistent set of importance scores. Obviously, π^* is the left eigenvector of $\mathbf{G}$ corresponding to the eigenvalue of 1:

$$\pi^{*T} = \pi^{*T}\mathbf{G}. \tag{3.4}$$

One can then normalize π^*: take $\pi_i^* / \sum_j \pi_j^*$ as the new value of π_i^*, and rank the entries in descending order, before outputting them on the search result webpage in that order. The matrix $\mathbf{G}$ is designed such that there is a unique solution to (3.4) and that (3.3) converges from any initialization.

However you compute π^*, taking (the normalized and ordered version of) π^* as the basis of ranking is called the **PageRank algorithm**. Compared with DPC for wireless networks in Chapter 1, the matrix $\mathbf{G}$ in PageRank is much larger, but we can afford a centralized computation.

3.3   Examples

Consider the network in Figure 3.4 with eight nodes and sixteen directional links. We have

$$\mathbf{H} = \begin{bmatrix} 0 & 1/2 & 1/2 & 0 & 0 & 0 & 0 & 0 \\ 1/2 & 0 & 0 & 0 & 1/2 & 0 & 0 & 0 \\ 0 & 1/2 & 0 & 0 & 0 & 0 & 0 & 1/2 \\ 0 & 0 & 1 & 0 & 0 & 0 & 0 & 0 \\ 0 & 0 & 0 & 1/2 & 0 & 0 & 0 & 1/2 \\ 0 & 0 & 0 & 1/2 & 1/2 & 0 & 0 & 0 \\ 0 & 0 & 0 & 1/2 & 0 & 1/2 & 0 & 0 \\ 1/3 & 0 & 0 & 1/3 & 0 & 0 & 1/3 & 0 \end{bmatrix}.$$

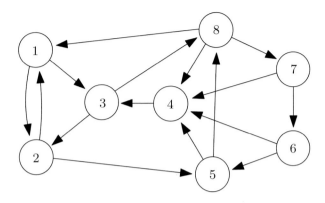

Figure 3.4 An example of the PageRank algorithm with eight webpages and sixteen hyperlinks. Webpage 3 is ranked the highest even though webpage 4 has the largest in-degree. Importance scores computed by PageRank can be quite different from node degrees.

Here $\hat{\mathbf{H}} = \mathbf{H}$ since there is no dangling node. Taking $\theta = 0.85$, we have

$$
\mathbf{G} = \begin{bmatrix}
0.0188 & 0.4437 & 0.4437 & 0.0188 & 0.0188 & 0.0188 & 0.0188 & 0.0188 \\
0.4437 & 0.0188 & 0.0188 & 0.0188 & 0.4437 & 0.0188 & 0.0188 & 0.0188 \\
0.0188 & 0.4437 & 0.0188 & 0.0188 & 0.0188 & 0.0188 & 0.0188 & 0.4437 \\
0.0188 & 0.0188 & 0.8688 & 0.0188 & 0.0188 & 0.0188 & 0.0188 & 0.0188 \\
0.0188 & 0.0188 & 0.0188 & 0.4437 & 0.0188 & 0.0188 & 0.0188 & 0.4437 \\
0.0188 & 0.0188 & 0.0188 & 0.4437 & 0.4437 & 0.0188 & 0.0188 & 0.0188 \\
0.0188 & 0.0188 & 0.0188 & 0.4437 & 0.0188 & 0.4437 & 0.0188 & 0.0188 \\
0.3021 & 0.0188 & 0.0188 & 0.3021 & 0.0188 & 0.0188 & 0.3021 & 0.0188
\end{bmatrix}.
$$

Initializing $\pi[0] = [1/8 \; 1/8 \; \ldots 1/8]^T$, iteration (3.3) gives

$$\pi[1] = [0.1073 \; 0.1250 \; 0.1781 \; 0.2135 \; 0.1250 \; 0.0719 \; 0.0542 \; 0.1250]^T$$
$$\pi[2] = [0.1073 \; 0.1401 \; 0.2459 \; 0.1609 \; 0.1024 \; 0.0418 \; 0.0542 \; 0.1476]^T$$
$$\pi[3] = [0.1201 \; 0.1688 \; 0.2011 \; 0.1449 \; 0.0960 \; 0.0418 \; 0.0606 \; 0.1668]^T$$
$$\pi[4] = [0.1378 \; 0.1552 \; 0.1929 \; 0.1503 \; 0.1083 \; 0.0445 \; 0.0660 \; 0.1450]^T$$
$$\pi[5] = [0.1258 \; 0.1593 \; 0.2051 \; 0.1528 \; 0.1036 \; 0.0468 \; 0.0598 \; 0.1468]^T$$
$$\pi[6] = [0.1280 \; 0.1594 \; 0.2021 \; 0.1497 \; 0.1063 \; 0.0442 \; 0.0603 \; 0.1499]^T$$

$$\vdots$$

and $\pi^* = [0.1286 \; 0.1590 \; 0.2015 \; 0.1507 \; 0.1053 \; 0.0447 \; 0.0610 \; 0.1492]^T$, to four decimal places. Therefore, the ranked order of the webpages are: 3, 2, 4, 8, 1, 5, 7, 6.

The node with the largest in-degree, i.e., with the largest number of links pointing to a node, is node 4, which is *not* ranked the highest. This is in part because its importance score is spread exclusively to node 3. As we will see again in Chapter 8, there are many more useful metrics measuring node importance than just the degree.

3.4   Advanced Material

3.4.1   Generalized PageRank and some basic properties

The Google matrix $\mathbf{G}$ can be generalized if the randomization ingredient is more refined. First, instead of the matrix $\frac{1}{N}\mathbf{1}\mathbf{1}^T$, we can add the matrix $\mathbf{1}\mathbf{v}^T$ (again, the outer product of two vectors), where $\mathbf{v}$ can be *any* probability distribution. Certainly, $\frac{1}{N}\mathbf{1}^T$ is a special case of that.

We can also generalize the dangling-node treatment: instead of adding $\frac{1}{N}\mathbf{w}\mathbf{1}^T$ to $\mathbf{H}$, where $\mathbf{w}$ is the indicator vector of dangling nodes, we can add $\mathbf{w}\mathbf{v}^T$. Again, $\frac{1}{N}\mathbf{1}$ is a special case of $\mathbf{v}$.

Now, the Google update equation can be written in the long form (not using the shorthand notation $\mathbf{G}$) as a function of the given webpage connectivity matrix $\mathbf{H}$, the vector $\mathbf{w}$ indicating the dangling webpages, and the two algorithmic parameters; scalar θ and vector $\mathbf{v}$:

$$\pi^T\mathbf{G} = \theta\pi^T\mathbf{H} + \pi^T(\theta\mathbf{w} + (1-\theta)\mathbf{1})\mathbf{v}^T. \tag{3.5}$$

You should verify that the above equation is indeed the same as (3.3).

There are many viewpoints to further interpret (3.3) and connect it to matrix theory, to Markov chain theory, and to linear systems theory. For example,

- π^* is the left eigenvector corresponding to the dominant eigenvalue of a positive matrix;
- it represents the so-called stationary distribution of a Markov chain whose transition probabilities are in $\mathbf{G}$; and
- it represents the equilibrium of an economic growth model according to $\mathbf{G}$ (more on this viewpoint is given later in this section).

The major operational challenges of running the seemingly simple update (3.3) are *scale* and *speed*: there are billions of webpages and Google needs to return the results almost instantly.

Still, the power method (3.3) offers many numerical advantages compared with a direct computation of the dominant eigenvector of $\mathbf{G}$. First, (3.3) can be carried out by multiplying a vector by the sum of $\mathbf{H}$ and two rank-1 matrices. This is numerically simple: $\mathbf{H}$ is very large but also very *sparse*: each webpage usually links to just a few other webpages, so almost all the entries in $\mathbf{H}$ are zero. Multiplying by rank-1 matrices is also easy. Furthermore, at each iteration, we only need to store the current π vector.

While we have not quantified the speed of convergence, it is clearly important to speed up the computation of π^*. As we will see again in Chapter 8, the convergence speed in this case is governed by the second-largest eigenvalue

$\lambda_2(\mathbf{G})$ of $\mathbf{G}$, which can be shown to be approximately θ here. So this parameter θ controls the tradeoff between convergence speed and the relevance of the hyperlink graph in computing the importance scores: smaller θ (closer to 0) drives the convergence faster, but also de-emphasizes the relevance of the hyperlink graph structure. This is hardly surprising: if you view the webpage importance scores more like random objects, it is easier to compute the equilibrium. Usually $\theta = 0.85$ is believed to be a pretty good choice. This leads to convergence in tens of iterations while still giving most of the weight to the actual hyperlink graph structure (rather than the randomization component in $\mathbf{G}$).

3.4.2 PageRank as the solution to a linear equation

PageRank sounds similar to distributed power control in Chapter 1. Both of them apply the power method to solve a system of linear equations. The solutions to those equations capture the right engineering configuration in the network, whether that is the relative importance of webpages in a hyperlink graph or the best transmit power vector in a wireless interference environment. This conceptual connection can be sharpened to an exact, formal parallel.

First, we can rewrite the characterization of π^* as the solution to the following linear equation, rather than as the dominant left eigenvector of matrix $\mathbf{G}$ (3.4) that has been our viewpoint so far:

$$(\mathbf{I} - \theta\mathbf{H})^T \pi = \mathbf{v}. \qquad (3.6)$$

We will soon prove that (3.6) is true. We can now compare (3.6) with the characterization of the optimal power vector in the distributed power control algorithm in Chapter 1:

$$(\mathbf{I} - \mathbf{D}\mathbf{F})\mathbf{p} = \mathbf{v}.$$

Of course, the vectors $\mathbf{v}$ are defined differently in these two cases: in terms of webpage viewing behavior in PageRank and receiver noise in power control. But we see a striking parallel: the self-consistent importance-score vector π and the optimal transmit-power vector $\mathbf{p}$ are both solutions to a linear equation with the following structure: identity matrix minus a scaled version of the network connectivity matrix.

In PageRank, the network connectivity is represented by the hyperlink matrix $\mathbf{H}$. This makes sense since the key factor here is the hyperlink connectivity pattern among the webpages. In power control, the network connectivity is represented by the normalized channel gain matrix $\mathbf{F}$. This makes sense since the key factor here is the strength of the interference channels.

In PageRank, the scaling is done by the single scalar θ. In power control, the scaling is done by many scalars in the diagonal matrix $\mathbf{D}$: the target SIR for each user. To make the parallel exact, we may think of a generalization of the Google matrix $\mathbf{G}$ where each webpage has its own scaling factor θ.

The general theme for solving these two linear equations can be stated as follows. Suppose you want to solve a system of linear equations $\mathbf{Ax} = \mathbf{b}$, but do not want to directly invert the square matrix $\mathbf{A}$. You might be able to split $\mathbf{A} = \mathbf{M} - \mathbf{N}$, where matrix $\mathbf{M}$ is invertible and its inverse $\mathbf{M}^{-1}$ can be much more easily computed than $\mathbf{A}^{-1}$.

The following **linear stationary iteration** ("linear" because the operations are all linear, and "stationary" because the matrices themselves do not vary over iterations) over times indexed by k:

$$\mathbf{x}[k] = \mathbf{M}^{-1}\mathbf{N}\mathbf{x}[k-1] + \mathbf{M}^{-1}\mathbf{b}$$

will converge to the desired solution:

$$\lim_{k\to\infty} \mathbf{x}[k] = \mathbf{A}^{-1}\mathbf{b},$$

from any initialization $\mathbf{x}[0]$, provided that the largest eigenvalue of $\mathbf{M}^{-1}\mathbf{N}$ is smaller than 1. Both DPC and PageRank are special cases of this general algorithm.

But we still need to show that (3.6) is indeed equivalent to (3.4): a π that solves (3.6) also solves (3.4), and vice versa. First, starting with a π that solves (3.6), we can easily show the following string of equalities:

$$\begin{aligned}
\mathbf{1}^T\mathbf{v} &= \mathbf{1}^T(\mathbf{I} - \theta\mathbf{H})^T\pi \\
&= \mathbf{1}^T\pi - \theta(\mathbf{H}\mathbf{1})^T\pi \\
&= \mathbf{1}^T\pi - \theta(\mathbf{1} - \mathbf{w})^T\pi \\
&= \pi^T(\theta\mathbf{w} + (1-\theta)\mathbf{1}),
\end{aligned}$$

where the first equality uses (3.6) and the third equality uses the fact that summing each row of $\mathbf{H}$ gives a vector of 1s (except those rows corresponding to dangling webpages). The other two equalities are based on simple algebraic manipulations.

But $\mathbf{1}^T\mathbf{v} = 1$ by design, so we know

$$\pi^T(\theta\mathbf{w} + (1-\theta)\mathbf{1}) = 1.$$

Now we can readily check that $\pi^T\mathbf{G}$, using its definition in (3.5) and the above equation, equals $\theta\pi^T\mathbf{H} + \mathbf{v}^T$.

Finally, using one more time the assumption that π satisfies (3.6), i.e., $\mathbf{v} = (\mathbf{I} - \theta\mathbf{H})^T\pi$, we complete the argument:

$$\pi^T\mathbf{G} = \theta\pi^T\mathbf{H} + \pi^T(\mathbf{I} - \theta\mathbf{H}) = \theta\pi^T\mathbf{H} - \theta\pi^T\mathbf{H} + \pi^T = \pi^T.$$

Therefore, any π solving the linear equation (3.6) is also a dominant left eigenvector of $\mathbf{G}$ that solves (3.4).

Vice versa, a π that solves (3.4) also solves (3.6), which can be similarly shown.

3.4.3 Scaling up and speeding up

It is not easy to adjust the parameters in PageRank computation. We discussed the role of θ before, and we know that, when θ is close to 1, PageRank results become very sensitive to small changes in θ, since the importance matrix $(\mathbf{I} - \theta\mathbf{H})^{-1}$ becomes very big.

There is also substantial research going into designing the right randomization vector $\mathbf{v}$, e.g., the entries of the $\mathbf{H}$ matrix: web surfers likely will not pick all of the hyperlinked webpages equally likely, and their actual behavior can be recorded to adjust the entries of $\mathbf{H}$.

The biggest challenge to running PageRank, however, is *scale*: how does Google scale up PageRank to really large matrices? How can we quickly compute and update the rankings? There are both storage and computational challenges. Many interesting solutions have been developed over the years, including the following five. The first one is a computational acceleration method. The other four are *approximations*, two of which change the notion of optimality and two of which restructure the graph of hyperlinked webpages.

1. *Decompose* $\mathbf{H}$. A standard triangular decomposition gives $\mathbf{H} = \mathbf{DL}$, where $\mathbf{D}$ is a diagonal matrix with entries being $1/O_i$, and $\mathbf{L}$ is a binary adjacency matrix. So only integers, instead of real numbers, need to be stored to describe $\mathbf{H}$. Suppose there are N webpages, and, on average, each webpage points to M webpages. N is huge; maybe in the billions, and M is very small; often 10 or less. Instead of NM multiplications, this matrix decomposition reduces it to just M multiplications.

2. *Relax the meaning of convergence.* It is not the values of importance scores that matter to most people, it is just the *order* of the webpages, especially the top ones. So once the computation of PageRank is sure about the order, there is no need to further improve the accuracy of computing π towards convergence.

3. *Differentiate among the webpages.* The PageRank algorithm as applied to most webpages quickly converges, and can be locked while the other webpages' PageRanks are refined. This is an approximation that works particularly well when the importance scores follow the power law that we will discuss in Chapter 10.

4. *Leave the dangling nodes out.* There are many dangling nodes out there, and their behaviors in the matrices and the computation involved are pretty similar. So they might be grouped together and not be updated in order to speed up the computation.

5. *Aggregation of webpages.* When many nodes are lumped together into a cluster, then *hierarchical* computation of PageRanks can be recursively computed, first treating each cluster as one webpage, then distributing the PageRank of that cluster among the actual webpages within that cluster. We will visit an example of this important principle of building hierarchy to reduce the computation (or communication) load in a homework problem.

3.4.4   Beyond the basic search

There is another player in the game of search: companies that specialize in increasing a webpage's importance scores, possibly pushing it into the top few search results, or even to the very top spot. This industry is called **Search Engine Optimization** (SEO). There are many proprietary techniques used by SEO companies. Some techniques enhance content-relevance scores, sometimes by adding bogus tags in the html files. Other techniques increase the importance score, sometimes by adding links pointing to a customer's site, and sometimes by creating several truly important webpages and then attaching many other webpages as their outgoing neighbors.

Google is also playing this game by detecting SEO techniques and then updating its ranking algorithm so that the artificial help from SEO techniques is minimized. For example, in early 2011, Google made a major update to its ranking algorithm to counter the SEO effects, and another significant update in May 2012 to further reduce the benefit of having not-so-important webpages pointing to a given webpage.

There are also many useful variants to the basic type of search we discussed, e.g., personalized search based on user's feedback on how useful she finds the top webpages in the search result. Multimedia search is another challenging area: searching through images, audios, and video clips requires very different ways of indexing, storing, and ranking the content than text-based search.

Summary

> **Box 3** PageRank computes webpages' importance scores
>
> We can view the hyperlinked webpages as a network, and use the connectivity patterns to rank the webpages by their importance scores. If many important webpages point to a given webpage, that webpage maybe also important. PageRank uniquely defines and efficiently computes a consistent set of importance scores. This set of scores can be viewed as the dominant eigenvector of a Google matrix that consists of a network connectivity part and a randomization part.

Further Reading

The PageRank algorithm is covered in almost every single book on network science these days. Some particularly useful references are as follows.

1. Back in 1998, the Google founders wrote the following paper explaining the PageRank algorithm:

S. Brin and L. Page, "The anatomy of a large-scale hypertextual Web search engine," *Computer Networks and ISDN Systems*, vol. 33, pp. 107–117, 1998.

2. The standard reference book devoted to PageRank is

A. N. Langville and C. D. Meyer, *Google's PageRank and Beyond*, Princeton University Press, 2006.

3. The following excellent textbook on the computer science and economics of networks has a detailed discussion of navigation on the web:

D. Easley and J. Kleinberg, *Networks, Crowds, and Markets: Reasoning About a Highly Connected World*, Cambridge University Press, 2010.

4. Dealing with non-negative matrices and creating linear stationary iterations are well documented, e.g., in the following reference book:

A. Berman and R. J. Plemmons, *Nonnegative Matrices in the Mathematical Sciences*, Academic Press, 1979.

5. Computational issues in matrix multiplication are treated in textbooks like this one:

G. Golub and C. F. van Van Loan, *Matrix Computations*, 3rd edn., The Johns Hopkins University Press, 1996.

Problems

3.1   *PageRank sink* ⋆

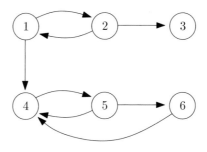

Figure 3.5 A simple network of webpages with a sink node.

Write out the **H** matrix of the graph in Figure 3.5. Iterate $\pi[k]^T = \pi[k-1]^T\mathbf{H}$, where $k = 0, 1, 2, \ldots$, and let the initialization be

$$\pi[0] = \begin{bmatrix} 1/6 & 1/6 & 1/6 & 1/6 & 1/6 & 1/6 \end{bmatrix}^T.$$

What problem do you observe with the converged π^* vector?

3.2   *Cyclic ranking* ★

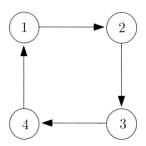

Figure 3.6 A simple network of webpages with a cycle.

Write out the $\mathbf{H}$ matrix of the graph in Figure 3.6. Iterate $\pi[k]^T = \pi[k-1]^T\mathbf{H}$, where $k = 0, 1, 2, \ldots$, and let the initialization be

$$\pi[0] = \begin{bmatrix} 1/2 & 1/2 & 0 & 0 \end{bmatrix}^T.$$

What happen to the vectors $\{\pi[k]\}$ as k becomes large? Solve for π^* such that $\pi^{*T} = \pi^{*T}\mathbf{H}$ and $\sum_i \pi_i^* = 1$.

3.3   *PageRank with different θ* ★★

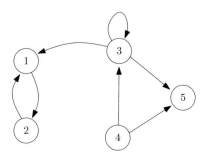

Figure 3.7 A simple example to try page PageRank with different θ.

Compute the PageRank vector π^* of the graph in Figure 3.7, for $\theta = 0.1, 0.3, 0.5$, and 0.85. What do you observe?

3.4   *Block aggregation in PageRank* ★★

Set $\theta = 0.85$ and start with any normalized initial vector $\pi[0]$.

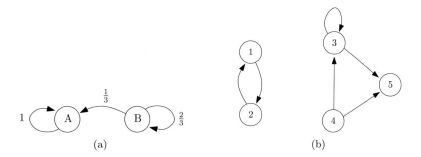

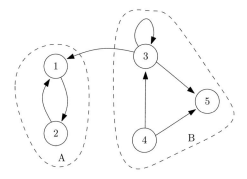

Figure 3.8 An example of hierarchical PageRank. Two different graphs that will be superimposed later.

Figure 3.9 Subgraphs A and B in Figures 3.8(a) and 3.8(b) are composed into a single graph.

(a) Compute the PageRank vector $\begin{bmatrix} \pi_A^* & \pi_B^* \end{bmatrix}^T$ of the graph in Figure 3.8(a) with

$$\mathbf{H} = \begin{bmatrix} 1 & 0 \\ 1/3 & 2/3 \end{bmatrix}.$$

Note the uneven splitting of link weights from node B. This will be useful later in the problem.

(b) Compute the PageRank vectors $\begin{bmatrix} \pi_1^* & \pi_2^* \end{bmatrix}^T$ and $\begin{bmatrix} \pi_3^* & \pi_4^* & \pi_5^* \end{bmatrix}^T$ of the two graphs in Figure 3.8(b).

(c) If we divide the graph in Figure 3.7 into two blocks as shown in Figure 3.9, we can approximate π^* in the previous question by

$$\tilde{\pi}^* = \begin{bmatrix} \pi_A^* \cdot \begin{bmatrix} \pi_1^* & \pi_2^* \end{bmatrix} & \pi_B^* \cdot \begin{bmatrix} \pi_3^* & \pi_4^* & \pi_5^* \end{bmatrix} \end{bmatrix}^T.$$

Compute this vector. Explain the advantage, in terms of computational load, of using this approximation instead of directly computing π^*.

3.5 *Personalized ranking (open-ended)*

How would you solicit and aggregate feedback from individual users to enable *personalized* ranking?

4 How does Netflix recommend movies?

We just saw three beautiful equations in the last three chapters, each used at least a billion times every single day:

$$p_i[t+1] = \frac{\gamma_i}{\mathrm{SIR}_i[t]} p_i[t],$$
$$U_i(\mathbf{b}) = v_i - p_i(\mathbf{b}),$$
$$\pi^{*T} = \pi^{*T} \mathbf{G}.$$

We continue with our first block of four chapters that present four fundamental algorithms: distributed power control, second price auction, PageRank, and now, collaborative filtering. These four chapters also introduce the basic languages of optimization, game, graph, and learning theories. A word of caution: as a chapter that introduces the basic ideas both in convex optimization and in machine learning, this chapter is among the longest in the book; you have to wait about 14 pages before we get to the most important idea on collaborative filtering for Netflix. This chapter is also mathematically more demanding than most others.

4.1 A Short Answer

4.1.1 Recommendation problem

Netflix started its DVD rental business in 1997: instead of going to rental stores, you can just wait for DVDs to arrive by mail. Instead of incurring a late fee for each day you hold the DVD beyond the return date, you can keep holding the DVD as long as you continue to pay the monthly subscription fee, but you cannot receive a new DVD without returning the old one. This is similar in spirit to the sliding window mechanism of congestion control in Chapter 14, or the tit-for-tat incentive mechanism of P2P in Chapter 15. Netflix also maintained an efficient inventory control and mail delivery system. It operated with great *scalability* (i.e., the per-customer cost is much lower as the number of customers goes up) and *stickiness* (i.e., users are reluctant to change the service). By 2008, there were about 9 million users in the USA and Canada.

Then Netflix moved on to the next mode of delivering entertainment. This time it was streaming movies and TV programs from video servers, through the Internet and wireless networks, to your Internet-connected devices: TVs, set-top boxes, games consoles, smartphones, and tablets. With its branding, choice of content, and aggressive pricing, Netflix's subscriber base nearly tripled to 23 million by April 2011. Netflix video streaming generated so much Internet traffic that over one in every four bits going through the Internet that month was Netflix traffic. In September 2011, Netflix announced that it would separate the DVD rental and online streaming businesses. Soon afterwards, Netflix reversed the decision, although the pricing for DVD rental and for online streaming became separated.

We will look at cloud-based video-distribution services, including Netflix, Amazon Prime, Hulu, HBO Go, etc., in Chapter 17, and how that is changing the future of both entertainment and networking. In this chapter, we instead focus on the social-network dimension used by Netflix: How does it recommend movies for you to watch (either by mail or by streaming)? It is like trying to read your mind and predict your movie rating. An effective **recommendation system** is important to Netflix because it enhances user experience, increases loyalty and volume, and helps with inventory control.

A recommendation system is a helpful feature for many applications beyond video distribution. Just like search engines in Chapter 3, recommendation systems give rise to structures in a "sea" of raw data and reduce the impact of information "explosion." Here are some representative systems of recommendation.

- You must have noticed how Amazon recommends products to you on the basis of your purchase and viewing history, adjusting its recommendation each time you browse a product. Amazon recommendation runs **content-based filtering**, in contrast to the **collaborative filtering** used by Netflix. (A related question is when can you trust the averaged rating of a product on Amazon? This is a different variant of the recommendation problem, and will be taken up in Chapter 5.)
- You must have also been swayed by recommendations on YouTube that followed each of the videos you watched. We will look at YouTube recommendation in Chapter 7.
- You may have used Pandora's online music selection, where recommendation is developed by experts of music selection. But you get to thumbs-up or thumbs-down the recommendation in the form of an explicit, binary feedback.

Netflix instead wants to develop a recommendation system that does *not* depend on any expert, but uses the rich history of *all* the user behaviors to profile *each* user's taste in movies. This system has the following inputs, outputs, and criteria of success.

- Among the inputs to this system is the history of star ratings across all the users and all the movies. Each data point consists of four numbers: (1) user ID, indexed by u, (2) movie title, indexed by i, (3) number of stars, $1-5$, in the rating, denoted as r_{ui}, and (4) date of the rating, denoted as t_{ui}. This is a really large data set: think of millions of users and tens of thousands of movies. But only a fraction of users will have watched a given movie, and only a fraction of that fraction actually bothered to rate the movie. Still, the size of this input is on the order of billions for Netflix. And the data set is also biased: knowing which users have watched and rated which movies already provides much information about people's movie taste. For a less popular service, there would also have been a cold-start problem: too little data to start with.

- The output is, first of all, a set of predictions $\hat{r}_{ui}$, one for each movie i that user u has not watched yet. These can be real numbers, not just integers like an actual rating r_{ui}. We can interpret a predicted rating of, say, 4.2 as saying that the user will rate this movie 4 stars with 80% probability and 5 stars with 20% probability. The final output is a short, rank-ordered list of movies recommended to each user u, presumably those movies receiving $\hat{r}_{ui} \geq 4$, or the top five movies with the highest predicted $\hat{r}_{ui}$.

- The real test of this mind-reading system is whether user u actually likes the recommended movies. This information, however, is hard to collect. So a proxy used by Netflix is the **Root Mean Squared Error** (RMSE), measured for those (u, i) pairs for which we have both the prediction and the actual rating. Let us say there are C such pairs. Each rating prediction's error is squared: $(r_{ui} - \hat{r}_{ui})^2$, and then averaged over all the predictions. Since the square was taken, to scale the numerical value back down, a square root is taken over this average:

$$\text{RMSE} = \sqrt{\sum_{(u,i)} \frac{(r_{ui} - \hat{r}_{ui})^2}{C}}.$$

The smaller the RMSE, the better the recommendation system. Netflix could have used the absolute value of the error instead of the squared error, but for our purposes we will stick to RMSE as the metric that quantifies the accuracy of a recommendation system. More importantly, regardless of the error metric it uses, in the end only the ranked order of the movies matters. Only the top few in that rank-ordered list are relevant as only they will be recommended to the user. The ultimate test is whether the user decides to watch the recommended movies, and whether she likes them or not. So, RMSE minimization is just a tractable approximation of the real problem of recommendation.

4.1.2 The Netflix Prize

Could recommendation accuracy be improved by 10% over what Netflix was using? That was the question Netflix presented to the research community in

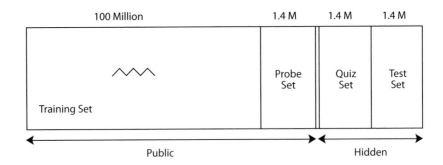

Figure 4.1 The Netflix Prize's four data sets. The training set and probe set were publicly released, whereas the quiz set and test set were hidden from the public and known only to Netflix. The probe, quiz, and test sets had similar statistical properties, but the probe set could be used by each competing team as often as they want, and the quiz set at most once a day. The final decision was based on comparison of the RMSE on the test set.

October 2006, through an open, online, international competition with a $1 million prize called the **Netflix Prize**.

The competition's mechanism is interesting in its own right. Netflix made available a set of over 100 million ratings, as part of its records from 1999 to 2005. That amount of data could fit in the memory of standard desktops in 2006, making it easy for anyone in the world to participate in the competition. The rating data came from more than 480000 users and 17770 movies. On average, each movie was rated by more than 5000 users and each user rated more than 200 movies. But those average numbers disguise the real difficulty here: many users rated only a few movies, and very few users rated a huge number of movies (one user rated over 17000 movies). Whatever recommendation system we use, it must work well for *all* users.

The exact distribution of the data is shown in Figure 4.1.

- A little fewer than 100 million ratings were made public as the *training set*.
- About 1.4 million additional ratings were also made public and they had similar statistical properties to the test set and the quiz set described next. This set of ratings was called the *probe set*, which competitors for the Netflix Prize could use to test their algorithms.
- About 1.4 million additional ratings were hidden from the competitors; this set was called the *quiz set*. Each competing team could submit an algorithm that would run on the quiz test, but not more than once a day. The RMSE scores continuously updated on the leaderboard of the Netflix Prize's website were based on this set's data.
- Another 1.4 million ratings, also hidden from the competitors, were called the *test set*. This was the real test. The RMSE scores on this set would determine the winner.

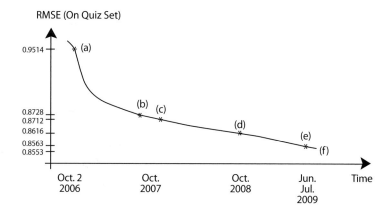

Figure 4.2 The Netflix Prize's timeline and some of the highlight events. It lasted for almost three years and the final decision came down to a 20-minute differential. The y-axis shows the progress towards reducing the RMSE on the quiz set data by 10% relative to the benchmark.

Each competing team first came up with a model for its recommendation system. Then it decided its model parameters' values by minimizing the RMSE between the known ratings in the training set and their model's predictions. Finally, it used this model with tuned parameters to predict the unknown ratings in the quiz set. Of course, Netflix knew the actual ratings in the quiz set, and could evaluate the RMSE between those ratings and the predictions from each team.

This was a smart arrangement. No team could reverse engineer the actual test set, since only scores on the quiz set were shown. It was also helpful to have a probe set on which the competing teams could run their own tests as many times as they wanted.

Netflix had its own algorithm called Cinematch that gave an RMSE of 0.9514 on the quiz set if its parameters were tuned by the training set. Improving the RMSE by even 0.01 could sometimes make a difference in the top ten recommendations for a user. If the recommendation accuracy could be improved by 10% over Cinematch, it would push RMSE to 0.8563 on the quiz set, and 0.8572 on the test set.

This Netflix Prize ignited the most intense and high-profile surge of activities in the research communities of machine learning, data mining, and information retrieval in recent years. To some researchers, the quality and sheer amount of the available data were as attractive as the hype and prize. Over 5000 teams worldwide entered more than 44000 submissions. Both Netflix and these research fields benefited from the three-year quest towards the goal of 10%. It turned out that setting the target as a 10% improvement was a really good decision. For the given training set and quiz set, getting an 8% improvement was reasonably easy, but getting a 11% would have been extremely difficult.

Here are a few highlights in the history of the Netflix Prize, also shown in the timeline in Figure 4.2.

- (a) Within a week of the start of the competition, Cinematch was beaten.
- (b) By early September, 2007, team BellKor made an 8.26% improvement over Cinematch, but the first place changed hands a couple of times, until
- (c) in the last hour before the first year of competition ended, the same team got 8.43% improvement and won the $50,000 annual progress prize for leading the pack during the first year of the competition.
- Then teams started merging. BellKor and BigChaos, two of the leading teams, merged and (d) received the 2008 progress prize for pushing the RMSE down to 0.8616. They further merged with Pragmatic Theory, and
- (e) in June 2009, the new team, BellKor's Pragmatic Chaos, became the first team to achieve more than 10% improvement, beating Cinematch by 10.06%, on the quiz set.
- Then the competition entered the "last call" period: all teams had 30 days to make their final submissions. (f) At the end of this period, two teams beat Cinematch by more than 10% on the quiz set: BellKor's Pragmatic Chaos had an RMSE of 0.8554, and The Ensemble had an RMSE of 0.8553, slightly better. The final winner was to be declared by comparing their RMSEs on the test set.
- Here is the grand finale: both teams beat Cinematch by more than 10% on the test set, and actually got the same RMSE on that set: 0.8567. But BellKor's Pragmatic Chaos submitted their algorithm 20 minutes earlier, and thus became the winner of the grand prize. A world-class science race lasting almost three years concluded with a 20-minute differential.

You must be wondering what algorithm BellKor's Pragmatic Chaos used in the final winning solution. The answer is documented in detail in a set of three reports, one from each component of this composite team, linked from the Netflix Prize website. But what you will find is that the winning solution was really a cocktail of many methods combined, with hundreds of ingredient algorithms blended together and thousands of model parameters fine-tuned specifically to the training set provided by Netflix. That was what it took to get that last 1% of improvement. But if you are interested only in the main approaches, big ideas, and getting $8 - 9\%$ improvement over Cinematch, there are actually just a few key methodologies. Those are what we will focus on in the rest of this chapter.

4.1.3   Key ideas

To start with, take a look at the table in Figure 4.3. We can also think of the table as a matrix $\mathbf{R}$, or as a *weighted* bipartite graph where the user nodes are on the left column and the movie nodes on the right. There is a link connecting user node u and movie node i if u rated i, and the value of the rating is the

Movies

		1	2	3	4	5	6	7	8
	1		5		2	4			
	2	4		3	1			3	
Users	3		5	4		5		4	
	4						1	1	2
	5	3			?		?	3	
	6		?	2		4		?	

Figure 4.3 Recommendation system's problem: predicting missing ratings from given ratings in a large yet sparse table. In this small example of six users and eight movies, there are eighteen known ratings as a training set, and four unknown ratings to be predicted. Real problems are much larger (with billions of cells in the table) and much sparser (only about 1% filled with known ratings).

weight of the link. In Chapter 8 we will discuss other matrices that describe the structure of different graphs.

Each column in this table is a movie (or an item in general), each row is a user, and each cell's number is the star rating by that user for that movie. Most cells are empty, since only a few users rated a given movie. You are given a large yet sparse table like this, and asked to predict some missing entries like those four indicated by question marks in the last two rows in this table.

There are two main types of techniques for any recommendation system: content-based filtering and collaborative filtering.

Content-based filtering looks at each row in isolation and attaches labels to the columns: if you like a comedy with Rowan Atkinson, you will probably like another comedy with Rowan Atkinson. This straightforward solution is often inadequate for Netflix.

In contrast, collaborative filtering exploits all the data in the entire table, across all the rows and all the columns, trying to discover *structures* in the pattern across the table. Drawing an imperfect analogy with search engines in Chapter 3, content-based filtering is like the relevance score of individual webpages, and collaborative filtering is like the importance score determined by the connections among the webpages.

In collaborative filtering, there are in turn two main approaches.

- The intuitively simpler one is the **neighborhood model**. Here, two users are "neighbors" if they share similar tastes in movies. If Alice and Bob both like "Schindler's List" and "Life is Beautiful," but not as much "E.T." and "Lion King," then knowing that Alice likes "Dr. Zhivago" would make us think Bob likes "Dr. Zhivago," too. In the neighborhood method, we first compute a similarity score between each pair of users. The larger the score, the closer these two users are in their taste for movies. Then for a given user whose opinion of a movie we would like to predict, we select, say, 50 of the most similar users who have rated that movie. Then take a weighted sum of these ratings and call that our prediction.

- The second approach is called the **latent-factor model**. It assumes that underneath the billions of ratings out there, there are only a few hundred key factors on which users and movies interact. Statistical similarities among users (or among movies) are actually due to some hidden, low-dimensional structure in the data. This is a big assumption: that there are many fewer *types* of people and movies than there are people and movies, but it sounds about right. It turns out that one way to represent a low-dimensional model is to factorize the table into two sets of *short vectors* of "latent factors."

Determining baseline predictors for the neighborhood model, or finding just the right short vectors in the latent-factor model, boils down to solving **least squares** problems, also called **linear regressions**.

Most of the mileage in the leading solutions to the Netflix Prize was obtained by combining variants of these two approaches, supplemented by a whole bag of tricks. Two of these supplementary ideas are particularly interesting.

One is **implicit feedback**. A user does not have to rate a movie to tell us something about her mind. Which movies she browsed, which ones she watched, and which ones she bothered to rate at all are all helpful hints. For example, it is useful to leverage the information in a binary table where each entry simply indicates whether this user rated that movie or not.

Another idea played an important role in pushing the improvement to 9% in the Netflix Prize: incorporating *temporal dynamics*. Here, the model parameters become time-dependent. This allows the model to capture changes in a person's taste and in trends of the movie market, as well as the mood of the day when a user rated movies, at the expense of dramatically increasing the number of model parameters to optimize. One interesting observation is that when a user rates many movies on the same day, she tends to give similar ratings to all of these movies. By discovering and discounting these temporal features, the truly long-term structures in the training set are better quantified.

In the next section, we will present baseline predictor training and the neighborhood method, leaving the latent-factor model to the Advanced Material.

4.2   A Long Answer

Before diving into specific predictors, let us take a look at the generic workflow consisting of two phases, as shown in Figure 4.4.

- *Training*: We put in a model (a mathematical representation of what we want to understand) with its parameters to work on the observed data, and

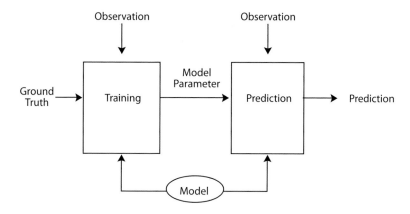

Figure 4.4 The main flow of two steps in building and using a model: first training and then prediction. The training module optimizes over model parameters using the known ground truth. Then the prediction module uses the models with optimized parameters in the model to make predictions.

then compare the predictions with the ground truth that we know in the training data. Then we tune the model parameters so as to minimize some error metric, like the RMSE, relative to the ground truth that we know.

- *Prediction*: Now we use the optimized model parameter to work on data observed beyond those in the training set. These predictions are then used in practice (or, in the case of the Netflix Prize, compared against a quiz set of ground truth that only Netflix knew).

4.2.1   Baseline predictor through least squares

Collaborative filtering extracts and leverages *structures* from the data set such as those in the table in Figure 4.3. But here is an easy method that does not even require any understanding of the structure: just use a simple averaging to compute the biases in the aggregate data. Let us start with this **baseline predictor**, and along the way introduce two important topics: the least squares problem and its solution, and time-dependent parameters that capture temporal shifts.

Suppose you look at the table of ratings in Figure 4.3 and decide *not* to study the user–movie interactions. Instead, you just take the average of all the ratings out there, denoted as $\bar{r}$, and use that as the predictor for all $\{\hat{r}_{ui}\}$. In the Netflix Prize's training set, $\bar{r} = 3.6$. That is an extremely lazy and inaccurate recommendation system.

So how about incorporating two parameters: b_i to model the quality of each movie i relative to the average $\bar{r}$, and b_u to model the bias of each user u relative to $\bar{r}$? Some movies are of higher quality than average, while some users tend to give lower ratings to all movies. For example, "The Godfather" might have

$b_i = 1.2$, but Alice tends to be a harsh reviewer with a $b_u = -0.5$. Then you might predict that Alice would rate "The Godfather" with $3.6 + 1.2 - 0.5 = 4.3$ stars.

If you think along this line, you will be using a model of baseline predictor

$$\hat{r}_{ui} = \bar{r} + b_u + b_i. \tag{4.1}$$

We could have used $b_u = (\sum_i r_{ui}/M_u) - \bar{r}$, where M_u is the number of movies rated by user u, and $b_i = (\sum_u r_{ui}/M_i) - \bar{r}$, where M_i is the number of users who rated movie i. But these averages might not minimize the RMSE. Instead, we choose the model parameters $\{b_u, b_i\}$ so that the resulting prediction's RMSE (for the training data) is minimized in the training phase. This is equivalent to

$$\text{minimize}_{\{b_u, b_i\}} \sum_{(u,i)} (r_{ui} - \hat{r}_{ui})^2, \tag{4.2}$$

where the sum is, of course, only over (u, i) pairs in the training set where user u actually rated movie i. The minimization is over all the $N + M$ parameters, where N is the number of users and M the number of movies in the training set. This type of optimization problem is called least squares. It has been studied extensively and will be encountered later.

Least squares minimizes a convex *quadratic* function with *no* constraints, where the objective function is the sum of squares of some linear function of the variables. That is exactly what we have in (4.2).

To simplify the notation, just consider a very small example with one user (user 1) rating two movies (A and B) in the training set: r_{1A} and r_{1B}, with the average rating $\bar{r} = (r_{1A} + r_{1B})/2$. The model parameters we have are b_1 (for b_u) and b_A and b_B (for b_i). The RMSE minimization boils down to minimizing the following convex, quadratic function

$$(b_1 + b_A + \bar{r} - r_{1A})^2 + (b_1 + b_B + \bar{r} - r_{1B})^2$$

over (b_1, b_A, b_B).

We can rewrite this minimization (4.2) in the standard form of a least squares problem, where $\mathbf{A}$ and $\mathbf{c}$ are a given constant matrix and vector, respectively: minimize the sum of squares of all the elements in a vector $\mathbf{Ab} - \mathbf{c}$, i.e., the square of the L-2 norm of $\mathbf{Ab} - \mathbf{c}$:

$$\|\mathbf{Ab} - \mathbf{c}\|_2^2,$$

where the subscript 2 denotes the L-2 norm. In this case, we have

$$\left\| \begin{bmatrix} 1 & 1 & 0 \\ 1 & 0 & 1 \end{bmatrix} \begin{bmatrix} b_1 \\ b_A \\ b_B \end{bmatrix} - \begin{bmatrix} r_{1A} - \bar{r} \\ r_{1B} - \bar{r} \end{bmatrix} \right\|_2^2.$$

More generally, the variable vector $\mathbf{b}$ contains all the user biases and movie biases, and is thus $N + M$ elements long. Each entry in the constant vector $\mathbf{c}$ is the difference between a known rating r_{ui} in the training set and $\bar{r}$, thus being C

elements long. The constant matrix $\mathbf{A}$ is $C \times (N + M)$, with each row containing exactly two 1s, corresponding to the current user and movie, and all other entries being 0.

You can take the first derivatives of this $\|\mathbf{A}\mathbf{b} - \mathbf{c}\|_2^2$ with respect to $\mathbf{b}$, and set them to zero. We know this is a minimizer because the objective function is convex (more on convexity in the next subsection).

From the definition of the L-2 norm, we have $\|\mathbf{x}\|_2^2 = \sum_i x_i^2$. So we can write $\|\mathbf{A}\mathbf{b} - \mathbf{c}\|_2^2$ as

$$(\mathbf{A}\mathbf{b} - \mathbf{c})^T (\mathbf{A}\mathbf{b} - \mathbf{c}) = \mathbf{b}^T \mathbf{A}^T \mathbf{A}\mathbf{b} - 2\mathbf{b}^T \mathbf{A}^T \mathbf{c} + \mathbf{c}^T \mathbf{c}.$$

Taking the derivative with respect to $\mathbf{b}$ and setting to 0 gives

$$2\left(\mathbf{A}^T \mathbf{A}\right) \mathbf{b} - 2\mathbf{A}^T \mathbf{c} = 0. \tag{4.3}$$

(You can also write out the quadratic function in long hand and take the derivative to verify the above expression.)

So the least-squares solution $\mathbf{b}^*$ is the solution to the following system of linear equations:

$$\left(\mathbf{A}^T \mathbf{A}\right) \mathbf{b} = \mathbf{A}^T \mathbf{c}. \tag{4.4}$$

There are many ways to numerically solve this system of linear equations to obtain the result $\mathbf{b}^*$. This is a useful fact: minimizing (convex) quadratic functions boils down to solving linear equations, because we take the derivative and set it to zero.

In the above example of estimating three parameters (b_1, b_A, b_B) from two ratings (r_{1A}, r_{1B}), this linear equation (4.4) becomes

$$\begin{bmatrix} 2 & 1 & 1 \\ 1 & 1 & 0 \\ 1 & 0 & 1 \end{bmatrix} \begin{bmatrix} b_1 \\ b_A \\ b_B \end{bmatrix} = \begin{bmatrix} r_{1A} + r_{1B} - 2\bar{r} \\ r_{1A} - \bar{r} \\ r_{1B} - \bar{r} \end{bmatrix}.$$

It so happens that there is an infinite number of solutions to the above linear equations. But in the realistic case where the number of model parameters is much fewer than the number of known ratings, like the one in the next section, we will not have that problem.

As explained in the Advanced Material, least squares solutions often suffer from the **overfitting** problem. It fits the known data in the training set so well that it loses the flexibility to adjust to a new data set. Then you have a super-refined explanatory model that loses its predictive power.

To avoid overfitting, a standard technique is called **regularization**. We simply minimize a weighted sum of (1) the sum of squared error and (2) the sum of squared parameter values:

$$\text{minimize}_{\{b_u, b_i\}} \sum_{(u,i)} (r_{ui} - \hat{r}_{ui})^2 + \lambda \left(\sum_u b_u^2 + \sum_i b_i^2 \right), \tag{4.5}$$

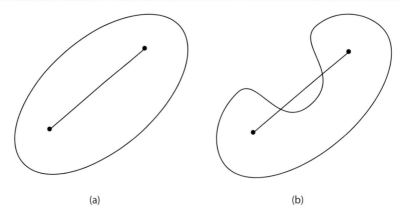

(a)   (b)

Figure 4.5 Here (a) is a convex set and (b) is not. A set is convex if for any two points in the set, the line segment in between them is also in the set. A key property of convex sets is that we can use a straight line (or a hyperplane in higher than two-dimensional spaces) to separate two (non-intersecting) convex sets. The constraint set of a convex optimization problem is a convex set. The domain of a convex function must be a convex set.

where $\hat{r}_{ui}$ is the baseline predictor (4.1), as a function of model parameters $\{b_u, b_i\}$. This balances the need to fit known data with the desire to use small parameters. The weight λ is chosen to balance these two goals. Bigger λ gives more weight to regularization and less to fitting the training data.

Since we picked the L-2 norm as the penalty function of $\{b_u, b_i\}$ in the regularization terms, (4.5) still remains a least squares (minimizing a convex quadratic function in the variables $\{b_u, b_i\}$), even though the objective function is different from the one in (4.2).

In the rest of the chapter, we will subtract the baseline predictors from the raw rating data:

$$\tilde{r}_{ui} = r_{ui} - \hat{r}_{ui} = r_{ui} - (\bar{r} + b_u + b_i).$$

After this bias removal, in matrix notation, we have the following error matrix:

$$\tilde{\mathbf{R}} = \mathbf{R} - \hat{\mathbf{R}}.$$

4.2.2   Quick detour: Convex optimization

Least squares is clearly *not* a linear programming problem as introduced in Chapter 1. But it is a special case of **convex optimization**: minimizing a **convex function** subject to a **convex set** of constraints. Detecting convexity in a problem statement takes some experience, but defining the convexity of a set and the convexity of a function is straight-forward.

We call a *set* convex if the following is true: whenever two points are in the set, the entire line segment connecting them is also in the set. So, in Figure 4.5, (a) is a convex set but (b) is not. An important property of convex sets is that you can use a straight line (or, a hyperplane in more than two dimensions) to separate two (non-intersecting) convex sets.

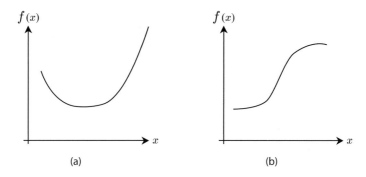

Figure 4.6 Here (a) is a convex function and (b) is not. A (twice differentiable) function is convex if its second derivative Hessian matrix is positive semidefinite. A function f is concave if $-f$ is convex. Some functions, like (b) above, have convex parts and concave parts. Linear functions are both convex and concave.

We call a (twice differentiable) *function* convex if its second derivative is positive (and the domain of the function is a convex set). In Figure 4.6, (a) is a convex function but (b) is not. In the case of a multivariate function f, the second derivative is a matrix, the **Hessian**, where the (i, j)th entry is the partial derivative of f with respect to the ith and jth arguments of the function.

For example, the Hessian of $f(x, y) = x^2 + xy + y^2$ is

$$\begin{bmatrix} 2 & 1 \\ 1 & 2 \end{bmatrix}.$$

For the functions we will deal with in this book, there is a simple test of convexity. If the Hessian's eigenvalues are all non-negative, it is called a **positive semidefinite matrix** (not to be confused with a positive matrix in Chapter 1); and we say the "second derivative" is "positive," and the multivariate function is convex. For example, $f(x, y) = x^2 + xy + y^2$ above is convex because its Hessian is a positive semidefinite matrix.

Least squares is a convex quadratic minimization, since the Hessian of $\|\mathbf{Ab} - \mathbf{c}\|_2^2$ is $2\mathbf{A}^T\mathbf{A}$ (just take the derivative of the left side of (4.3) with respect to $\mathbf{b}$), which is always positive semidefinite no matter what $\mathbf{A}$ is. This is analogous to $x^2 \geq 0$ even if $x < 0$.

We can also define convexity for functions that are not smooth enough to have second derivatives. Roughly speaking, convex functions curve *upwards*. But since all functions we will see have first and second derivatives, let us stick to this easily verifiable, second-derivative definition.

It turns out that convex optimization problems are easy to solve, both in theory and in practice, almost as easy to solve as linear programming problems. One justification is that, for a convex optimization problem, any locally optimal solution (i.e., no worse than other feasible solutions in a small neighborhood of this solution) is also a globally optimal solution (i.e., no worse than any other feasible solution). We will see several more justifications in later chapters. The "watershed" between easy and hard optimization is *convexity*, rather than linearity. This has been a very useful realization in the optimization theory community over the past two decades.

4.2.3   Quick detour: Baseline predictor with temporal models

If you compute the model parameters $\{b_u, b_i\}$ of the baseline predictor through the above least squares (4.5) using the training data $\{r_{ui}\}$, the accuracy of the prediction in the probe set or the quiz set will not be that impressive. But here is an idea that can substantially reduce the RMSE: incorporate temporal effects into the baseline model parameters.

Movies went in and out of fashion over the period of more than five years in the Netflix Prize's datasets. So b_i should not be just one number for user i, but a function that depends on what day it is. These movie trends do not shift substantially from one day to the next, but if we bin all the days in five years into thirty bins, each about ten weeks long, we might expect some shift in b_i across the bins. Let us denote that shift, whether positive or negative, as $b_{i,bin(t)}$, where time t is measured in days but then binned into ten-week periods. Then we have thirty additional model parameters for each movie i, in addition to the time-independent b_i:

$$b_i(t) = b_i + b_{i,bin(t)}.$$

Each user's taste also changes over time. But taste is trickier to model than a movie's temporal effect. There is a continuous component $\sigma_u(t)$: user u's rating standard changes over time, and this deviation can be measured in several ways. There is also a *discrete* component $b_{u,t}$: big fluctuations on each day that a user rates movies. Why is that so? One reason is that a user account on Netflix is often shared by a family, and the actual person giving the rating might be a different one on any given day. Another is the "batch-rating effect:" when a user decides to rate many movies on the same day, these ratings often fall into a much narrower range than they would have if not batch processed. This was one of the key insights among the hundreds of tricks used by the leading teams in the Netflix Prize. Putting these together, we have the following time-dependent bias term per user:

$$b_u(t) = b_u + \sigma_u(t) + b_{u,t}.$$

Now our baseline predictor becomes time-dependent:

$$\hat{r}_{ui}(t) = \bar{r} + b_i(t) + b_u(t).$$

The least squares problem in (4.5) remains the same if we substitute $\hat{r}_{ui}$ by $\hat{r}_{ui}(t)$, and add regularization terms for all the additional time-dependent parameters defining $b_i(t)$ and $b_u(t)$.

We just added a lot of model parameters to the baseline predictor, but doing so was well worth it. Even if we stop right here with just the baseline predictor,

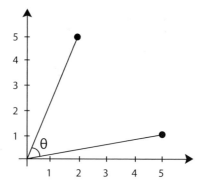

Figure 4.7 Given two points (in this case, in two-dimensional space), we can measure how close they are in terms of the size of the angle between the lines connecting each point to the origin. In the case of movie recommendation, The cosine of this angle θ is the cosine similarity measure if we view these two points as two users' ratings and one axis for each movie.

without any discovery and exploitation of user–movie interactions in the training set, we can already achieve an RMSE of 0.9555, not far from the benchmark, Cinematch's RMSE of 0.9514. Further, the two families of methods we will describe next can also benefit from temporal-effect models.

4.2.4 Neighborhood method: Similarity measure and weighted prediction

We have said nothing about collaborative filtering yet. Everything so far is about taking averages in a row or in a column of the rating matrix. What about the structures in the network of interactions among movies and users? We now move on to extract such structures in the table shown in Figure 4.3. The neighborhood method is one of the two main approaches. It relies on *pairwise statistical correlation.*

- *User–user correlation*: Two users with similar ratings of movies in the training set are row-wise "neighbors:" they are likely to have similar ratings for a movie in the quiz set. If one of the users rates movie i with 4 stars, the other user is likely to rate it with 4 stars too.
- *Movie–movie correlation*: Two movies that got similar ratings from users in the training set are column-wise "neighbors:" they are likely to have similar ratings by a user in the quiz set. If one of the movies is rated 4 stars by user u, the other movie is likely to be rated 4 stars by this user too.

Both arguments sound intuitive. We mentioned user–user correlation in the first section, and we will focus on movie–movie correlation now.

Since there are so many (u, i) pairs in the training data, of course we want to leverage more than just one neighbor. "Neighbor" here refers to closeness in movies' styles. Given a movie i, we want to pick its L nearest neighbors, where "distance" among movies is measured by a **similarity metric**. A standard similarity metric is the **cosine coefficient** illustrated in Figure 4.7, where we view each pair of columns in the $\tilde{\mathbf{R}}$ matrix as two vectors $\mathbf{r}_i$ and $\mathbf{r}_j$ in the Euclidean space, and the cosine of their angle is

$$d_{ij} = \frac{\mathbf{r}_i^T \mathbf{r}_j}{\|\mathbf{r}_i\|_2 \|\mathbf{r}_j\|_2} = \frac{\sum_u \tilde{r}_{ui} \tilde{r}_{uj}}{\sqrt{\sum_u (\tilde{r}_{ui})^2 \sum_u (\tilde{r}_{uj})^2}}, \tag{4.6}$$

where the summation is of course only over those users u that rated both movies i and j in the training set. We can now collect all the $\{d_{ij}\}$ in an $M \times M$ matrix $\mathbf{D}$ that summarizes all pairwise movie–movie similarity values.

Now, for a given movie i, we rank all the other movies, indexed by j, in descending order of $|d_{ij}|$. Then we pick those top L movies as the neighbors that will count in neighborhood modeling. L can be some integer between 1 and $M - 1$. Call this set of neighbors $\mathcal{L}_i$ for movie i.

We say the predicted rating is simply the baseline predictor plus a weighted sum of the ratings from these neighbor movies (and normalized by the weights). There can be many choices for these weights w_{ij}. One natural, though suboptimal, choice is to simply let $w_{ij} = d_{ij}$. There is actually no particular reason why the similarity measure must also be the prediction weight, but for simplicity let us stick with that.

Then we have the following predictor, where N in $\hat{r}_{ui}^N$ denotes "neighborhood method:"

$$\hat{r}_{ui}^N = (\bar{r} + b_u + b_i) + \frac{\sum_{j \in \mathcal{L}_i} d_{ij} \tilde{r}_{uj}}{\sum_{j \in \mathcal{L}_i} |d_{ij}|}. \tag{4.7}$$

This equation is the most important one for solving the Netflix recommendation problem in this chapter. So we pause a little to examine it.

- The three terms inside the brackets simply represent the estimate before taking advantage of similarities of users or movies.
- The quotient term is the weighted sum of the intelligence we gathered from collaborative filtering. The weights $\{d_{ij}\}$ can be positive or negative, since it is just as helpful to know two movies are very different as to know they are very much alike. Of course, the magnitude-normalization term in the denominator needs to take the absolute value of $\{d_{ij}\}$ to avoid a cancellation of the effects.

There are quite a few extensions to the neighborhood predictor shown above. Choosing an appropriate L and generalizing the choice of weights w_{ij} are two of those extensions. We can also throw away those neighbors that have very few overlaps, e.g., if very few users rated both movies 1 and 2, then we might not want to count movie 2 as a neighbor of movie 1 even if d_{12} is large. We also want to discount those users that rate too many movies as all high or all low. These users tend to be not so useful in recommendation systems.

Now we can again collect all these predictors into a matrix $\hat{\mathbf{R}}^N$ and use that instead of just the baseline predictors in the matrix $\hat{\mathbf{R}}$.

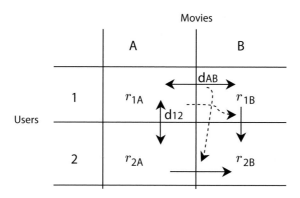

Movies

Figure 4.8 Two methods of collaborative filtering: user-user correlation vs. movie-movie correlation. Suppose we want to predict user 2's rating of movie B. We can either exploit the similarity between movies A and B and r_{2A}, or exploit the similarity between users 1 and 2 and r_{1B}.

We have seen two styles of neighborhood models: user–user and movie–movie. But, as you might suspect, these two are really two sides of the same coin, creating different "chains" of connection in the table of ratings. This is illustrated in Figure 4.8.

4.2.5 Summary

The procedure developed so far can be summarized into five steps:

1. Train a baseline predictor by solving a least squares (4.2).
2. Obtain the baseline prediction matrix $\hat{\mathbf{R}}$, and shift $\mathbf{R}$ by $\hat{\mathbf{R}}$ to get $\tilde{\mathbf{R}}$.
3. Compute the movie–movie similarity matrix $\mathbf{D}$.
4. Pick a neighborhood size L to construct a neighborhood of movies $\mathcal{L}$ for each movie i.
5. Compute the sum of the baseline predictor and the neighborhood predictor (4.7) as the final prediction for each (u, i) pair. This gives us the neighborhood-prediction matrix $\hat{\mathbf{R}}^N$.

At the end of the above steps, Netflix can pick the top few unwatched and highly rated (by its prediction) movies to recommend to each user u. Or, as in the case of the Netflix Prize, it can compute the RMSE against the ground truth in quiz and test sets to determine the prize winner.

4.3 Examples

We illustrate the baseline predictor and the neighborhood predictor with a simple example. The matrix $\mathbf{R}$ in this example consists of forty hypothetical ratings from ten users on five movies. The five columns are labeled from A to E, and the ten rows from 1 to 10. This is an 80% dense matrix, *much* denser than a real system's data, but we are constrained by the size of a matrix that can be written out on one page. We will use thirty randomly chosen ratings as the training set, and the remaining ten, in boldface, as the test set. The other ten missing ratings are denoted by "–". The average rating of the training data is $\bar{r} = 3.83$.

$$
\mathbf{R} = \begin{bmatrix}
5 & 4 & 4 & - & \mathbf{5} \\
- & 3 & 5 & \mathbf{3} & 4 \\
5 & 2 & - & \mathbf{2} & 3 \\
- & \mathbf{2} & 3 & 1 & 2 \\
4 & - & \mathbf{5} & 4 & 5 \\
\mathbf{5} & 3 & - & 3 & 5 \\
3 & \mathbf{2} & 3 & 2 & - \\
5 & \mathbf{3} & 4 & - & 5 \\
4 & 2 & \mathbf{5} & 4 & - \\
\mathbf{5} & - & 5 & 3 & 4
\end{bmatrix}
$$

4.3.1   Baseline predictor

The baseline predictor minimizes the sum of squares of all the elements in the vector $\mathbf{Ab} - \mathbf{c}$, where the dimension of $\mathbf{A}$ is 30×15 since there are 30 training data points and 15 variables (for 10 movies and 5 users):

$$
\begin{bmatrix}
1 & 0 & 0 & \cdots & 0 & 1 & 0 & \cdots & 0 \\
0 & 0 & 1 & \cdots & 0 & 1 & 0 & \cdots & 0 \\
\vdots & & & \ddots & & & & & \vdots \\
0 & 0 & 0 & \cdots & 1 & 0 & 0 & \cdots & 1
\end{bmatrix}
\begin{bmatrix}
b_1 \\ b_2 \\ \vdots \\ b_{10} \\ b_A \\ b_B \\ \vdots \\ b_E
\end{bmatrix}
-
\begin{bmatrix}
r_{1A} - \bar{r} \\ r_{3A} - \bar{r} \\ \vdots \\ r_{10E} - \bar{r}
\end{bmatrix}.
$$

Solving the above system of linear equations with 30 equations and 15 variables, we find the optimal user bias,

$$
\mathbf{b}_u^* = [0.62\ 0.42\ -0.28\ -1.78\ 0.52\ 0.49\ -1.24\ 0.45\ 0.40\ 0.23]^T,
$$

and the optimal movie bias

$$
\mathbf{b}_i^* = [0.72\ -1.20\ 0.60\ -0.60\ 0.33]^T.
$$

These values quantify the intuition from what we observe from the training data. For example:

1. users 1, 2, 5, 6, and 8 tend to give higher ratings;
2. users 4 and 7 tend to give lower ratings;

3. movies A and C tend to receive higher ratings;
4. movies B and D tend to receive lower ratings.

We clip any predicted rating lower than 1 to 1 and any higher than 5 to 5, since it is ranking data we are dealing with. The rating matrix estimated by the baseline predictor (after clipping) is as follows:

$$
\hat{\mathbf{R}} =
\begin{bmatrix}
5.00 & 3.09 & 4.90 & - & \mathbf{4.62} \\
- & 2.89 & 4.69 & \mathbf{3.49} & 4.42 \\
4.10 & 2.19 & - & \mathbf{2.78} & 3.71 \\
- & \mathbf{1.00} & 2.49 & 1.29 & 2.22 \\
4.90 & - & \mathbf{4.79} & 3.58 & 4.51 \\
\mathbf{4.88} & 2.96 & - & 3.56 & 4.48 \\
3.15 & \mathbf{1.23} & 3.03 & 1.82 & - \\
4.84 & \mathbf{2.92} & 4.72 & - & 4.44 \\
\mathbf{4.78} & 2.87 & 4.67 & 3.46 & - \\
\mathbf{4.61} & - & 4.49 & 3.29 & 4.22
\end{bmatrix}.
$$

We compute the RMSE between $\mathbf{R}$ and $\hat{\mathbf{R}}$ above. It is 0.51 for the training set and 0.58 for the test set.

The baseline predictor with parameters trained through least squares suffers from the overfitting problem. Take the predicted rating for movie B by user 4 as an example: $\hat{r}_{4B} = \bar{r} + b_4 + b_B = 3.67 - 1.78 - 1.20 = 0.69$, significantly lower than the real rating $r_{4B} = 2$. Indeed, the test data's RMSE is higher than training data's. We will see a solution for this problem in the Advanced Material.

4.3.2 Neighborhood model

The neighborhood model goes one step further to extract the user–movie interactions. We start with the differences between the raw ratings of the training set and the corresponding biases captured by the (unregularized) baseline predictor:

$$
\tilde{\mathbf{R}} = \mathbf{R} - \hat{\mathbf{R}} =
\begin{bmatrix}
0 & 0.91 & -0.90 & - & ? \\
- & 0.11 & 0.31 & ? & -0.42 \\
0.90 & -0.19 & - & ? & -0.71 \\
- & ? & 0.51 & -0.29 & -0.22 \\
-0.90 & - & ? & 0.42 & 0.49 \\
? & 0.040 & - & -0.56 & 0.52 \\
-0.15 & ? & -0.031 & 0.18 & - \\
0.16 & ? & -0.72 & - & 0.56 \\
? & -0.87 & 0.33 & 0.54 & - \\
? & - & 0.51 & -0.29 & -0.22
\end{bmatrix}.
$$

We used ? to indicate test-set data, and − to denote unavailable data (e.g., user 1 never rated movie D).

We use the cosine coefficient to measure the similarity between movies represented in $\tilde{\mathbf{R}}$. Take the calculation of the similarity between movie B and movie C as an example. According to the training data in $\tilde{\mathbf{R}}$, users 1, 2, and 9 rated both movies. Therefore,

$$
\begin{aligned}
d_{BC} &= \frac{\tilde{r}_{1B}\tilde{r}_{1C} + \tilde{r}_{2B}\tilde{r}_{2C} + \tilde{r}_{9B}\tilde{r}_{9C}}{\sqrt{(\tilde{r}_{1B}^2 + \tilde{r}_{2B}^2 + \tilde{r}_{9B}^2)(\tilde{r}_{1C}^2 + \tilde{r}_{2C}^2 + \tilde{r}_{9C}^2)}} \\
&= \frac{(0.91)(-0.90) + (-0.11)(0.31) + (-0.87)(0.33)}{\sqrt{(0.91^2 + 0.11^2 + 0.87^2)(0.90^2 + 0.31^2 + 0.33^2)}} \\
&= -0.84.
\end{aligned}
$$

Similarly, we can calculate the entire similarity matrix, a 5×5 symmetric matrix (where the diagonal entries are not of interest since they concern the same movie):

$$
\mathbf{D} = \begin{bmatrix}
- & -0.20 & -0.45 & -0.97 & -0.75 \\
-0.20 & - & -0.84 & -0.73 & 0.51 \\
-0.45 & -0.84 & - & -0.22 & -0.93 \\
-0.97 & -0.73 & -0.22 & - & 0.068 \\
-0.75 & 0.51 & -0.93 & 0.068 & -
\end{bmatrix}.
$$

With the above pairwise movie–movie similarity values, we can carry out the following procedure to compute $\hat{r}_{ui}$.

1. Find $L = 2$ movie neighbors with the largest absolute similarity values, $|d_{ik}|$ and $|d_{il}|$.

2. Check whether user u has rated both movies k and l. If so, use both, as in the formula below. If the user rated only one of them, then just use that movie. If the user rated neither, then do not use the neighborhood method.

3. Calculate the predicted rating $\hat{r}_{u,i} = (\bar{r} + b_u + b_i) + \frac{d_{ik}\tilde{r}_{uk} + d_{il}\tilde{r}_{ul}}{|d_{ik}| + |d_{il}|}$.

Take $\hat{r}_{3D}$ for an example. The two nearest neighbors for movie D are movie A and movie B, whose cosine coefficients are -0.97 and -0.73 respectively. User 3 has rated both movie A and movie B. Hence, we compute $\hat{r}_{3D}$ as

$$
(\bar{r} + b_3 + b_D) + \frac{d_{DA}\tilde{r}_{3A} + d_{DB}\tilde{r}_{3B}}{|d_{DA}| + |d_{DB}|} = 2.78 + \frac{-0.97 \times 0.90 + (-0.73) \times (-0.19)}{0.97 + 0.73} = 2.35.
$$

In the baseline predictor, $\hat{r}_{3D} = 2.78$. So the neighborhood predictor reduces the error compared to the ground truth of $r_{3D} = 2$.

Similarly, the closest neighbors for movie B are movies C and D. From the training set, we know that user 2 rated movie C but not movie D. Hence $\hat{r}_{2B} = (\bar{r} + b_2 + b_B) + d_{BC}\tilde{r}_{2C}/|d_{BC}| = 2.89 - 0.31 = 2.58$.

The predictions by the neighborhood model (after clipping) are given below:

$$
\hat{\mathbf{R}}^N =
\begin{bmatrix}
5.00 & 3.99 & 3.99 & - & \mathbf{5.00} \\
- & 2.58 & 4.86 & \mathbf{3.38} & 4.11 \\
4.81 & 2.19 & - & \mathbf{2.35} & 2.81 \\
- & \mathbf{1.00} & 2.71 & 1.29 & 1.71 \\
4.46 & - & \mathbf{4.30} & 4.49 & 5.00 \\
\mathbf{4.97} & 3.52 & - & 3.52 & 4.48 \\
2.97 & \mathbf{1.16} & 3.03 & 1.97 & - \\
4.28 & \mathbf{3.64} & 4.16 & - & 4.77 \\
\mathbf{4.25} & 2.44 & 5.00 & 4.33 & - \\
\mathbf{4.87} & - & 4.71 & 3.29 & 3.71
\end{bmatrix}
$$

The training error and test error using the neighborhood model are 0.32 and 0.54, respectively, compared with 0.51 and 0.58 using the baseline predictor. This represents a 7% improvement in RMSE for the test set.

We could also take a look at the errors in terms of *Hamming distance*

$$
\sum_{(u,i)} |r_{ui} - \hat{r}_{ui}|.
$$

If we just look at the wrong predictions (after rounding up or down to integers), it turns out all of them are off by one star. But the baseline predictor gets 17 wrong, whereas the neighborhood predictor gets 8 wrong, a 53% reduction in error.

Even in this extremely small example, the movie-based neighborhood method manages to take advantage of the statistical relationship between the movies and outperform the baseline predictor both for the training and for the test data sets. In much larger and more realistic cases, neighborhood methods have a lot more information to act on and can reduce the RMSE further relative to a baseline predictor.

4.4 Advanced Material

4.4.1 Regularization: Robust learning without overfitting

Learning is both an exercise of *hindsight* and one of *foresight*: it should generate a model that fits training data, but also make predictions that match unseen data. There is a tradeoff between the two. You need reasonable hindsight to have foresight, but perfect hindsight often means you are simply re-creating history. This is called overfitting, and is a particularly common problem when there is a lot of noise in your data (you end up modeling the noise), or there are too many parameters relative to the amount of training data (you end up creating a model tailor-made for the training data). Eventually it is the foresight that matters.

Overfitting is intuitively clear: you can always develop a model that fits training data perfectly. For example, my recommendation system for Netflix might be as follows: Alice rates 4 stars to a comedy starring Julia Roberts if she watches it on April 12, and 5 stars to a drama starring Al Pacino if she watches on March 25, etc. I can make it fit the training data perfectly by adding enough model parameters, such as lead actors and actresses, date of the user watching the movie, etc. But then, the model is no longer simple, robust, or predictive.

We will go through an example of overfitting in learning in a homework problem. There are several ways to avoid overfitting. Regularization is a common one: add a penalty term that reduces the sensitivity to model parameters by rewarding smaller parameters. We measure size by some norm, and then we add this penalty term to the objective function that quantifies learning efficiency. In our case, that efficiency is captured by the RMSE. To maintain the least squares structure of the optimization over model parameters, it is common practice to add the L-2 norm of the parameter values as the penalty term to control hindsight. The optimization problem becomes

$$\text{minimize}_{\{\text{model parameters}\}} \; (\text{Squared error term}) + \lambda \, (\text{Parameter size squared}).$$

This problem is called a **regularized least squares**, and it can be efficiently solved, just like the original least squares. The theme in regularization is to use parameters with smaller magnitudes to avoid "over-optimization." This goal can also be achieved by using fewer parameters.

In the baseline predictor for Netflix recommendation, adding the regularization term $\lambda(\sum_u b_u^2 + \sum_i b_i^2)$ limits the magnitudes of the user and movie biases, which in turn helps bring the test error more in line with the training error. For the same example as in the last section, Figure 4.9 shows how the training and test errors vary with respect to different values of λ. The test error first falls with more regularization and then starts to rise once λ exceeds an optimal value, which happens to be about 1 in this example.

After regularization with $\lambda = 1$, $\hat{\mathbf{R}}$ becomes

$$\hat{\mathbf{R}} = \begin{bmatrix}
4.70 & 3.23 & 4.61 & - & \mathbf{4.40} \\
- & 3.05 & 4.44 & \mathbf{3.38} & 4.23 \\
4.01 & 2.53 & - & \mathbf{2.85} & 3.71 \\
- & \mathbf{1.47} & 2.86 & 1.80 & 2.65 \\
4.67 & - & \mathbf{4.58} & 3.52 & 4.38 \\
\mathbf{4.54} & 3.07 & - & 3.39 & 4.25 \\
3.37 & \mathbf{1.90} & 3.28 & 2.22 & - \\
4.66 & \mathbf{3.18} & 4.57 & - & 4.36 \\
\mathbf{4.49} & 3.02 & 4.40 & 3.34 & - \\
\mathbf{4.45} & - & 4.36 & 3.29 & 4.15
\end{bmatrix},$$

The resulting RMSE is 0.56 for the training set and 0.50 for the test set. Contrast this with 0.51 and 0.58 for the training and test sets, respectively, without regularization.

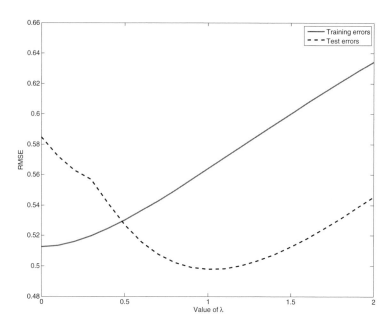

Figure 4.9 The effect of regularization on overfitting for the baseline predictor. As the regularization weight parameter λ increases, the training set's RMSE increases but the test set's RMSE drops before λ becomes too large. In hindsight, we see that $\lambda = 1$ turns out to be the right regularization weight in this numerical example. Setting the right λ in each regularized least squares can be tricky.

The tricky part is to set the weight λ just right so as to strike the balance between hindsight and foresight. There are several techniques to do that, such as **cross validation**, where we divide the available training data into K parts. In each of K rounds of cross-validation, we leave out one of the K parts as test data. Of course, a larger K provides more opportunities for validation, but leaves a smaller test data sample per round.

4.4.2 Latent-factor method: matrix factorization and alternating projection

There are two major methods in collaborative filtering. We have seen one, and will now turn to the other. The latent-factor method relies on *global structures* underlying the table in Figure 4.3. One of the challenges in recommendation system design is that the table is both *large* and *sparse*. In the case of the Netflix Prize, the table has about $N = 480000$ times $M = 17770$ (i.e., more than 8.5 billion cells), yet only slightly more than 1% of those are occupied with ratings.

But we suspect there may be structures that can be captured by two, much smaller matrices. We suspect that the similarities among users and movies are not just a statistical fact, but are actually induced by some *low-dimensional* structures hidden in the data. Therefore, we want to build a low-dimensional model for these high-dimensional data.

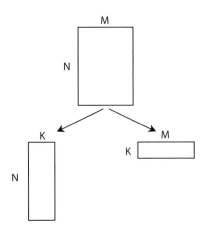

Figure 4.10 Factorize matrix $\mathbf{R}$ into a skinny matrix $\mathbf{P}$ and a thin matrix $\mathbf{Q}$. In the case of the Netflix Prize, $N = 480000$ and $M = 17770$, both of which are very large, whereas the latent factor dimension is between $K = 10$ and 200, much smaller than either N or M.

How about using a K-dimensional vector $\mathbf{p}_u$ to explain each user u's movie taste. And for each movie i, we use a K-dimensional vector $\mathbf{q}_i$ explaining the movie's appeal. The inner product between these two vectors, $\mathbf{p}_u^T \mathbf{q}_i$, is the prediction $\hat{r}_{ui}$.

Typical numbers of K for the Netflix Prize were between 10 and 200. Let us consider a toy example where K is 2. One dimension is whether the movie is romantic or action, and the other is long or short. Let us say Alice likes romantic movies but is not too much into long movies: $\mathbf{p}_u = [1, \ 0.7]^T$. Let us say "Gone With the Wind" is pretty romantic and long: $\mathbf{q}_i = [2.5, \ 2]^T$. Then their inner product is $\mathbf{p}_u^T \mathbf{q}_i = 1 \times 2.5 + 0.7 \times 2 = 3.9$.

Now our job in the latent factor method is to fix K, the number of dimensions, and then optimize over all these latent factor vectors $\{\mathbf{p}_u, \mathbf{q}_i\}$ to minimize the Mean Squared Error (MSE) on the basis of the given ratings $\{r_{ui}\}$ in the training set:

$$\text{minimize}_{\mathbf{P},\mathbf{Q}} \sum_{(u,i)} \left(r_{ui} - \mathbf{p}_u^T \mathbf{q}_i\right)^2. \tag{4.8}$$

Here, we collect the $\mathbf{p}_u$ vectors across all the users to form an $N \times K$ matrix $\mathbf{P}$, and we collect the $\mathbf{q}_i$ vectors across all the movies to form a $K \times M$ matrix $\mathbf{Q}$. Of course, we should also add quadratic regularization terms, penalizing the sum of squares of all the entries in $(\mathbf{P}, \mathbf{Q})$.

On solving (4.8) we obtain $\mathbf{P}^*$ and $\mathbf{Q}^*$, and the resulting latent-factor prediction matrix is

$$\hat{\mathbf{R}}^L = \mathbf{P}^* \mathbf{Q}^*.$$

How many model parameters are we talking about here? If $K = 200$, then we have $200 \times (480000 + 17770)$, a little under 100 million parameters. That is almost the same as the number of the training set's rating data points. We can view $(\mathbf{P}, \mathbf{Q})$ as a matrix factorization of the much larger $N \times M$ matrix $\mathbf{R}$, as illustrated in Figure 4.10. By picking a K value so that $K(N + M)$ is about the same as the number of entries in the large, sparse matrix $\mathbf{R}$, we turn the sparsity in the high-dimensional data (given to us) into structures in a low-dimensional model (constructed by us).

How do we solve problem (4.8), or its regularized version? It might look like a least squares, but it is not. *Both* $\mathbf{p}_u$ and $\mathbf{q}_i$ are optimization variables now. In fact, it is not even a convex optimization. However, if we hold $\mathbf{P}$ as constants and vary over $\mathbf{Q}$, or vice versa, it reduces to a least squares problem that we know how to solve. So a standard way to solve (4.8) is by **alternating projections**: hold $\mathbf{P}$ as constants and vary over $\mathbf{Q}$ in a least squares, then hold $\mathbf{Q}$ as constants and vary over $\mathbf{P}$ in another least squares, and repeat until convergence. Since this is a nonconvex optimization problem, the converged point $(\mathbf{P}^*, \mathbf{Q}^*)$ might not be globally optimal.

Coming back to those latent factors, what exactly are they? Many models may be able to fit given observations, but we hope some of them can also be explained through intuition. In the neighborhood method, we have an intuitive explanation of statistical correlation. But in the latent-factor method, precisely where we expect to see some meaning of these latent factors, we actually do not have intuitive labels attached to all the K dimensions. If the prediction works out, these K dimensions are telling us something, but we might not be able to explain what that "something" is. Some people view this as a clever strength: extracting structures even when words fail us. Others consider it an unsatisfactory explanation of a "trick."

In any case, a straightforward implementation of the latent-factor model, with the help of temporal effect modeling, can give an RMSE of 0.8799. That is about an 8% improvement over Cinematch, and enough to get you close to the leading position in the 2007 progress prize.

The difference between the neighborhood model and the latent-factor model lies not just in statistical vs. structural understanding, but also in the fact that the neighborhood model utilizes only pairwise (local) interactions, whereas the latent-factor model leverages global interactions. But we can expand the neighborhood model to incorporate global interactions too. In fact, it has recently been shown that neighborhood models become equivalent to a particular matrix factorization model.

By combining the neighborhood method and latent-factor method, with the help of temporal effect models and implicit feedback models, and after blending in other tricks and carefully tuning all the parameters with the training set, about two years after the 8% improvement was achieved, we finally saw the 10% mark barely reached by the two teams in July 2009.

Summary

Box 4 Recommendation by collaborative filtering

A recommendation system makes predictions about each user's preference over a set of products. Collaborative filtering methods leverage the structures across all the users and the products, such as the similarities among users or among products. Minimizing RMSE as the prediction error metric sometimes lead to least squares, a special case of convex optimization. Regularizing a least squares helps avoid overfitting.

Further Reading

The Netflix Prize was widely reported in popular science magazines from 2006 to 2009.

1. The original announcement by Netflix in October 2006 can be found here, along with links to the winning team's algorithms:
   The Netflix Prize, `http://www.netflixprize.com`

2. A non-technical survey of the first two years of progress can be found at
   R. Bell, J. Bennett, Y. Koren, and C. Volinsky, "The million dollar programming prize," *IEEE Spectrum*, vol. 46, no. 5, pp. 28–33, May 2009.

3. A more technical survey can be found in the following book chapter:
   Y. Koren and R. Bell, "Advances in collaborative filtering," *Recommender Systems Handbook*, Springer, 2011.

4. Another survey written by part of the winning team of the Netflix Prize focuses on factorization methods, which we followed in the Advanced Material:
   Y. Koren, R. Bell, and C. Volinsky, "Matrix factorization techniques for recommender systems," *IEEE Computer*, vol. 42, vol. 8, pp. 30–77, August 2009.

5. This chapter introduces the basic notion of convex optimization. An excellent textbook of convex optimization and applications is
   S. Boyd and L. Vandenberghe, *Convex Optimization*, Cambridge University Press, 2004.

Problems

4.1 *Baseline predictor* ⋆

Compute the baseline predictor $\hat{\mathbf{R}}$ based on the following raw data matrix $\mathbf{R}$:

$$\mathbf{R} = \begin{bmatrix} 5 & - & 5 & 4 \\ - & 1 & 1 & 4 \\ 4 & 1 & 2 & 4 \\ 3 & 4 & - & 3 \\ 1 & 5 & 3 & - \end{bmatrix}.$$

(Hint: This involves a least squares with sixteen equations and nine variables. Feel free to use any programming language. For example, the backslash operator or `pinv()` in Matlab can be helpful. If there are multiple solutions to the least squares problem, take any one of those.)

4.2 *Neighborhood predictor* ★ ★ ★

Using the given $\mathbf{R}$ and the computed $\hat{\mathbf{R}}$ from the previous question, compute the neighborhood predictor $\hat{\mathbf{R}}^N$ with $L = 2$. Computer neighbors across the columns (movies).

4.3 *Least squares* ★★

(a) Solve for $\mathbf{b}$ in the following least squares problem, by hand or using any programming language:

$$\text{minimize}_{\mathbf{b}} \quad \|\mathbf{Ab} - \mathbf{c}\|_2^2,$$

where

$$\mathbf{A} = \begin{bmatrix} 1 & 0 & 2 \\ 1 & 1 & 0 \\ 0 & 2 & 1 \\ 2 & 1 & 1 \end{bmatrix} \quad \text{and} \quad \mathbf{c} = \begin{bmatrix} 2 \\ 1 \\ 1 \\ 3 \end{bmatrix}.$$

(b) Solve the above least squares problem again with regularization. Vary the regularization parameter λ for $\lambda = 0, 0.2, 0.4, \ldots, 5.0$, and plot both $\|\mathbf{Ab} - \mathbf{c}\|_2^2$ and $\|\mathbf{b}\|_2^2$ against λ.

(Hint: take the derivative of

$$\|\mathbf{Ab} - \mathbf{c}\|_2^2 + \lambda\|\mathbf{b}\|_2^2$$

with respect to $\mathbf{b}$ to obtain a system of linear equations.)

4.4 *Convex functions* ★

Determine whether the following functions are convex, concave, both, or neither:

(a) $f(x) = 3x + 4$, for all real x;

(b) $f(x) = 4\ln(x/3)$, for all $x > 0$;

(c) $f(x) = e^{2x}$, for all real x;

(d) $f(x, y) = -3x^2 - 4y^2$, for all real x and y;

(e) $f(x, y) = xy$, for all real x and y.

4.5　*Log-Sum-Exp and geometric programming* ⋆⋆

(a) Is $\exp(x + y)$ convex or concave in (x, y)?

(b) Is $\exp(x + y) + \exp(2x + 5y)$ convex or concave in (x, y)?

(c) Is $\log(\exp(x+y)+\exp(2x+5y))$ convex or concave in (x, y)? This log-sum-exp function is heavily used in a class of convex optimization called **geometric programming** in fields like statistical physics, chemical engineering, communication systems, and circuit design.

(d) Can you turn the following problem in variables $x, y, z > 0$ into convex optimization?

$$\begin{array}{ll} \text{minimize} & xy + xz \\ \text{subject to} & x^2 yz + xz^{-1} \le 10 \\ & 0.5x^{-0.5}y^{-1} = 1. \end{array}$$

(Hint: try a log change of variables.)

5   When can I trust an average rating on Amazon?

In this and the next three chapters, we will walk through a remarkable landscape of intellectual foundations. But sometimes we will also see significant gaps between theory and practice.

5.1   A Short Answer

We continue with the theme of recommendation. Webpage ranking in Chapter 3 turns a graph into a rank-ordered list of nodes. Movie ranking in Chapter 4 turns a weighted bipartite user–movie graph into a set of rank-ordered lists of movies, with one list per user. We now examine the aggregation of a vector of rating scores by reviewers of a product or service, and turn that vector into a scalar for each product. These scalars may in turn be used to rank order a set of similar products. In Chapter 6, we will further study aggregation of many vectors into a single vector.

When you shop on Amazon, likely you will pay attention to the number of stars shown below each product. But you should also care about the number of reviews behind that averaged number of stars. Intuitively, you know that a product with two reviews, both 5 stars, might not be better than a competing product with one hundred reviews and an average of 4.5 stars, especially if these one hundred reviews are all 4 and 5 stars and the reviewers are somewhat trustworthy. We will see how such intuition can be sharpened.

In most online review systems, each review consists of three fields:

1. rating, a numerical score often on the scale of 1–5 stars (this is the focus of our study),
2. review, in the form of text, and
3. review of review, often a binary up or down vote.

Rarely do people have time to read through all the reviews, so a summary review is needed to aggregate the individual reviews. In particular, we need to aggregate a vector of rating numbers into a single number, so that a ranking of similar products can be generated from these ratings. What is a proper aggregation? That is the subject of this chapter.

Reviews are sometimes not very trustworthy, yet they are important in so many contexts, from peer reviews in academia to online purchases of every kind. The hope is that the following two approaches can help.

First, we need methods to ensure some level of accuracy, screening out the really "bad" ratings. Unlimited and anonymous reviews have notoriously poor quality, because a competitor may enter many negative reviews, the seller herself may enter many positive reviews, or someone who has never even used the product or service may enter random reviews. So before anything else, we should first check the mechanism used to enter reviews. Who can enter reviews? How strongly are customers encouraged, or even rewarded, to review? Do you need to enter a review of reviews before you are allowed to upload your own review? Sometimes a seemingly minor change in formatting leads to significant differences: Is it a binary thumbs-up or thumbs-down, followed by a tally of up vs. down votes? What is the dynamic range of the numerical scale? It has been observed that the scale of 1–10 often returns 7 as the average, with a bimodal distribution around it. A scale of 1–3 gives a very different psychological hint to the reviewers compared to a scale of 1–5, or a scale of 1–10 compared to −5 to 5.

Second, the review population size needs to be large enough. But *how* large is large enough? And can we run the raw ratings through some signal processing to get the most useful aggregation?

These are tough questions with no good answers yet, and are not even well-formulated problem statements. The first question depends on the nature of the product being reviewed. Movies (e.g., on IMDb) are very subjective, whereas electronics (e.g., on Amazon) are much less so, with hotels (e.g., on tripadvisor) and restaurants (e.g., on opentable) somewhere in between. It also depends on the quality of the review, although reputation of the reviewer is a difficult metric to quantify in its own right.

The second question depends on the metric of "usefulness." Each user may have a different metric, and the seller of the product or the provider of the service may use yet another one. This lack of clarity in what should be optimized is the crux of the ill-definedness of the problem at hand.

With these challenges, it may feel like opinion aggregation is unlikely to work well. But there have been notable exceptions recorded. A famous example is Galton's 1906 observation on a farm in Plymouth, UK, where 787 people in a festival there participated in a game of guessing the weight of an ox, each writing down a number *independently* of others. There was also no common bias; everyone could take a good look at the ox. While the estimates by each individual were all over the places, the average was 1197 pounds. It turned out the ox weighed 1198 pounds. Just a simple averaging worked remarkably well. For the task of guessing the weight of an ox, 787 was more than enough to get the right answer (within a margin of error of 0.1%).

But in many other contexts, the story is not quite as simple as Galton's experiment. There were several key factors here that made simple averaging work so well.

- The task was relatively easy; in particular, there was an objective answer with a clear numerical meaning.
- The estimates were both unbiased and independent of each other.
- There were enough people participating.

More generally, three factors are important in aggregating individual opinions.

- *Definition of the task*: Guessing a number is easy. Consensus formation in social choice is hard. Reviewing a product on Amazon is somewhere in between. Maybe we can define "subjectivity" by the size of the review population needed to reach a certain "stabilization number."
- *Independence of the reviews*: As we will see, the **wisdom of crowds**, if there is one to a degree we can identify and quantify, stems not from having many smart individuals in the crowd, but from the independence of each individual's view from the rest. Are Amazon reviews independent of each other? Kind of. Even though you can see the existing reviews before entering your own, usually your rating will not be significantly affected by the existing ratings. Sometimes, reviews are indeed entered as a reaction to recent reviews posted on the website, either to counter-argue or to reinforce points made there. This influence from the sequential nature of review systems will be studied in Chapter 7.
- *Review population*: For a given task and the degree of independence, there is correspondingly a minimum number of reviews, a threshold, needed to give a target confidence of trustworthiness to the average. If these ratings pass through some signal-processing filters first, then this threshold may be lowered.

What kind of signal processing do we need? For text reviews, there need to be tools from natural language processing to detect inconsistencies or extreme emotions in a review and to discount it. We in academia face this problem in each decision on a peer-reviewed paper, a funding proposal, a tenure-track position interview, and a tenure or promotion case.

For rating numbers, some kind of weighting is needed, and we will discuss a particularly well-studied one soon. In Chapter 6, we will also discuss voting methods, including majority rule, pairwise comparison, and positional counting. These voting systems require each voter to provide a complete ranking, and sometimes on a numerical rating scale. Therefore, we will have more information, perhaps too much information, as compared with our current problem in this chapter.

5.1.1   Challenges of rating aggregation

Back to rating aggregation. Here are several examples illustrating three of the key challenges in deciding when to trust ratings on Amazon.

Philips 22PFL4505D/F7 22-Inch 720p LED LCD HDTV, Black
Buy new: $349.99 $229.99
20 new
3 used from $210.00
Get it by **Tuesday, Sep 20** if you order in the next 71 hours and choose one-day shipping.
☆☆☆☆☆ ☑ (121)
Eligible for **FREE** Super Saver Shipping and **FREE** Returns.

Panasonic VIERA TC-L32C3 32-Inch 720p LCD HDTV
Buy new: $399.95 Click for product details
6 Used & new from $308.00
Get it by **Tuesday, Sep 20** if you order in the next 72 hours and choose one-day shipping.
☆☆☆☆☆ ☑ (55)
Eligible for **FREE** Super Saver Shipping and **FREE** Returns.

Figure 5.1 The tradeoff between review population and average rating score. Should a product with fewer reviews (55) but a higher average rating (4.5 stars) be ranked higher than a competing product with more reviews (121) but a lower average rating (4 stars)?

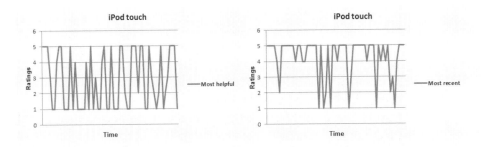

Figure 5.2 How to view the aggregated ratings: should it be based on helpful ratings or on the latest trend? The same set of iPod touch ratings on Amazon is used to extract two different subsets of ratings, and their values are quite different.

Example 1. Many online rating systems use a naive averaging method for their product ratings. Given that different products have different numbers of reviews, it is hard to determine which product has a better quality. For example, as in Figure 5.1, one day in 2011 on Amazon, Philips 22PFL4504D HDTV has 121 ratings with a mean of 4, while Panasonic VIERA TC-L32C3 HDTV has 55 ratings with a mean of 4.5. So the customer is faced with a tradeoff between choosing a product with a lower average rating and a larger number of reviews versus one with a higher average rating and a smaller number of reviews.

Example 2. Consider two speaker systems for home theater on Amazon. However, RCA RT151 and Pyle Home PCB3BK have comparable mean scores around 4. On the one hand, 51.9% of users gave RCA RT151 a rating of 5 stars while 7.69% gave 1 star. On the other hand, 54.2% of users gave 5 stars to Pyle Home PCB3BK while 8.4% gave 1 star. So Pyle Home PCB3BK has not only a higher percentage of people giving it 5 stars, but also a higher percentage of people giving it 1 star. There is a larger *variation* in the ratings of Pyle Home PCB3BK. Does that make the average rating more trustworthy or less?

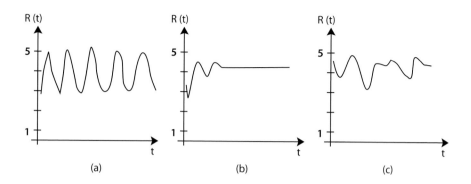

Figure 5.3 Three time series with the same long-term average rating but very different stabilization behaviors. Suppose the time-axis scale is on the order of weeks. Then curve (a) shows continued cyclic fluctuations of ratings over time; curve (b) shows a clear convergence; and curve (c) shows signs of convergence, but it is far from clear that the ratings have converged.

Example 3. In Figure 5.2, we compare the ratings of the top 60 "most helpful" reviews of iPod3 Touch (32 GB) on Amazon with those from the 60 most recent ratings. The mean of the most recent ratings is 1.5 times greater than the mean of the most helpful reviews. Does this reflect a "real" change or just normal fluctuations? What should the timescale of averaging be?

At the heart of these problems is the challenge of turning *vectors* into *scalars*, which we will meet again in Chapter 6. This can be a "lossy compression" with very different results depending on how we run the process of scalarization.

5.1.2   Beyond basic aggregation of ratings

We may run a time-series analysis to understand the dynamics of rating. In Figure 5.3, the three curves of ratings entered over a period of time give the same average, but "clearly" some of them have not converged to a stable average rating. What kind of *moving-window size* should we use to account for cumulative averaging and variance over time?

We may consider detecting anomalous ratings and throwing out the highly suspicious ones. If we detect a trend change, that may indicate a change of ownership or generational upgrade. And if such detection is accurate enough, we can significantly discount the outdated ratings. For ratings on the scale of $1 - 5$, the coarse granularity makes this detection more difficult.

We may consider zooming into particular areas of this vector of ratings, e.g., the very satisfied customers and the very dissatisfied ones, although it is often the case that those who care enough to enter ratings are either extremely satisfied or reasonably dissatisfied. There might be a bimodal distribution in the underlying customer satisfaction for certain products, but for many products there is often another bimodal distribution on the biased sampling since only those who cared enough to write reviews.

Across all these questions, we can use the cross-validation approach from Chapter 4 to train and test the solution approach. Or, if we can stand back one year and predict the general shape of ratings that have unfolded since then; that would be a strong indicator of the utility of our signal-processing method.

But these questions do not have well-studied answers yet, so we will now focus instead on some simpler questions as proxies to our real questions: Why does simple averaging sometimes work, and what should we do when it does not?

5.2   A Long Answer

5.2.1   Averaging a crowd

We start from a significantly simplified problem. Take the Galton example, and say the number that a crowd of N people wants to guess is x, and each person i in the crowd makes a guess y_i:

$$y_i(x) = x + \epsilon_i(x),$$

i.e., the true value plus some error ϵ_i. The error depends on x but not on other users j; it is independent of other people's errors. This error can be positive or negative, but we assume that it averages across different x to be 0; it has no bias. In reality, errors are often neither independent nor unbiased. Sequential estimates based on publicly announced estimates made by others may further exacerbate the dependence and bias. We will see examples of such information cascades in Chapter 7.

We measure error by the metric of mean squared error (MSE), just like what we did for Netflix recommendation in Chapter 4. We want to compare the following two quantities:

- the average of individual guesses' errors, and
- the error of the averaged guess.

The average of errors and the error of the average are not the same, and we will see how much they differ. Since x is a number that can take on different values with different probabilities, we should talk about the *expected MSE*, where the expectation $\mathbf{E}_x$ is the averaging procedure over the probability distribution of x.

The average of (expected, mean-squared) errors, denoted by AE, by definition, is

$$E_{AE} = \frac{1}{N} \sum_{i=1}^{N} \mathbf{E}_x \left[\epsilon_i^2(x) \right]. \tag{5.1}$$

On the other hand, the (expected, mean-squared) error of the average, denoted by EA, is

$$E_{EA} = \mathbf{E}_x \left[\left(\frac{1}{N} \sum_{i=1}^{N} \epsilon_i(x) \right)^2 \right] = \frac{1}{N^2} \mathbf{E}_x \left[\left(\sum_{i=1}^{N} \epsilon_i(x) \right)^2 \right], \tag{5.2}$$

since the error term is now

$$\frac{1}{N} \sum_i y_i - x = \frac{1}{N} \left(\sum_i y_i - Nx \right) = \frac{1}{N} \left(\sum_i (y_i - x) \right) = \frac{1}{N} \left(\sum_i \epsilon \right).$$

It looks like (5.1) and (5.2) are the same, but they are not: the sum of squares and the square of the sum are different. Their difference is a special case of Jensen's inequality on convex quadratic functions. There are many terms in expanding the square in (5.2): some are ϵ_i^2, and others are $\epsilon_i \epsilon_j$, where $i \neq j$. For example, if $N = 2$, we have one cross-term, $2\epsilon_1 \epsilon_2$:

$$(\epsilon_1 + \epsilon_2)^2 = \epsilon_1^2 + \epsilon_2^2 + 2\epsilon_1 \epsilon_2.$$

These cross-terms $\{\epsilon_i \epsilon_j\}$ take on different values depending on whether the estimates $\{y_i\}$ are independent or not. If they are independent, we have

$$\mathbf{E}_x[\epsilon_i(x)\epsilon_j(x)] = 0, \quad \forall i \neq j.$$

In that case, all the cross-terms in expanding the square are zero, and we have

$$E_{EA} = \frac{1}{N} E_{AE}. \tag{5.3}$$

If you take the square root of the MSE to get the RMSE, the scaling in (5.3) is then $1/\sqrt{N}$.

This may appear to be remarkable. In such a general setting and using such an elementary derivation, we have mathematically crystallized (a type of) the wisdom of crowds in terms of efficiency gain: the error is reduced by a factor as large as the size of the crowd if we average the estimates first, provided that the estimates are *independent* of each other. But we have to caution you on several grounds. This result holds for a crowd of two as much as it holds for a crowd of a thousand. Some people think there must be something beyond this analysis, which is essentially the Law of Large Numbers at work: variance is reduced as the number of estimates goes up. There should be some type of the wisdom of crowds that shows up only for a large enough crowd. Furthermore, we have so far assumed there is no systematic bias; averaging will not help reduce any bias that is in everyone's estimate.

The $1/N$ factor is only one dimension of the wisdom of crowds, what we refer to as "multiplexing gain" from independent "channels." We will later see "diversity gain," symbolically summarized as $1 - (1 - p)^N$, where p is, say, the probability of some undesirable event.

What if the estimates are completely *dependent*? Then the averaged estimate is just the same as each estimate, so the error is exactly the same, and it does not matter how many reviews you have:

$$E_{EA} = E_{AE}.$$

In most cases, the estimates are somewhere in-between completely indepen-dent and completely dependent. We have seen that each person in the crowd

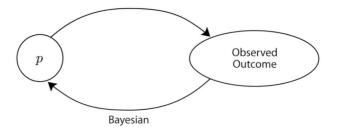

Figure 5.4 The key idea in Bayesian estimation. Not only does the underlying probability p of an event impact what we observe, but the probability p itself can be better estimated by tallying the observation.

can be quite wrong. What is important is that they are wrong in independent ways. Of course, if we could identify which member in the crowd have the correct estimates, we should just use their estimates. So the wisdom of crowds we discussed is more about achieving robustness arising out of independent randomization than getting it right by identifying the more trustworthy estimates (like in boosting in the Advanced Material).

5.2.2   Bayesian estimation

Let us move beyond a simple averaging now. **Bayesian estimation** can help us quantify the intuition that the number of ratings, N, should matter. Let us first get a feel for the Bayesian view with a simple, illustrative example. We will then rank products in the Bayesian way.

Suppose you run an experiment that returns a number, and you run it n times (the same experiment and independent runs). Suppose that s times it returns an outcome of 1. What do you think is the chance that the result from the next experiment, the $(n+1)$th one, will return an outcome of 1 too? Without going into the foundation of probability theory, the answer is the intuitive one:

$$\frac{s}{n}.$$

Now if you know the experiment is actually a flip of a biased coin, what do you think is the chance that the next experiment will give a head and return 1? Hold on, is that the same question?

Actually, it is not. Now you have some *prior* knowledge: you know there are two possible outcomes, one with probability p for heads, and the other $1-p$ for tails. That prior knowledge changes the derivation, and this is the essence of the Bayesian reasoning.

We first write down the probability distribution of p given that s out of n flips showed heads. Intuitively, the bigger s is, the more likely the coin is biased towards heads, and the larger p is. As illustrated in Figure 5.4. This is the gist of the Bayesian view: having more observations makes the model better.

Now, if p were *fixed*, the probability of observing s heads and $n - s$ tails would follow the Binomial distribution:

$$\binom{n}{s} p^s (1 - p)^{n-s}. \tag{5.4}$$

So the probability distribution of p must be *proportional* to (5.4). This was the key step in Laplace's work that turned Bayes' insights into a systematic mathematical language in the eighteenth century.

This step is perhaps less straightforward that it may sound. We are "flipping the table" here. Instead of looking at the probability of observing s out of n heads for a given p, we are looking at the probability distribution of p that gave rise to this observation in the first place, since we know the observation but not p.

Once the above realization is internalized in your brain, the rest is easy. The probability distribution of p is proportional to (5.4), but we need to divide it by a normalization constant so that it is between 0 and 1. Knowing $p \in [0, 1]$, the normalization constant is simply

$$\int_0^1 \binom{n}{s} p^s (1 - p)^{n-s} \, dp.$$

Using the beta function to get the above integral, we have

$$f(p) = \frac{\binom{n}{s} p^s (1 - p)^{n-s}}{\int_0^1 \binom{n}{s} p^s (1 - p)^{n-s} \, dp} = \frac{(n+1)!}{s!(n-s)!} p^s (1 - p)^{n-s}.$$

Finally, since the *conditional* probability of seeing a head given p is just p, the *unconditional* probability of seeing a head is simply the whole integral

$$\int_0^1 p f(p) dp,$$

which evaluates to

$$\frac{s + 1}{n + 2}.$$

A remarkable and remarkably simple answer, this is called the **rule of succession** in probability theory. It is perhaps somewhat unexpected. The intermediate step to understanding p's distribution is not directly visible in the final answer, but that was in the core of the inner-workings of Bayesian estimation. It is similar in spirit to the latent-factor model in Chapter 4, and to other hidden-factor models like the hidden-Markov models used in voice recognition and portfolio optimization.

Why is it *not* s/n? One intuitive explanation is that if you know the outcome must be success or failure, it is as if you have already seen two experiments "for free", one success and another failure. If you incorporate the prior knowledge in this way, then the same intuition on the case without prior knowledge does indeed give you $(s + 1)/(n + 2)$ now.

Figure 5.5 A Bayesian-adjusted rating $\tilde{r}_i$ sits somewhere between the raw rating r_i and the overall average rating R. The exact location of $\tilde{r}_i$ on this sliding bar differs depending on the weighting formula used.

5.2.3   Bayesian ranking

So how is Bayesian estimation related to rating averages on Amazon? Because ratings' population size matters. Back to our motivating question: should a product with only two reviews, even though both are 5 stars, be placed higher than a competing product with a hundred reviews that averages 4.5 stars? Intuitively, this would be wrong. We should somehow weight the raw rating scores with the population sizes, just like we weighted a node's importance by the in-degree in PageRank in Chapter 3. Knowing how many reviews there are gives us prior knowledge, just like knowing a coin shows up heads 100 times out of 103 flips is a very different observation than knowing it shows up heads 10 times out of 13 flips.

More generally, we can think of a "sliding ruler" in Figure 5.5 between the average rating of all the products, R, and the averaged rating of brand i, r_i. The more reviews there are for brand i, call that n_i, relative to N, the total number of reviews for all the brands, the more trustworthy r_i is relative to R. The resulting Bayesian rating for brand i is

$$\tilde{r}_i = \frac{NR + n_i r_i}{N + n_i}. \tag{5.5}$$

We may also want to put an upper bound on N, for otherwise as time goes by and N monotonically increases, the dynamic range of the above ratio can only shrink.

Quite a few websites adopt Bayesian ranking. The Internet Movie Database (IMDb)'s top 250 movies ranking follows (5.5) exactly. So the prior knowledge used is the average of all the movie ratings.

Beer Advocate's beer ranking, `http://beeradvocate.com/lists/popular`, uses the following formula:

$$\frac{N_{min}R + n_i r_i}{N_{min} + n_i},$$

where N_{min} is the minimum number of reviews needed for a beer to be listed there. Perhaps a number in-between N and N_{min} would have been a better choice, striking a tradeoff between following the Bayesian ranking exactly and avoiding the saturation effect (when some beers get a disproportionately large numbers of reviews).

All of the above suffer from a drawback in their assuming that there is a single, "true" value of a product's ranking, as the mean of some Gaussian distribution.

But some products simply create bipolar reactions: some love it and some hate it. The idea of Bayesian ranking can be extended to a multinomial model and the corresponding Dirichlet prior.

Of course, this methodology applies only to adjusting the relative ratings of each brand within a comparable family of products, so that proper ranking can be achieved on the basis of $\{\tilde{r}_i\}$. It cannot adjust ratings without this backdrop of a whole family of products that provides the *scale* of relative trustworthiness of ratings. It is good for *ranking*, but not for *rating* refinement.

5.3 Examples

5.3.1 Bayesian ranking changes order

Consider Table 5.1, a compilation of ratings and review populations for Mac-Books on Amazon. The items are listed in descending order of their average rating. Following (5.5), the Bayesian rankings for each of the five items can be computed. First, we compute the product $NR = \sum_i n_i r_i$ as follows:

$$10 \times 4.920 + 15 \times 4.667 + 228 \times 4.535 + 150 \times 4.310 + 124 \times 4.298 = 2332.752$$

Then, with $N = \sum_i n_i = 527$, we apply (5.6) to each of the items:

$$\tilde{r}_1 = \frac{2332.752 + 10 \times 4.920}{527 + 10} = 4.436,$$

$$\tilde{r}_2 = \frac{2332.752 + 15 \times 4.667}{527 + 15} = 4.433,$$

$$\tilde{r}_3 = \frac{2332.752 + 228 \times 4.535}{527 + 228} = 4.459,$$

$$\tilde{r}_4 = \frac{2332.752 + 150 \times 4.310}{527 + 150} = 4.401,$$

$$\tilde{r}_5 = \frac{2332.752 + 124 \times 4.298}{527 + 124} = 4.402.$$

These calculations and the Bayesian rankings are shown in Table 5.1. All of the MacBook's ranking positions change after the adjustment is applied, because the Bayesian ranking takes into account the number of reviews as well as the average rating for each item. The third MacBook (MB402LL/A) rises to the top because the first and second were rated by far fewer people, and their Bayesian ratings drop.

MacBook	No. Ratings	Ave. Rating	Rank	Bayesian Rating	Bayesian Rank
MB991LL	10	4.920	1	4.436	2
MB403LL	15	4.667	2	4.433	3
MB402LL	228	4.535	3	4.459	1
MC204LL	150	4.310	4	4.401	5
MB061LL	124	4.298	5	4.402	4

Table 5.1 An example where average ratings and Bayesian-adjusted ratings lead to entirely different rankings of the items. For instance, though the first listed MacBook (MB991LL/A) has the highest average, this average is based on a small number of ratings, which lowers its Bayes ranking by two places.

5.3.2   Bayesian ranking quantifies subjectivity

Sometimes, Bayesian ranking does not alter the ranked order of a set of comparable products, but we can still look at the differences between the original average ratings and the Bayesian ratings across these products, and then take the L-2 norm of this vector of differences.

For example, the Bayesian ratings for a set of digital cameras and those for a set of women's shoes on Amazon are computed in Table 5.2 and Table 5.3, respectively. This L-2 distance between the two vectors of ratings, normalized by the total number of products in each product category, is 1.5% for digital cameras and 1.7% for shoes. This difference of 12% is a quantified indicator about the higher subjectivity and stronger dependence on review population size for fashion goods than for electronic goods.

Digital Camera	No. reviews n_i	Ave. Rating r_i	Bayesian Rating $\tilde{r}_i$
Canon Powershot	392	4.301	4.133
Nikon S8000	163	3.852	4.008
Polaroid 10011P	168	3.627	3.965

Table 5.2 Bayesian ratings for three digital cameras, with a total of 723 ratings. The L-2 distance between the vector of average ratings and that of Bayesian ratings is 1.5%.

5.3.3   What does Amazon do?

On Amazon, each individual product rating shows only the raw ratings. When it comes to ranking similar products, Amazon actually follows some secret formula that combines the average reviews with three additional elements:

- the Bayesian adjustment by review population,
- the recency of the reviews, and
- the reputation score of the reviewer (or quality of review, as reflected in review of review).

Women's Shoes	No. reviews n_i	Ave. Rating r_i	Bayesian Rating $\tilde{r}_i$
Easy Spirit Traveltime	150	3.967	4.182
UGG Classic Footwear	148	4.655	4.289
BearPaw Shearling	201	4.134	4.204
Sketchers Shape-Ups	186	4.344	4.245
Tamarac Slippers	120	3.967	4.189

Table 5.3 Bayesian ratings for five women's fashion shows, with a total of 805 ratings. The L-2 distance between the vector of average ratings and that of Bayesian ratings is 1.7%.

The exact formula is not known outside of Amazon. In fact, even the reviewer reputation scores, which lead to a ranking of Amazon reviewers, follow some formula that apparently has been changed three times in past years and remains a secret. See www.amazon.com/review/top-reviewers-classic for the hall of fame of Amazon reviewers. Obviously, how high a reviewer is ranked depends on how many yes/useful votes (say, x) and no/not-useful votes (say, y) are received by each review she writes. If x is larger than a threshold, either x itself or the fraction $x/(x+y)$ can be used to assess this reviewer's quality. And the reviewer reputation changes as some moving-window average of these review-quality measures changes over time. The effect of fan vote, where some people always vote yes on a particular reviewer's reviews, is then somehow subtracted. We will see in Chapter 6 that Wikipedia committee elections also follow some formula that turns binary votes into a ranking.

Now back to ranking on Amazon. Maybe we can reverse-engineer how these rankings are generated through an example. Let us consider the list of the top 20 LCD HDTVs of size $30 - 34$ inches in April 2012. It can be obtained from Amazon by applying the following sequence of filters: Electronics > Television & Video > Televisions > LCD > 30 to 34 inches.

There are two rank-ordered lists, as we have alluded to previously.

- The first is the ranking by Amazon, which orders the list in Table 5.4.
- The second is the ranking by averaged ratings, which leads to a different order:
  1, 7, 12, 14, 2, 5, 4, 8, 3, 9, 15, 16, 20, 18, 6, 10, 11, 13, 17, 19.

We have to make an educated guess as to the reasons behind the disconnect between the two rank-ordered lists.

The "average rating" of a product is it's average number of stars, shown in Table 5.4. Rather than showing this number explicitly, Amazon rounds it and gives an "average customer review". Average rating has many differences from Amazon's ranking: 7, 12, 14, 15, 16, 20, 18, and 6 all seem to be strangely out of order. Let us try to reverse-engineer what factors might have contributed to the actual ranking.

HDTV	No. reviews	5 star	4 star	3 star	2 star	1 star	"Ave. review"	Ave. rating
1	47	37	8	1	1	0	4.7	4.723
2	117	89	19	0	3	6	4.6	4.556
3	315	215	61	19	9	11	4.5	4.460
4	180	116	47	9	2	6	4.5	4.472
5	53	36	12	3	1	1	4.5	4.528
6	111	71	19	6	6	9	4.2	4.234
7	22	16	4	2	0	0	4.6	4.636
8	56	43	5	3	1	4	4.5	4.464
9	130	89	22	8	4	7	4.4	4.400
10	155	96	26	11	9	13	4.2	4.181
11	231	135	48	17	15	16	4.2	4.173
12	8	5	3	0	0	0	4.6	4.625
13	116	55	35	9	5	12	4.0	4.000
14	249	175	60	3	3	8	4.6	4.570
15	8	5	1	2	0	0	4.4	4.375
16	34	20	8	4	0	2	4.3	4.294
17	47	20	14	6	5	2	4.0	3.957
18	44	20	20	1	1	2	4.2	4.250
19	56	24	17	4	5	6	3.9	3.857
20	7	3	3	1	0	0	4.3	4.286

Table 5.4 A list of the top twenty $30 - 34$ inch LCD HDTVs on Amazon.

1. *Bayesian adjustment*: The population sizes of the ratings matter. The raw rating scores must be weighted with the population size in some way.

2. *Recency of the reviews*: Perhaps some of the reviewers rated their HDTVs as soon as they purchased them, and gave them high ratings because the products worked initially. But, especially with electronics, sometimes faulty components cause the products to stop working over time.

3. *Quality of the reviewers or reviews*: (a) Reputation score of the reviewer: Reviewers with higher reputations should be given more "say" in the average customer review of a product. (b) Quality of review: The quality of a review can be measured in terms of its length or associated keywords in the text. (c) Review of review: Higher review scores indicate that customers found the review "helpful" and accurate. (d) Timing of reviews: Review spamming from competing products can be partially detected from review timing.

First, Bayesian adjustment uses (5.6), so we need to know N and R. Here, R is the averaged rating of all the products (here assumed to be the top 20), and N is either the total number of reviews, the *minimum* number of reviews necessary for a product to be listed (as in the Beer Advocate's website), or possibly some number in between. From Table 5.4, we can compute $R = \sum_i n_i r_i / \sum_i n_i = 4.36$.

Now, what should we choose for N? We compare the Bayesian rankings for $N_{min} = 7$, which is the lowest number of reviews for any product, $N_{max} = 315$, which is the highest number of reviews for any product, $N_{ave} = 99$, which is the average of the reviews across the products, and $N_{sum} = 1986$, the total number of reviews entered for the top 20 products. They lead to different Bayesian rankings:

- N_{min}: 1, 7, 14, 2, 5, 12, 4, 3, 8, 9, 15, 20, 16, 18, 6, 10, 11, 13, 17, 19.
- N_{max}: 14, 2, 3, 1, 4, 5, 7, 8, 9, 12, 15, 20, 16, 18, 6, 17, 10, 19, 11, 13.
- N_{ave}: 14, 1, 2, 3, 4, 5, 7, 8, 9, 12, 15, 20, 16, 18, 6, 10, 11, 17, 19, 13.
- N_{sum}: 14, 3, 2, 4, 1, 5, 7, 8, 9, 12, 15, 20, 16, 18, 6, 17, 10, 19, 11, 13.

Clearly, having N too large or too small results in rankings that are far out of order. Both N_{max} and N_{ave} give results closer to Amazon's ranking, at least in terms of grouping clusters of products together in the rank-ordered list. Still, there are some outliers in each case: 14, 15, 20, 6, 10, and 11 are considerably out of order.

We notice that products 12, 15, and 20 all have very few ratings, specifically 8, 8, and 7, respectively. But, even so, why would they be placed so far apart in the top 20? To answer this, we take into account the review of reviews: In the case of product 12, for instance, the "most helpful" review had twenty-six people find it helpful, whereas in the cases of products 15 and 20 the numbers were six and three, respectively. In addition, we can look at the recency of the reviews: The "most helpful" review for product 12 was made in November of 2011. Product 15's "saving grace" is that its corresponding review was more recent, made in December of 2011, which would push it closer to product 12 in the ranking. On the other hand, Amazon may have deemed that product 20's review in July of 2011 was too outdated. Finally, product 12 had an extremely high quality of review in its descriptive listing of the pros and cons in each case. This may have pushed its ranking higher.

On the other hand, why would Amazon decide to rank an item such as 6 so high, given that the Bayesian ranking places it around the 15th position? Well, when we look at item 6, we see that its "most helpful" review had 139 out of 144 people find it helpful, and similar percentages exist for all reviews below it. Further, the reviewers all have high ratings, and one of them is an Amazon "top reviewer." Quality of reviews overrides the quantity.

The final point of discussion is why product 14 is ranked so low. This can be explained by two factors: (1) the most helpful review was from 2010, which was extremely outdated; and (2) the eight reviewers who gave it 1 star all said that the TV had stopped working after a month, many of these reviews were high up in the "helpful" rankings. These reviewers dramatically increased the spread of the ratings and opinions for this product.

To summarize, the following set of guidelines is inferred from this (small) sample on how Amazon comes up with its ranking.

1. An initial Bayesian ranking is made, with N chosen to be somewhere around N_{max} or N_{ave}.
2. Products that have small numbers of reviews or low recency of their most helpful reviews are ranked separately amongst themselves, and re-distributed in the top 20 at lower locations (e.g., products 12, 15, and 20).
3. Products that have very-high-quality, positive reviews from top reviewers are bumped up in the ranking (e.g., product 6).
4. Products for which there could be a major issue, like the possibility of faulty electronics (e.g., product 14), are severely demoted in the ranking.

5.4   Advanced Material

As we just saw, reviews of reviews can be quite helpful: a higher score mean more confidence in the review, and naturally leads to a heavier weight for that review. If a review does not have any reviews, we may take some number between the lowest review score for a review and the average review score. More generally, we may take the Bayesian approach to determine the trustworthiness of a review by observation of the reviews it receives.

While a systematic study of the above approach for rating analytics is still ongoing, this idea of weighting individual estimates by its effectiveness has been studied in a slightly different context of statistical learning theory, called **boosting**. If we view each estimator as a person, boosting shows how they can collaborate, through sequential decision making, to make the resulting collective estimate much better than any individual estimate can be.

5.4.1   Averaging sequentially-trained estimators

Ada Boost, short for adaptive boosting, captures the idea that by sequentially training estimators we can make the average estimator more accurate. It is like an experience many students have while reviewing class material before an exam. We tend to review those points that we already know well (since that makes us feel better), while the right approach is exactly the opposite: to focus on those points that we do not know very well yet.

As in Figure 5.6, consider N estimators $y_i(\mathbf{x})$ that each map an input $\mathbf{x}$ into an estimate, e.g., the number of stars in an aggregate rating. The final aggregate rating is a weighted sum: $\sum_i \alpha_i y_i$, where $\{\alpha_i\}$ are scalar weights and $\{y_i\}$ are functions that map vector $\mathbf{x}$ to a scalar. The question is how to select the right weights α_i. Of course, those y_i that are more accurate deserve a larger weight α_i. But how much larger?

Let us divide the training data into M sets indexed by j: $\mathbf{x}_1, \mathbf{x}_2, \ldots, \mathbf{x}_M$. For each training set, there is a right answer t_j, $j = 1, 2, \ldots, M$, known to us since we are training the estimators. So now we have N estimators and M datasets. As each estimator y_i gets trained by the datasets, some datasets are

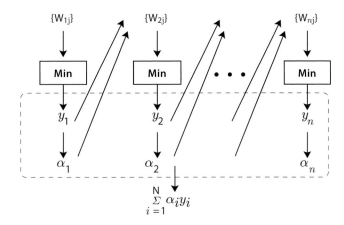

Figure 5.6 Ada Boosting. There are N estimators, indexed by i, $\{y_i\}$, and M training datasets, indexed by j. Training is done to minimize weighted errors, where the weights w_{ij} are sequentially chosen from estimator i to estimator $i+1$ according to how well each training dataset has been learned so far. The final estimator is a weighted sum of individual estimators, where the weights $\{\alpha_i\}$ are also set according to the error of each estimator y_i.

well handled while others are less so. We should *adapt* accordingly, and give challenging datasets, those leading to poor performance thus far, heavier weight w in the next estimator's parameter training.

In summary, both the training weights $\{w_{ij}\}$ and the estimator combining weights $\{\alpha_i\}$ are determined by the performance of the estimators on the training sets.

We start by initializing the training weights w_{1j} for estimator 1 to be even across the datasets:

$$w_{1j} = \frac{1}{M}, \quad j = 1, 2, \ldots, M.$$

Then, *sequentially* for each i, we train estimator y_i by minimizing

$$\sum_{j=1}^{M} w_{ij} 1_{y_i(\mathbf{x}_j) \neq t_j},$$

where 1 is an indicator function, which returns 1 if the subscript is true (the estimator is wrong and should be included in the penalty being minimized) and 0 otherwise (the estimator is correct).

After this minimization, we get the resulting estimator, leading to an error indicator function denoted by 1_{ij}: the optimized error indicator of estimator i being wrong on dataset j.

The (normalized and weighted) sum of these error terms $\{1_{ij}\}$ becomes

$$\epsilon_i = \frac{\sum_j w_{ij} 1_{ij}}{\sum_j w_{ij}},$$

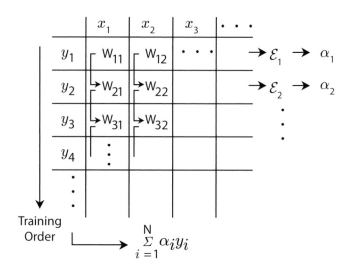

Figure 5.7 A detailed view of the sequence of the steps in the iterations of adaptive boosting. Each column represents a training dataset. Each row represents an estimator. Sequential training takes place down each column and across the rows. Performance per row is summarized by α_i, which defines the linear combination y as the final estimator derived from "the wisdom of crowds."

for estimator i. Let the estimator-combining weight α_i for estimator i be the (natural) log-scaled version of this error term:

$$\alpha_i = \log(1/\epsilon_i - 1).$$

Now we update the training weights for the next estimator $i + 1$:

$$w_{i+1,j} = w_{ij}e^{\alpha_i 1_{ij}}.$$

Obviously, $w_{i+1,j}$ is still w_{ij} if estimator i gets dataset j right, and $w_{ij}(1/\epsilon_i - 1)$ otherwise. If the jth training dataset is poorly estimated, then $w_{i+1,j} > w_{i,j}$ in the next estimator, just as we desired.

Now we repeat the above loop for all the estimators i to compute all the α_i. Finally, the aggregate estimator is

$$y(\mathbf{x}) = \sum_{i=1}^{N} \alpha_i y_i(\mathbf{x}).$$

Clearly, we put more weight on estimators that are more accurate; the smaller ϵ_i, the bigger α_i. This Ada Boost method is illustrated in Figure 5.7. It has also been connected to repeated games like one of the homework problems in Chapter 1. It turns out that both boosting through batch-training of estimators and certain methods in online learning can be viewed as special cases of repeated games.

Summary

Box 5 Bayesian ranking

Averaging ratings scalarizes a vector and leads to a ranking of products. The number of ratings in each product should matter, and Bayesian ranking provides a way to do that. Other factors such as the timing and quality of the review associated with each rating can also be taken into account. In simpler tasks, a factor-of-N multiplexing gain in the wisdom of crowds follows from independent and unbiased individual inputs.

Further Reading

There is a gap between the rich theory of signal processing, statistical learning, and Bayesian estimation on the one hand, and the simple but tricky issue of when I can trust a product's rating average on Amazon.

1. The following blog provides a concise summary of Bayesian ranking:
`http://www.thebroth.com/blog/118/bayesian-rating`
The next blog adds further details:
`http://andrewgelman.com/2007/03/bayesian-sortin/`
And this one too, especially on using a Dirichlet prior for a multinomial model:
`http://masanjin.net/blog/how-to-rank-products-based-on-user-input`

2. Here is a standard textbook on machine learning, including different ways to combine estimators:
C. M. Bishop, *Pattern Recognition and Machine Learning*, Springer, 2006.

3. The following recent book summarizes two decades of research results on the subject of boosting, with connections to game theory, optimization theory, and other topics in machine learning:
R. E. Schapire and Y. Freund, *Boosting: Foundations and Algorithms*, The MIT Press, 2012.

4. For a graduate-level, comprehensive treatment on the subject of the Bayesian approach to data analysis, the following is one of the standard choices:
A. Gelman, J. B. Carlin, H. S. Stern, and D. B. Rubin, *Bayesian Data Analysis*, 2nd edn., Chapman and Hall/CRC, 2004.

5. The following is a unique overview that organizes the methodologies of analyzing patterned structures in signals around six prominent applications:

D. Mumford and A. Desolneux, *Pattern Theory: The Stochastic Analysis of Real-World Signals*, A. K. Peters, 2010.

Problems

5.1 *Bayesian ranking* ⋆

Suppose there are 50 ratings for all printers, with an average rating of 4. Given ratings 5, 5, and 5 for a printer by Canon, and ratings 4.5, 5, 4, 5, 4, 5, 3, 5, 4.5, and 3.5 for another by HP, calculate the ranking by average rating and the Bayesian ranking, respectively.

5.2 *Analyzing Amazon data* ⋆⋆⋆

Take a look at the rating data in worksheet "raw" of the following file:
http://www.network20q.com/hw/amazon_data.xls
for the iPod touch, address the following.

(a) Compute the mean and the adjusted mean. Here, we define the adjusted mean as $\sum$(rating×number of helpful reviews)/total number of helpful reviews.

(b) Plot the monthly mean.

(c) How does the temporal pattern of the monthly mean influence your perception of the product quality?

(d) Which metric (raw mean, adjusted mean, or monthly mean) is the most accurate in your opinion? Would a certain combination of the metrics be more useful?

5.3 *Averaging a crowd* ⋆

Suppose the estimation errors $\{\epsilon_i\}$ are independent and identically distributed random variables that takes values 1 and -1 with an equal probability. X is a uniform random variable that takes on 100 possible values ($P(X = x) = 1/100$, $x = 1, 2, \ldots, 100$). Calculate E_{AE} and E_{EA} for $N = 100$. Does the relationship between E_{AE} and E_{EA} hold?

Now let $N = 1000$ and let X take on 1000 possible values uniformly. Plot the histogram of $\frac{1}{N} \sum_{i=1}^{N} \epsilon_i(x)$ over all trials of X. What kind of distribution is the histogram? What is the variance? How does this relate to E_{EA}?

5.4 *Averaging a dependent crowd* ⋆⋆

Consider three people making dependent estimates of a number, with the following expectations of errors and correlations of errors:

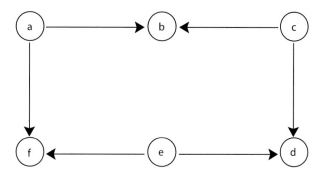

Figure 5.8 A belief network with six random variables. Their dependence relationships are represented by directed links.

$$\mathbf{E}[\epsilon_1^2] = 1773,$$
$$\mathbf{E}[\epsilon_2^2] = 645,$$
$$\mathbf{E}[\epsilon_3^2] = 1796,$$
$$\mathbf{E}[\epsilon_1\epsilon_2] = 1057,$$
$$\mathbf{E}[\epsilon_1\epsilon_3] = 970,$$
$$\mathbf{E}[\epsilon_2\epsilon_3] = 708.$$

Compute the average of errors and the error of the average in this case.

5.5 *Independence of random variables* ★★

Many types of logical relationships can be visualized using a **directed acyclic graph**. It is a graphical model with directed links and no cycles (a cycle is a path that ends at the same node as that where it starts). Here is an important special case. A **belief network** visualizes a probability distribution of the following form:

$$p(x_1, x_2, \ldots, x_n) = \prod_{i=1}^{n} p(x_i | Parent(x_i)),$$

where $Parent(x)$ is the set of parental (conditioning set of) random variables of random variable x. For example, we can write a generic joint distribution among three random variables as

$$p(x_1, x_2, x_3) = p(x_3 | x_1, x_2) p(x_2 | x_1) p(x_1).$$

We can now represent dependence among random variables in a directed acyclic graph. Each node i in this graph corresponds to a random variable. A (directed) link in this graph represents a parental relationship. Look at the graph in Figure 5.8. Is (a, c) independent of e?

6   Why does Wikipedia even work?

Now we move from recommendation to influence in social networks. We start with consensus formation from conflicting opinions in this chapter before moving on to a collection of influence models in the next two.

But first, let us compare the four different "consensus" models we covered in Chapters 3 – 6, as visualized in Figure 6.1.

- Google's PageRank turns a graph of webpage connections into a single rank-ordered list according to their importance scores.
- Netflix's recommendation turns a user–movie rating matrix into many ranked order lists, one list per user, based on the predicted movie ratings for each user.
- Amazon's rating aggregation turns a vector of rating scores into a single scalar for each product.
- Voting systems turn a set of rank-ordered lists into a single rank-ordered list, as we will see in this chapter.

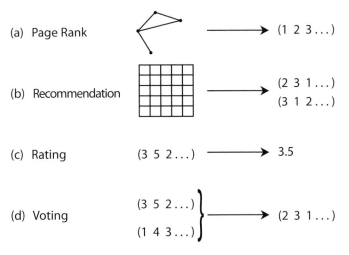

Figure 6.1 Comparison of four types of "consensus-formation" mechanisms with their inputs and outputs. The first three mechanisms have been covered in the last three chapters, and the last one will be part of this chapter.

6.1   A Short Answer

Crowdsourcing knowledge representation with unpaid and possibly anonymous contributors is very tricky. It faces many challenges for this idea to "work." For example, how do we create incentives for people to keep contributing; and how do we handle disagreements among the contributors?

Launched in 2001, Wikipedia represented a convergence of three forces that had been gathering momentum: (1) wikis for online collaboration among people, (2) the free- and open-software movement, and (3) the appearance of online encyclopedias. Within a decade, Wikipedia has generated 4 million articles in the USA and 27 million articles worldwide. It has become one of the most popular sources of information online. For certain fields, like medicine, the quality of Wikipedia articles is consistently high; and for many fields, if you google a term, a Wikipedia entry will likely come up in the top few search results. It is quite amazing that Wikipedia actually "worked" as well as it did. As we have seen in Chapter 1 and will see again in Chapter 11, when people interact with each other, there is often the risk of the "tragedy of the commons". How does Wikipedia turn that into effective collaboration? This is the driving question for this chapter.

Of course, there are also limitations to Wikipedia in its capacity as an encyclopedia.

- *Misinformation*: Sometimes information on Wikipedia is plainly wrong, especially in articles with a small audience. But Wikipedia provides an effective self-correcting mechanism: it is open to edits from anyone.
- *Mistakes*: There are also honest mistakes, but, again, anyone can edit an article, and the edit will stay there as long as no other contributor can present a stronger case otherwise.
- *Missing information*: No encyclopedia can be truly *complete* to everyone's liking, not even the largest encyclopedia in history.

Due to these limitations, there have been some high-profile cases of abuse in Wikipedia. Still, Wikipedia stands as a major success of online collaboration.

There had been other efforts aimed at creating free, open, online encyclopedias before, and the unique success of Wikipedia is often attributed to a "good-faith collaboration" environment within the Wikipedia contributor community. If we count the number of pairwise links in a fully connected graph with n nodes, we have on the order of n^2 such links. But if we examine the number of opinion configurations, we have 2^n possibilities even if each person has just two choices. This n^2 vs. 2^n tension exemplifies the positive and the negative sides of the network effect. Converging on one of these 2^n configurations is difficult, and Wikipedia mostly follows the principle of "rough consensus." In this chapter, we will study the process of reaching a rough consensus from voting theory (even though it does not explicitly involve voting through rank-ordered lists) and from bargaining theory.

Wikipedia is free, open, dynamic, interactive, and extensively linked. There are natural pros and cons associated with such a model of an encyclopedia that complements other forms of encyclopedia. Let us consider three distinct features of Wikipedia.

- It is free. How can people be motivated to contribute? Incentives do not have to be financial; the ability to influence others is a reward in its own right to most people. This requires the Wikipedia audience to be very large.
- Anyone can write or add to an article, including non-experts, anonymous writers, and people with conflicts of interest. The key is *check and balance*. Precisely because anyone can contribute, Wikipedia has a large body of writers who check others' writing frequently through a mechanism for debates and updates. Sometimes, however, a contributor or an IP address may be blocked if it is detected as a regular source of deliberate misinformation.
- Any subject may be contributed, including controversial ones. Sometimes, however, certain articles can be "protected" from too frequent edits to give time for the community of contributors to debate. How does Wikipedia avoid unbalanced treatment or trivial subjects? It turns out that there are Policies and Guidelines, and there are mechanisms for conflict resolution by editors.

The first and second features above provide Wikipedia with a strong, positive network effect: a larger audience leads to more contributors, which in turn leads to more audience, provided that the quality of contributions is kept high.

This brings us to the third feature above. How does Wikipedia enforce quality and resolve conflicting contributions? (Before addressing this question, we must bear in mind the obvious fact that Wikipedia is not a sovereign state with the power of a government. So issues such as voting, decision-making, and free speech do not have the same context.)

To start with, there are three core Policies on Wikipedia to help ensure reliability and neutrality of the articles as much as possible.

- *Verifiability (V)*: each key point and all data in an article must be externally verifiable, with a link to the primary source for verification by readers.
- *No Original Research (NOR)*: this is to prevent people from using Wikipedia as a publication venue of their new results.
- *Neutral Point of View (NPOV)*: the basic rule is that a reader must not be able to tell the bias of the author in reading through a Wikipedia article. It is particularly important for controversial topics, but also the most difficult to use exactly in those cases, e.g., contentious political, social, and religious topics. Unlike the above two policies, it is harder to enforce this one since "neutrality" is subjective.

Wikipedia has also installed several mechanisms for debates and updates. One is the use of the history page and the talk page, which are available for public view through tags on top of each article's page. All previous versions and all the changes made are recorded.

Furthermore, there is a reputation system for contributors, similar to the reviewer rating system on Amazon. Each article can be rated on a 1–6 scale. For those who do not reveal their names, it is a reputation system of the IP addresses of the devices from which contributions are sent. In addition, links across article pages are analyzed in a manner similar to Google's PageRank.

But perhaps the ultimate mechanism still boils down to people negotiating. Depending on the stage of the article, expert and non-expert contributors may join the discussion. There is a hierarchy of Wikipedia communities, and debates among contributors who cannot come to an agreement will be put forward to a group of the most experienced editors. This committee of editors acts like a jury, listening to the various sides of the debate, and then tries to decide by "rough consensus."

How do we model the process of reaching a "rough consensus" through "good faith collaboration?" Not easy. We will see in this chapter two underlying theories: voting theory and bargaining theory. But much translation is needed to connect either of these theories to the actual practice of Wikipedia.

- In a voting model, each contributor has some partially ordered list of preferences, and a *threshold* on how far away from her own preferences the group decision can be before she vetoes the group decision and thereby preventing the consensus from being reached. (Sometimes a decision is actually carried out by explicit votes in the arbitration committee. And the committee members are also elected through a voting system.) Dealing with partial ordering, quantifying the distance between two ordered lists, and modeling each editor's veto-decision threshold are still under-explored in the study of group-interaction dynamics in Wikipedia.
- In a bargaining model, the contributors need to reach a compromise, otherwise there would be no agreement. Each contributor's utility function, and the default position in the case of no agreement, need to reflect the goodwill typically observed in Wikipedia collaboration.

In contrast to coordination through well-defined pricing feedback signals, which we will see in several later chapters, coordination though bargaining or voting is much harder to model. In the next section, we will present the basics of the rich theory of voting and social choice. Then, in the Advanced Material, we will briefly discuss the mathematical language for bargaining and cooperative games.

6.2   A Long Answer

Wikipedia consensus formation illustrates important issues in reaching consensus among a group of individuals that is binding for everyone. This is different from presenting the rank-ordered list for each person to evaluate individually (like Netflix recommendation in Chapter 4). It is different from coordinating individual actions through pricing (like ad space auction in Chapter 2 and Internet congestion control in Chapter 14). Ranking preferences is also different from ranking objects (like ranking webpages in Chapter 3 or ranking products in Chapter 5).

Voting is obviously essential for elections of those who will make and execute binding laws to be imposed on those casting the votes. It is also useful for many other contexts, from talent competitions to jury decisions. It is a key component of the **social choice theory** that studies how individual preferences are collected and summarized.

Voting theory studies how to aggregate vectors, where the vectors are collections of individual preferences. We will see an axiomatic treatment of voting methods in this section. Later, in Chapter 20, we will see another axiomatic treatment of scalarizing a vector of resource allocations. In both cases, turning a not-well-ordered input into an ordered output must satisfy certain intuitive properties, and it is often tricky to accomplish that.

6.2.1   Major types of voting systems

A **voting system** is a function that maps a set of voters' preferences, called a **preference profile**, to an **voting outcome**, just as illustrated at the bottom of Figure 6.1. There are N voters and M candidates. A preference profile is a collection of rank-ordered lists, one list per voter that lists all the candidates. A voting outcome is a single rank-ordered list.

The requirement of *complete* rank-ordered lists as inputs can be too stringent. When there are many candidates, often a coarser granularity is used. For example, the members of Wikipedia's arbitration committee are elected by a binary voting system. Each voter divides the list of candidates into just two parts: those she votes "for" and those she votes "against." Then the percentage of "for" votes, out of all the votes received by each candidate, is used to rank order the candidates. The tradeoff between user-interface simplicity and the voting result's consistency and completeness is interesting to explore, but this topic does not have as many results as voting systems with complete rank-ordered input lists.

A voting system *aggregates* N lists into one list, like squeezing a three-dimensional ball into a two-dimensional "paste." Naturally, some information in the original set of N lists will be lost after this mapping, and that sometimes leads to results not conforming to our (flawed) intuition. We will focus on three commonly used voting systems to illustrate the key points.

Perhaps the most popular voting system is **plurality voting**. We simply count the number of voters who put a candidate j in the first position in their lists. Call these numbers $\{V_j\}$, and the candidate j with the largest V_j wins and is put on the first position of the list in the outcome. Then put the candidate with the second largest V_j in the second position, and so on. To simplify the discussion, we will assume that there are no ties. There is also a variant called the **Kemeny**

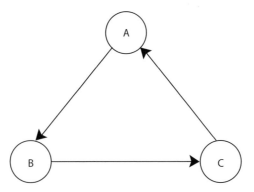

Figure 6.2 Condorcet voting can lead to pairwise comparisons that aggregate into a cyclic result. A pointing to B means A is more preferred to B, denoted also as A>B. It is crucial and tricky to avoid cyclic preferences in both the inputs and the output of a voting system.

rule, where we count how many "least-liked" votes a candidate receives, and rank in reverse order of those numbers. There are other voting systems that try to determine the least objectionable candidate to help reach consensus. Plurality voting sounds reasonable and is often practiced, but there are many preference profiles that lead to counter-intuitive voting outcomes.

A generalization of plurality voting is **positional voting**. Looking at a voter's rank-ordered list, we assign some numbers to each candidate on the basis of its position in the list. The most famous, and the only "right" positional voting that avoids fundamental dilemmas in voting, is **Borda count**. It is named after the French mathematician who initiated the scientific study of voting systems in the eighteenth century. By Borda count, the first-position candidate in each list gets $M - 1$ points, the second-position one gets $M - 2$ points, and so on, and the last-position one gets no points, since being the last one in a list that must be complete carries no information about a voter's preference at all. Then, the candidates are ranked according to their total points across all the voters.

Yet another method is **Condorcet voting**, named after another French mathematician, who founded the research field of voting theory shortly after Borda. It is an aggregation of binary results from *pairwise* comparisons. All voting paradoxes must have at least three candidates. When $M = 2$, the result is always clear-cut, since each voter's preference can be characterized by one number, and aggregating scalars is unambiguous. So how about looking at each possible pair of candidates (A, B), and seeing how many voters think one is better than the other? This unambiguously decides the winner out of that pair: if more voters think A is better than B, denoted as A>B, then A is placed higher than B in the aggregated rank-ordered list. Now, if the pairwise decisions generate a consistent rank-ordered list, we are done. But this might not happen, for example, when the preference profile is: A>B>C, B>C>A, and C>A>B. We can see that the pairwise comparison between A and B is A>B, similarly B>C and C>A as in Figure 6.2. But that is *cyclic*, and thus logically inconsistent. There is no Condorcet voting output that is self-consistent in this case.

6.2.2   A counter-intuitive example

Suppose the editors of Wikipedia need to vote on a contentious line in an article on "ice cream" about which flavor is the best: chocolate (C), vanilla (V), or strawberry (S); with $M = 3$ candidates and $N = 9$ voters. There are six positional possibilities for three candidates, and it turns out that half of these receive zero votes, while the other three possibilities receive the following votes:

- C>V>S: 4 votes,
- S>V>C: 3 votes,
- V>S>C: 2 votes.

What should the aggregated rank-ordered list look like?

By plurality vote, the aggregation is clear: C>S>V. But something does not sound right. Those who favor strawberry over chocolate outnumber those who favor chocolate over strawberry. So how could C>S>V be right?

Well, let us look at Condorcet voting then.

- C vs. V? Pairwise comparison shows V wins.
- S vs. V? Pairwise comparison shows V wins again.
- C vs. S? Pairwise comparison shows S wins.

The above three pairwise comparisons are all we need in this case, and aggregation is clear: V wins over both C and S, so it should come out first. Then S wins over C. So the outcome is clear: V>S>C. But wait, this is *exactly* the opposite of the plurality vote's outcome.

How about the Borda count? V gets 11 points, C 8 points, and S 8 points too. So V wins and C and S tie.

Three voting methods gave three different results. This is strange. You may object: "But these preference profiles are synthesized artificially. Real-world ones will not be like this."

Well, first of all, this synthesized input is indeed designed to highlight that our intuition of what constitutes an accurate aggregation of individual preferences is incomplete at best.

Second, there are many more paradoxes like this. In fact, we will go through a method that can generate as many paradoxes as we like. This is not an isolated incident.

Third, how do you define "real-world cases"? Maybe through some intuitively correct statements that we will take as true to start the logical reasoning process. We call those **axioms**. But some seemingly innocent axioms are simply not compatible with each other. This is the fundamental, negative result of **Arrow's Impossibility Theorem** in social choice theory.

6.2.3   Arrow's impossibility result

Let us look at the following five statements that sound very reasonable about any voting system. We will consider them as axioms that any voting system must satisfy. Two of them concern a basic property called **transitivity**, a logical self-consistency requirement: if there are three candidates (A, B, C), and in a list A is more preferred than B, and B is more preferred than C, then A is more preferred than C. Symbolically, we can write this as A>B>C $\Rightarrow$ A>C. Had this not been true, we would have a cyclic, thus inconsistent preference: A>B>C>A.

Now the five axioms proposed by Arrow are as follows:

1. Each input list (in the preference profile) is complete and transitive.
2. The output list (in the voting outcome) is complete and transitive.
3. The output list cannot just be the same as one input list no matter what the other input lists are.
4. *Pareto Principle.* If all input lists prefer candidate A over candidate B, the output list must do so too.
5. *IIA Principle.* If between a given pair of candidates (A, B), each input list's preference does not change, then, even if their preferences involving other candidates change, the output list's preference regarding A and B does not change.

The last statement above is called the **Independence of Irrelevant Alternatives** (IIA). As we will see, these alternatives are actually not irrelevant after all.

You might think there should be a lot of voting systems that satisfy all the five axioms above. Actually, as soon as we have $M = 3$ candidates or more, there are *none*. (If the surprise factor is a measure of a fundamental result's elegance, this impossibility theorem by Arrow in his Ph.D. thesis in 1950 is among the most elegant ones we will see.)

How could that be? Something is wrong with the axioms. Some of them are not as innocent as they might seem to be at first glance. The first two axioms are about logical consistency, so we have to keep them. The third one is the underlying assumption of social choice, without which aggregation becomes trivial. So it must be either the Pareto axiom or the IIA axiom.

Usually in an axiomatic system, the axiom that takes the longest to describe is the first suspect for undesirable outcomes of the axiomatic system. And IIA looks suspicious. Actually, how A and B compare with each other in the *aggregate output list* should depend on other options, such as candidate C's ranking in the *individual input lists*. To assume otherwise actually opens the possibility that transitive inputs can lead to cyclic output. When a voter compares A and B, there may be a C in between, or not, and that in turn determines whether the outcome is transitive or not. IIA prohibits the voting system from differentiating between those input lists that lead to only transitive outputs and those that may lead to cyclic outputs.

This will be clearly demonstrated in the next section, where we will see that a group of input rank lists, each transitive, can be just the same as a group of input rank lists where some of them are cyclic. Clearly, when an input is cyclic, the output may not be transitive. In the language of axiomatic construction, if axiom 5 can block axiom 2, no wonder this set of axioms is not self-consistent. The negative result is really a positive highlight on the importance of maintaining logical consistency and on the flawed intuition in IIA.

In hindsight, Arrow's impossibility theorem states that when it comes to ranking three or more candidates, pairwise comparisons are inadequate. Then the next question naturally is what additional information do we need? It turns out that a numerical scale, rather than just relative order, will lead to a "possibility result."

6.2.4   Possibility results

What is the *true intention* of the voters? Well, the answer is actually obvious: the entire preference profile *itself* is the true intention. Voting can be no more universally reflective of the "true intent" of voters than two points, say $(1, 4)$ and $(3, 2)$, on a two-dimensional plane can be compared to decide which is bigger.

In particular, voting in political systems, despite the occurrence of counter-intuitive results stemming from our flawed intuition, is a universal right that provides the basic bridge between individuals and the aggregate, an effective means to provide checks and balances against absolute power, and the foundation of consent from the governed to the government. No voting system is perfect, but it is better than the alternative of no voting. Moreover, some voting systems *can* achieve a possibility result.

For example, by replacing IIA with the **Intensity form of IIA** (IIIA), there are voting systems that can satisfy all the axioms. What is IIIA? When we write A>B, we now also have to write down the number of other candidates that sit in between A and B; this is the *intensity*. If there are none, the intensity is zero. IIIA then states that, in the outcome list, the ranking of a pair of candidates depends only on the pairwise comparison *and* the intensity.

While the original five axioms put forward by Arrow are not compatible, it turns out that the modified set of axioms, with IIA replaced by IIIA, is. Borda count is a voting system that satisfies all five axioms now. This stems from a key feature of Borda count: the point spread between two adjacent positions in a rank list is the same no matter which two positions we are looking at. In essence, we need to *count* (the gaps between candidates) rather than just *order* (the candidates).

We have not, however, touched on the subject of untruthful voting, such as manipulation and collusion based on information or estimates about others' votes. For example, Borda count can be easily manipulated if people do not vote according to their true rank-ordered list.

6.3   Examples

6.3.1   Sen's impossibility result

Another fundamental impossibility theorem was developed by Sen in the 1970s. This time, it turns out that the following four axioms are incompatible:

1. Each input list is complete and transitive.
2. The output list is complete and transitive.
3. If all input lists prefer candidate A over candidate B, the output list must too.
4. There are at least two "decisive voters."

The first three are similar to the axioms by Arrow. The last one concerns a **decisive voter**: a voter who can decide (at least) one pair of candidates' relative ranking for the whole group of voters, i.e., other voters' preferences do not matter for this pairwise comparison. Obviously, we need more than one decision voter for the system to qualify as real voting.

Just like Arrow's impossibility theorem, the problematic axiom, in this case axiom 4, precludes the guarantee of transitivity of the output list. What axiom 4 implies is actually the following: one voter can impose strong negative externality on all other voters. This is illustrated next.

6.3.2   Constructing any counter-example you want

All examples of Sen's result can be constructed following a procedure illustrated in a small example.

Suppose there are $N = 5$ candidates and $M = 3$ voters. Voter 1 is the decisive voter for the (A, B) pairwise comparison, voter 2 for the (C, D) pair, and voter 3 for the (E, A) pair. These will be marked in bold in tables that follow. We will show how decisive voters can preclude transitivity in the output list.

Let us start with a cyclic ranking for every voter: A>B>C>D>E>A, as shown in Table 6.1.

	A, B	B, C	C, D	D, E	E, A
1	**A>B**	B>C	C>D	D>E	E>A
2	A>B	B>C	**C>D**	D>E	E>A
3	A>B	B>C	C>D	D>E	**E>A**

Table 6.1 Step 1 of constructing examples showing inconsistency of Sen's axioms. Each row represents the draft preferences of a voter. The columns contain a subset of the pairwise comparisons of five candidates. Pairwise preferences in bold indicate that they come from decisive voters.

We do not want input rankings to be cyclic, so we need to flip the pairwise comparison order at least at one spot for each voter. But we are guaranteed to find, for each voter, two spots where flipping the order will not change the outcome. Those are exactly the two pairwise comparisons where some other voter is the decisive voter. So flip the order at those two spots and you are guaranteed a transitive ranking by each voter. The resulting output, however, remains cyclic, as shown in Table 6.2. This means that Axiom 2 can be beaten by the other axioms.

	A, B	B, C	C, D	D, E	E, A
1	**A>B**	B>C	"D>C"	D>E	"A>E"
2	"B>A"	B>C	**C>D**	D>E	"A>E"
3	"B>A"	B>C	"D>C"	D>E	**E>A**
Outcome	A>B	B>C	C>D	D>E	E>A

Table 6.2 Step 2 of constructing examples showing the inconsistency of Sen's axioms. Pairwise preferences in quotation marks are those which have been flipped from the draft version in Table 6.1 to turn input rankings transitive without destroying the cyclic nature of the outcome ranking.

This example demonstrates not only how easy it is to generate examples illustrating Sen's negative result, but also that each decisive voter is actually destroying the knowledge of the voting system on whether transitivity is still maintained. If a decisive voter ranks A>B, and another voter ranks not just B>A, but also B > k other candidates > A, then we say the decisive voter imposes a k-*strong negative externality* on that voter. In cyclic ranking in Sen's system, each voter suffers strong negative externality from some decisive voter.

This example highlights again the importance of keeping track of the *position* of candidates in the *overall* ranking list by each voter, motivating the use of Borda count. This is yet another example in this book where we see how negative externality is internalized. We simply cannot consolidate ranking lists by extracting some portion of each list in isolation. "Sensible voting" is still possible if we avoid that kind of compression of voters' intentions.

6.3.3   Connection to prisoner's dilemma

In fact, the prisoner's dilemma we saw back in Chapter 1 is a special case of Sen's negative result, and similar dilemmas can be readily constructed now. Recall that there are four possibilities: both prisoners 1 and 2 do not confess (A), 1 confesses and 2 does not (B), 1 does not confess and 2 does (C), and both confess (D).

Prisoner 1 is the decisive voter on the (A, B) and (C, D) pairs. Prisoner 2 is the decisive voter on the (A, C) and (B, D) pairs. Together with the obvious consensus A>D, we have an outcome that contains two cyclic rankings: D>B>A>D

	A, B	B, D	A, D	C, D	A, C
Prisoner 1	**B>A**	–	A>D	**D>C**	–
Prisoner 2	–	**D>B**	A>D	–	**C>A**
Outcome	B>A	D>B	A>D	D>C	C>A

Table 6.3 Prisoner's dilemma as a special case of Sen's impossibility result. Think of the two prisoners as two voters, and the four possible outcomes as four candidates. Four pairs of candidates are compared since they are the four individual actions afforded to the two prisoners. Those pairwise preferences that do not matter are marked in "–" since the other prisoner is the decisive voter for that pair. An additional pairwise comparison is obvious: both do not confess (candidate A) is better than both confess (candidate D).

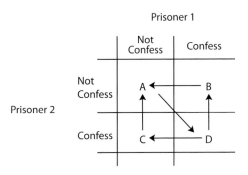

Figure 6.3 The prisoner's dilemma produces two cycles of ranking orders: D>B>A>D and A>D>C>A. This leads to a special case of Sen's impossibility result.

and A>D>C>A, as shown in Figure 6.3. This presence of cyclic output rankings in the outcome is another angle to understand the rise of the socially suboptimal D as the Nash equilibrium.

6.4   Advanced Material

The second conflict-resolution mechanism, in a Wikipedia editorial decision as well as in many other contexts, is **bargaining**. Each bargaining party has a selfish motivation and yet all the parties want to achieve some kind of agreement. If no agreement is achieved, then each party goes back to its own **disagreement point**. This interaction is studied in **cooperative game theory**.

We will first follow the more intuitive approach developed by Rubenstein in the 1980s before turning to Nash's axiomatic approach in his Ph.D. thesis in 1950, almost at the same time as Arrow's thesis. We will see an IIA-style axiom too. But this time it is a possibility theorem that followed; in fact, a unique function modeling bargaining that satisfies all the axioms.

6.4.1   Bargaining: Interactive offers

Suppose there are two people, A and B, bargaining over how to divide a cake of size 1. This cake-cutting problem will be picked up again in Chapter 20 in our study of fairness, with extensions like "one cuts, the other chooses."

For now, consider the following procedure. At the end of each of the discrete timeslots with duration T, each person takes a turn to offer the other person how to share the cake. It is essentially a number $x_1 \in [0, 1]$ for A and $x_2 = 1 - x_1$ for B. This iterative procedure starts with the initial offer at time 0 from A to B. If the offer is accepted, an agreement is reached. If it is rejected, the other person makes another offer in the next timeslot.

But wait a minute. This bargaining process can go on forever. Why would either person be motivated to accept an offer (other than one that gives her the whole cake)? There must be a price to pay for disagreeing. In Rubenstein's model, the price to pay is *time*. If an agreement is reached at the kth iteration, a person's payoff is

$$u_i = x_i e^{-r_i k T}, \quad i = 1, 2,$$

where r_i is a positive number capturing "the tolerance to wait and keep bargaining." So the payoff depends both on the deal itself (x_i) and on when you seal the deal (k), with the second dependence sensitive to each person's bargaining power: the one with a larger r_i has more to lose by hanging on to keep bargaining for and rejecting offers.

We will not go into the details of the equilibrium properties of this procedure. But it is intuitively clear that if waiting for the next round of negotiation gives me the same payoff as accepting this round's offer, I might as well accept the offer. Indeed, it can be shown that the equilibrium offers (x_1^*, x_2^*) from each side satisfy the following two equations simultaneously:

$$1 - x_1^* = x_2^* e^{-r_2 T},$$
$$1 - x_2^* = x_1^* e^{-r_1 T}.$$

There is a unique solution to the above pair of equations:

$$x_1^* = \frac{1 - e^{-r_2 T}}{1 - e^{-(r_1 + r_2)T}},$$

$$x_2^* = \frac{1 - e^{-r_1 T}}{1 - e^{-(r_1 + r_2)T}}.$$

As the bargaining rounds get more efficient, i.e., $T \to 0$, the exponential penalty of disagreement becomes linear: $\exp(-r_i T) \to 1 - r_i T$, and the solution simplifies to the following approximation for small T:

$$x_1^* = \frac{r_2}{r_1 + r_2},$$

$$x_2^* = \frac{r_1}{r_1 + r_2}.$$

This proportional allocation makes sense: a bigger r_2 means a weaker hand of B, and thus a bigger share goes to A at equilibrium.

6.4.2   Bargaining: Nash bargaining solution

Iterative bargaining is just one mechanism of bargaining. The model can be made agnostic to the mechanism chosen.

Let the payoff function U_i map from the space of allocation $[0, 1]$ to some real number. Assume these are "nice" functions: strictly increasing and concave. If no agreement is reached, a disagreement point will be in effect: (d_1, d_2). Assume disagreement is at least as attractive as accepting the worst agreement: $d_1 \geq U_1(0)$ and $d_2 \geq U_2(0)$.

The set of possible agreement points is obviously

$$\mathcal{X} = \{(x_1, x_2) : x_1 \in [0, 1], x_2 = 1 - x_1\}.$$

Therefore, the set of possible utility pairs is

$$\mathcal{U} = \{(u_1, u_2) : \text{there is some } (x_1, x_2) \in \mathcal{X} \text{ such that } U_1(x_1) = u_1, U_2(x_2) = u_2\}.$$

Nash showed that the following four reasonable statements about a payoff point $\mathbf{u}^* = (u_1^*, u_2^*)$ can be taken as axioms that lead to a unique and useful solution.

1. *Symmetry*: If two players, A and B, are identical ($d_1 = d_2$ and U_1 is the same as U_2), the payoffs received are the same too: $u_1^* = u_2^*$.
2. *Affine Invariance*: If utility functions or disagreement points are scaled and shifted (an affine transformation), the resulting $\mathbf{u}^*$ is scaled and shifted in the same way (invariant).
3. *Pareto Efficiency*: There cannot be a strictly better payoff pair than $\mathbf{u}^*$.
4. *IIA*: Suppose A and B agree on a point $\mathbf{x}^*$ that leads to $\mathbf{u}^*$ in $\mathcal{U}$. Then, if in a new bargaining problem, the set of possible utility pairs is a strict subset of $\mathcal{U}$, and $\mathbf{u}^*$ is still in this subset, the new bargaining's payoffs remain the same.

The first axiom on symmetry is the most intuitive. The second one on affine invariance says changing your unit of payoff accounting should not change the bargaining result. The third one on Pareto effiency prevents clearly inferior allocation. The fourth one on IIA is again the most controversial one. But at least it does not preclude other axioms here, and indeed the four axioms are consistent.

Nash proved that there is one and only one solution that satisfies the above axioms, and it is the solution to the following maximization problem, which maximizes the product of the gains (over the disagreement point) by both A and B, over the set of payoffs that is feasible and no worse than the disagreement point itself:

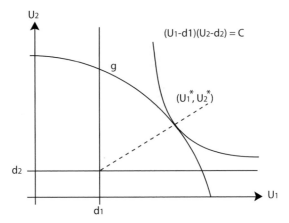

Figure 6.4 Nash bargaining solution illustrated on the two-player payoff plane. It is the intersection of the g curve (which captures the feasible tradeoff between the two players' payoffs) and the straight line normal to g and originating from the disagreement point (d_1, d_2).

$$\begin{aligned} \text{maximize} \quad & (u_1 - d_1)(u_2 - d_2) \\ \text{subject to} \quad & (u_1, u_2) \in \mathcal{U} \\ & u_1 \geq d_1 \\ & u_2 \geq d_2 \\ \text{variables} \quad & u_1, u_2. \end{aligned}$$

This solution (u_1^*, u_2^*) is called the **Nash Bargaining Solution** (NBS).

Obviously there is a tradeoff between possible u_1 and possible u_2. If A gets payoff u_1, what is the payoff for B? Using the above definitions, it is

$$u_2 = g(u_1) = U_2(1 - U_1^{-1}(u_1)).$$

We just defined a mapping, g, from player 1's payoff value to player 2's. This allows us to draw g as a curve in the payoff plane as in Figure 6.4. It is a similar visualization to the SIR feasibility region in Chapter 1.

If this function g is differentiable, we can differentiate the objective function

$$(u_1 - d_1)(g(u_1) - d_2)$$

with respect to u_1, set it to zero, and obtain

$$(u_1 - d_1)g'(u_1) + (g(u_1) - d_2) = 0.$$

Since $g(u_1) = u_2$, we have

$$g'(u_1) = -\frac{u_2 - d_2}{u_1 - d_1}.$$

This has a geometric interpretation, as illustrated in Figure 6.4. The NBS (u_1^*, u_2^*) is the unique point on the graph of g where the line from the disagreement point

(d_1, d_2) intersects g perpendicular to the slope of g. This clearly illustrates that the bigger the ratio d_1/d_2, i.e., the bigger A's bargaining power relative to B's, the more favorable will u_1 be relative to u_2 at the NBS.

For example, if both payoff functions are linear (which actually violates the strictly increasing property we assumed of U_i), then the cake-cutting NBS has a simple solution: each person first takes the disagreement-point allocation away, and then evenly splits the rest of the cake:

$$x_1^* = d_1 + 0.5(1 - d_1 - d_2),$$
$$x_2^* = d_2 + 0.5(1 - d_1 - d_2).$$

There is another way to model bargaining power: turn the objective function to

$$(u_1 - d_1)^\theta (u_2 - d_2)^{1-\theta},$$

where θ and $1 - \theta$, for $\theta \in [0, 1]$, are the normalized bargaining power exponents for A and B, respectively. It turns out that there is an interesting connection between the iterative bargaining solution and this axiomatically developed NBS. In the limit of $T \to 0$, the iterative bargaining solution is the same as the NBS solution with asymmetric bargaining power $\theta = \frac{r_2}{r_1} + r_2$ and disagreement point $(0, 0)$.

Summary

Box 6 Consensus formation

Wikipedia's success depends on the positive network effect and good-faith collaboration. Forming consensus for collaboration can be modeled through voting or bargaining. In trying to compress many rank-ordered lists into a single one, a voting system may not satisfy some intuitive conditions such as the Independence of Irrelevant Alternatives. The positional voting method of Borda count can resolve some of these issues.

Further Reading

There is very little mature work on the mathematical models of Wikipedia, but a rich research literature both on voting theory and on bargaining theory.

1. A comprehensive survey of the features in Wikipedia, including the policies, guidelines, and editorial procedures, can be found in the following book:

P. Ayers, C. Matthews, and B. Yates, *How Wikipedia Works*, No Starch Press, 2008.

2. Arrow's impossibility theorem was part of his Ph.D. dissertation, and originally published in 1950 and then in his book in 1951:

K. Arrow, *Social Choice and Individual Values*, Yale University Press, 1951.

3. One of the foremost researchers in voting theory today is Saari, who published several books interpreting and overcoming the negative results of Arrow and of Sen. Our treatment of IIIA, construction of examples of Sen's results, and connection to the prisoner's dilemma all follow Saari's books, such as the one below:

D. Saari, *Dethroning Dictators, Demystifying Voting Paradoxes*, Cambridge University Press, 2006.

4. The Nash bargaining solution was axiomatically constructed as part of Nash's Ph.D. dissertation, and originally published in the following paper in 1950:

J. Nash, "The bargaining problem," *Econometrica*, vol. 18, no. 2, pp. 155–162, April 1950.

5. Among the many books devoted to bargaining theory since 1950 is the following concise and rigorous survey:

A. Muthoo, *Bargaining Theory with Applications*, Cambridge University Press, 1999.

Problems

6.1 *Differences between Borda count, Condorcet voting and plurality voting* ⋆

Consider an election involving $N = 31$ voters and $M = 3$ candidates A, B, and C, with their preference profiles summarized as follows:

Preference	Votes
$C > A > B$	9
$A > B > C$	8
$B > C > A$	7
$B > A > C$	5
$C > B > A$	2
$A > C > B$	0

What is the voting result by (a) plurality voting, (b) Condorcet voting, and (c) Borda count?

6.2 *List's list* ⋆

A three-member faculty committee need to determine whether a student should be advanced to Ph.D. candidacy or not, based on the student's performance on both the oral and written exams. The following table summarizes the evaluation result of each faculty member:

Professor	Written	Oral
A	Pass	Pass
B	Fail	Pass
C	Pass	Fail

(a) Suppose the student's advancement is determined by a majority vote of all the faculty members, and a professor will agree on the advancement if and only if the student passes both the oral and written exams. Will the committee agree on advancement?

(b) Suppose the student's advancement is determined by whether she passes both the oral and written exams. And whether the student passes an exam or not is determined by a majority vote of the faculty members. Will the committee agree on advancement? How does this compare with the result in (a)?

6.3 *Anscombe's paradox* ⋆⋆

Suppose there are three issues where a "yes" or "no" vote indicates a voter's support or disapproval. There are two coalitions of voters, the majority coalition $A = \{A_1, A_2, A_3\}$ and the minority coalition $B = \{B_1, B_2\}$. The preference profile is summarized as follows:

Voter	Issue 1	Issue 2	Issue 3
A_1	Yes	Yes	No
A_2	No	Yes	Yes
A_3	Yes	No	Yes
B_1	No	No	No
B_2	No	No	No

(a) What is the majority-voting result of each issue?

(b) For each member in the majority coalition A, how many issues out of three does she agree with in the voting result?

(c) Repeat (b) for the minority coalition $\mathcal{B}$.

(d) Suppose the leader in coalition $\mathcal{A}$ enforces a "party discipline" on all members: they first vote internally to achieve agreement. Then on the final vote where coalition $\mathcal{B}$ is present, all members in coalition $\mathcal{A}$ will vote according to their internal agreement. What happens then to the final voting result?

6.4　*Nash Bargaining Solution* ★★

Alice has an alarm clock (good A_1) and an apple (good A_2). Bob has a bat (good B_1), a ball (good B_2), and a box (good B_3). Their utilities for these goods are summarized as follows:

Owner	Goods	Utility to Alice	Utility to Bob
Alice	Alarm Clock (A_1)	2	4
Alice	Apple (A_2)	2	2
Bob	Bat (B_1)	6	3
Bob	Ball (B_2)	2	1
Bob	Box (B_3)	4	2

What is the Nash bargaining result between Alice and Bob?

6.5　*Wikipedia articles (open-ended question)*

Take a look at the history pages and the discussion pages of two Wikipedia articles: "Abortion" and "Pythagorean Theorem." Summarize three key (qualitative) differences you can see between them.

7 How do I viralize a YouTube video and tip a Groupon deal?

A quick recap of where we have been so far in the space of online services and web 2.0. In Chapter 3, we discussed the recommendation of webpages with an objective metric computed by Google from the graph of hyperlinked webpages. In Chapter 4, we discussed the recommendation of movies with subjective opinions estimated by Netflix from movie–user bipartite graphs.

Then we investigated the wisdom of crowds. In Chapter 5, we discussed *aggregation* of opinion in (more or less) independent ratings on Amazon. In Chapter 6, we discussed *resolution* of opinion conflicts in Wikipedia.

In this chapter, we will talk about *dependence* of opinions, taking a macroscopic, topology-agnostic approach, and focusing on the viral effect in YouTube and tipping in Groupon. Then in the next chapter, we will talk about the effect of network topology on the dependence of opinion.

As will be further illustrated in this and the next chapters, network effects can be positive or negative. They can also be studied as *externalities* (e.g., coupling in the objective function or the constraint functions, where each user's utility or constraint depends on other users' actions), or as *information dependence* (e.g., information cascades or product diffusion as we will see in this chapter).

7.1 A Short Answer

7.1.1 Viralization

YouTube is a "viral" phenomenon itself. In the space of user-generated video content, it has become the dominant market leader, exhibiting the "winner takes all" phenomenon. More recently it has also featured movies for purchase or rental, and commissioned professional content, to compete against Apple's iTunes and the studios.

YouTube started in February 2005 and was acquired by Google in 2006. Within several years people watched videos on YouTube so much that it became the second largest search engine with 2.6 billion searches in August 2008, even though we normally would not think of YouTube as a search engine. Its short video-clip format, coupled with its recommendation page, is particularly addictive. In summer 2011, more than 40% of Internet videos were watched on YouTube, with over 100 million unique viewers each month just in the USA. Each day over a billion video plays were played, and each minute more than 24 hours of new video clips were uploaded.

There are interesting analytic engines like "YouTube Insight" that highlight the aggregate behavior of YouTube watching. Some videos have gone viral, the most extreme example being "Charlie bit my finger–again," a less-than-one-minute clip that had generated over 465 million views as of July 2012. If you just look at the viewer percentage across all the videos on YouTube, it exhibits the long-tail distribution that we will discuss in Chapter 10.

There has been a lot of social media and web 2.0 marketing research on how to make your YouTube video go viral, including practical advice on the four main paths that lead a viewer to a YouTube clip:

- web search,
- referral through email or twitter,
- subscription to a YouTube channel, and
- browsing through the YouTube recommendation page.

We have seen how tags, subscription, and recommendation play a bigger role than the counts of likes and dislikes in the rise of popularity of a video. It is also interesting to see that YouTube does not use a PageRank-style algorithm for ranking the video clips, since linking the videos by tags is too noisy. Nor does it use the sophisticated recommendation engine such as Netflix Prize solutions, since viewing data for short clips is too noisy and YouTube videos often have short lifecycles. Instead, YouTube recommendation simply leverages video association through **co-visitation count**: how often each pair of videos is watched together by a viewer over, say, 24 hours. This gives rise to a set of related videos for each video, a link relationship among the videos, and thus a graph with the videos as nodes. From the set of videos in k hops from a given video, together with matching of the keywords in the video title, tags, and summary, YouTube then generates a top-n recommendation page. It has also been observed that often only those videos with a watch-count number similar to, or slightly higher than, that of the current video are shown in the recommendation page. This is a version of "preferential attachment" that we will discuss in Chapter 10. It makes it easier for widely watched videos to become even more widely watched, possibly becoming viral.

Now, how do you even define "viral"? There is no commonly accepted definition, but probably the notion of "viral" means that the rate-of-adoption curve should exhibit three features, like curve (c) shown in Figure 7.1:

- high peak,
- large volume, i.e., the adoption lasts long enough in addition to having a high peak, and
- a short time to rise to the peak.

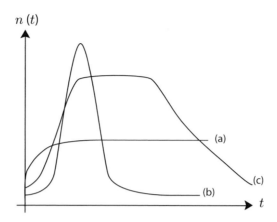

$n(t)$

(a)

(c)

(b)

t

Figure 7.1 A few typical shapes of adoption trajectory $n(t)$ over time t: curve (a) stays at a low level; curve (b) rises very quickly but then dies out rapidly too; and curve (c) has a reasonably sharp rise to a high level and a large area under the curve. Of course, we can also have combinations of these curves, e.g., a curve that has a sharp rise to a high level and stays there.

7.1.2   Tipping

Another web 2.0 sensation that has gone viral is the daily deal service, such as Groupon and LivingSocial. Groupon was formed in 2008, and after two years was generating over $300 million annual revenue from more than 500 cities. It went pubic in November 2011.

In a daily deal, a supplier of some goods or services announces a special discount, which must have a large enough number of users signed up within a 24-hour period. If the number of users exceeds the target threshold, the deal is tipped, and each user has, say, 3 months to redeem the coupon. The supplier's hope is that the discount is in part compensated for by the high volume, and in part the possibility of repeat customers who will return in the future and pay the full price. This is the **power of crowds** in action. More specifically, a daily-deal tipping needs a sufficiently large number of people to make the same decision within a sufficiently small window of time.

The effectiveness of Groupon (and the like) for the suppliers and the consumers is still somewhat under-studied. In a detailed survey by Dhulakia in 2011 with 324 businesses in 23 US market, some interesting results emerged: close to 80% of deal users were new customers, but only 36% of them spent beyond the deal's face value, and only 20% returned to buy at full price later. On the supplier side, 55% made money from the deals, but 27% lost money, and less than half of them expressed interest in participating in the future, restaurants and salons being particularly negative. Across hundreds of daily-deals websites, there are few differentiation factors at this point. The cut into the deals' revenues by these websites will have to be lowered in the future as the industry consolidates.

This chapter presents models that can be used to characterize and understand phenomena such as viralization, tipping, and synchronization observed for YouTube videos and Groupon deals.

7.2   A Long Answer

There are two distinct reasons for popularity of a product.

1. *Intrinsic value*: You may enjoy certain music or buy a particular product just because you like it, whether the rest of the world agrees or not.
2. *Network effect*: Your decision depends on what others do, either because (a) the fact that others like a product gives you information, possibly leading to what we call information cascades, an example of the fallacy of crowds, or (b) because the value of the service or product actually depends on the number of people who use it, like the fax machine or Wikipedia, an example of positive externality.

We will first discuss models of 2(a), such as **information cascade** studied in the political-economy literature and **tipping** from an unstable equilibrium towards a stable equilibrium. Then, in the Advanced Material, we will discuss the combination of 1 and 2(b), called the intrinsic factor and the imitation factor in **diffusion** models studied in the marketing literature, as well as a **synchronization** model.

All the models in this chapter are population-based and agnostic of the actual topologies (although our example of information cascade implicitly assumes a linear topology). In the next chapter, we will focus on topology-based models in the study of influence.

The models in both chapters are summarized in Table 7.2. Except for the synchronization and random-walk models, all assume the nodes (the people) have discrete, often binary, "states of mind." We will also see that some of these models become similar when generalized a little.

Which model to use really depends on what we are trying to model. Many people acting at the same time? Or a few early adopters changing others' minds? Or one more person carrying the system over the threshold and triggering a change in others' behavior? Each of these models is motivated by a different type of influence, and has its use and limitation.

Viralizing a YouTube video, tipping a Groupon deal, and influencing via Facebook and Twitter posts are three particular examples in these two chapters. But, for these emerging phenomena involving human-psychology factors, we still do not know much about which of these models and their extensions, if any, fit reality sufficiently well to render predictive power.

Across these models, the following issues are raised and some of them have been addressed:

	Deterministic-interaction Model	Random-interaction Model
Population-based	Information cascade (Sec. 7.2.1) Synchronization (Sec. 7.4.1)	Tipping (Sec. 7.2.2) Diffusion (Sec. 7.4.2)
Topology-dependent	Contagion (Sec. 8.2.1) Random walk (Sec. 8.4.1)	Infection (Sec. 8.2.2–8.2.5)

Table 7.1 Influence models in Chapters 7 and 8. Tipping and contagion models are highly related, as are information cascade and random walk, and diffusion and infection. The precise forms of deterministic or random interaction models will be presented as we go through these two chapters.

- distribution of the nodes across different states at equilibrium,
- the amount of time it takes to reach an equilibrium,
- more generally, transient behavior before an equilibrium is reached.

In addressing these issues, we first must observe how individual decisions and local interactions lead to a global property. Then, the general workflow of our modeling process may run as follows: pick a curve of adoption evolving over time, like one of those in Figure 7.1, reverse-engineer it through some mathematical languages (differential equation, dynamic systems, sequential decision-making, selfish optimization, etc.), and, finally, hope it will also shed light on forward engineering such as the design of social media viral marketing. The last step is where the gap between theory and practice lies. Having said that, let us see what kinds of explanatory models have been created.

7.2.1 Information cascade

One of the large-scale, controlled experiments that resemble the YouTube experience was run by Salganik, Dodds, and Watts in 2005. Each of the 14341 participants rated each of the $S = 48$ (unknown) songs (from unknown bands) from 1 to 5 stars. There were four different conditions tested, depending on

- whether the songs were presented at random, or in descending order of the current download counts; and
- whether the current download numbers were hidden, or shown next to each song.

When the current download numbers were shown, social influence was present. And when the order of the songs followed the download numbers, this influence became even stronger.

How do we measure the *spread* of the download numbers of the songs? Let m_i be the percentage of downloads, out of all the downloads, received by song i, for $i = 1, 2, \ldots, S$. We can examine the sum of the differences: $\sum_{i,j} |m_i - m_j|$, which the experimental data showed to be always larger under social influence than under the independent condition.

In addition, a heavier social influence also increased the *unpredictability* of a song's popularity. Really good songs (as defined by download popularity under no social influence) rarely do badly even under social influence, and really bad ones hardly ever do very well, but for the majority of the songs in between, stronger social influence significantly increases their range of popularity.

This experiment, together with many other controlled experiments or empirical social data analyses, demonstrates what we often feel intuitively: we are influenced by others' opinions even when we do not know the underlying reasons driving their opinions.

On YouTube, if many others before us watched a video, we are more likely to watch it too. We might abort the viewing if we realize we actually do not like it, but this still counts towards the viewing number shown next to the video and partially drives its place on the recommendation page.

On a street corner, if one person looks up to the sky, you may think she has a nose bleed or just turned philosophical, and you would pass by. But if ten people look up to the sky together, you may think there is something wrong up there, and stop and look up to the sky too. Now that makes the crowd even bigger. Until someone shouts "Hey, these two guys have nose bleeds, that is why they stopped and tilted their heads!" Then you think everyone else, just like you, was misled, and decide to leave the crowd and keep on walking. Suddenly, the whole crowd disperses.

These examples, and many others in our lives, from stock market bubbles and fashion fads to the emergence of pop stars and the collapse of totalitarian regimes, illustrate several key observations about information cascade as a model for influence in **sequential decision making**.

- Each person gets a *private signal* (my nose starts bleeding) and releases a *public action* (let me stop walking and tilt my head to the sky). Subsequent users can observe the public action but *not* the private signal.
- When there are enough public actions of the same type, at some threshold point, all later users start ignoring their own private signals and simply follow what others are doing. A cascade starts, and the *independence assumption* behind the wisdom of crowds in Chapter 6 breaks down.
- A cascade can start *easily* if people are ready to rely on others' public actions in their reasoning; it can accumulate to a *large size* through positive feedback; and it can be *wrong*.
- But a cascade is also *fragile*. Even if a few private signals are released to the public, the cascade can quickly disappear or even reverse direction, precisely because people have little faith in what they are doing even when there are many of them doing the same thing.

There are many dimensions to understanding the above observations. We focus in this subsection on a particular model from 1992 for sequential decision making

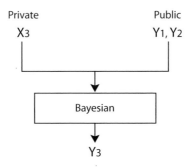

Figure 7.2 The third person in the binary-number-guessing model of information cascade. She has two public actions before her and receives her private signal. If $Y_1 = Y_2$, a Bayesian thinking process leads to the onset of an information cascade.

that leads to information cascades. The simplest version of this model is a binary-number-guessing process.

Consider a set of people lined up in sequential order. Each person takes a turn observing a private signal $X_i = \{0, 1\}$ and trying to guess whether the correct number is 0 or 1. To simplify the notation a little, we will assume that the correct number is equally likely to be 0 or 1, although that does not need to be the case for what we will see.

The chance that the private signal is the correct number is $p_i > 0.5$, and with probability $1 - p_i$ it is not. It is important that $p_i > 0.5$. We will assume for simplicity that all $p_i = p$. In addition, the private signal is conditionally independent of other people's signals, conditioned on the underlying correct number.

Upon receiving her private signal, each user writes down her guess $Y_i = \{0, 1\}$ on a blackboard and walks away. That is her public action. Since it is a binary guess, she assesses that one number is more likely than the other, and she writes down her guess accordingly. Every user can see the guesses made by people before her, but not their actual private signals.

If you are the first person, what should you do? If you see 1, you will guess 1, since the chance of that being the correct guess is $p > 0.5$. Similarly, if you see 0, you will guess 0.

If you are the second person, what should you do? You have one private signal X_2 and one public action recorded from the first person Y_1. You know how person 1 *reasoned*: X_1 must be equal to Y_1, even though you cannot see her X_1. So, as the second user, you actually know the first user's private signal: $X_1 = Y_1$. Now equipped with two private signals $\{X_1, X_2\}$, you will decide as follows: if both X_1 and X_2 are 0, then obviously the correct number is more likely to be 0, and you guess 0. If both are 1, then guess 1. If one is 0 and the other is 1, then flip a coin and randomly choose 0 or 1.

Now comes the first chance of an information cascade starting. If you are the third person, what should you do? As shown in Figure 7.2, you have one private signal, X_3, and two public actions Y_1 and Y_2.

- If $Y_1 \neq Y_2$, by reasoning through the first two persons' reasoning, you know the public actions by prior users collectively do not tell you anything, and you should just rely on your own private signal. In this case, you are in exactly the same shoes as the first user. And the fourth user will be in the same shoes as the second user.

- However, if $Y_1 = Y_2$, and they are both the same as X_3, you will obviously pick that number. But even if your private signal X_3 differs, you still will pick what two public signals suggest. This follows the same *Bayesian reasoning* as in Chapter 5. Say $Y_1 = Y_2 = 1$ and $X_3 = 0$. What is the probability that the correct number is 1 given this sequence of $(1,1,0)$? By Bayes' rule, this probability, denoted as $P[1|(1,1,0)]$, equals

$$\frac{P[1]P[(1,1,0)|1]}{P[(1,1,0)]},$$

where $P[1]$ and $P[0]$ denote the probabilities that the correct number is 1 and 0, respectively. We can easily see this because the joint probability that the correct number is 1 and the sequence is (1,1,0) can be written in two different ways: $P[1]P[(1,1,0)|1]$ and $P[(1,1,0)]P[1|(1,1,0)]$.

Furthermore, $P[(1,1,0)] = P[1]P[(1,1,0)|1] + P[0]P[(1,1,0)|0]$, where $P[(1,1,0)|1]$ and $P[(1,1,0)|0]$ denote the conditional probabilities that $Y_1 = 1, Y_2 = 1$, and $X_3 = 0$ given that the correct number is 1 and 0, respectively. This means

$$P[1|(1,1,0)] = \frac{P[1]P[(1,1,0)|1]}{P[(1,1,0)]} = \frac{P[1]P[(1,1,0)|1]}{P[1]P[(1,1,0)|1] + P[0]P[(1,1,0)|0]}. \tag{7.1}$$

Plugging the numbers into (7.1), we have

$$\frac{0.5(p^2(1-p) + 0.5p(1-p)^2)}{0.5(p^2(1-p) + 0.5p(1-p)^2) + 0.5((1-p)^2p + 0.5(1-p)p^2)}$$

since the sequence $Y_1 = 1, Y_2 = 1, X_3 = 0$ could have come from either $X_1 = 1, X_2 = 1, X_3 = 0$ or $X_1 = 1, X_2 = 0, X_3 = 0$, but the second person flips the coin and so happens to choose $Y_2 = 1$. Now, is this expression bigger than half? Certainly. After cancelling $p(1-p)$ and multiplying both the numerator and the denominator by 4, we are dividing $2p + (1-p)$ by $2p + (1-p) + 2(1-p) + p$, which is $(1+p)/3$. Since $p > 0.5$, that ratio is indeed bigger than $1/2$. So person 3 guesses that the correct number is 1 even when her private signal is 0.

In summary, once an odd-numbered user and then an even-numbered user have shown the same public action in a row, the next user will just follow, no matter what her private signal is. An information cascade starts.

What about the probability of no cascade after two people have received their private signals and made their public actions? That is equal to the probability that the first two public signals are different. If the correct number is 1, the probability that $X_1 = Y_1 = 1$ and $X_2 = 0$ is $p(1-p)$, and, with probability $1/2$, user 2 will choose $Y_2 = 0$. So the probability of $Y_1 = 1$ and $Y_2 = 0$ is $p(1-p)/2$. The probability that $Y_1 = 0$, $X_2 = 1$, and $Y_2 = 1$ is $p(1-p)/2$ too, so the probability of $Y_1 \neq Y_2$, i.e., no cascade, is $p(1-p)$. By symmetry, the answer is the same when the correct number is 0. In conclusion, the probability of having no cascade is

$$\text{Prob}_{no} = p(1-p).$$

By symmetry, the probability of a cascade of 1s (call that an "up" cascade) and that of a cascade of 0s (call that a "down" cascade) are the same, each taking half of $1 - \text{Prob}_{no}$:

$$\text{Prob}_{up} = \text{Prob}_{down} = \frac{1 - \text{Prob}_{no}}{2} = \frac{1 - p(1-p)}{2}.$$

Following a similar argument, we see that, after an even number, $2n$, of people have gone through the process, we have

$$\text{Prob}_{no} = (p(1-p))^n,$$

and

$$\text{Prob}_{up} = \text{Prob}_{down} = \frac{1 - (p(1-p))^n}{2},$$

since no cascades happen if the members of each pair of people $(1, 2)$, $(3, 4)$, $(5, 6)$... take different public actions.

Therefore, intuitively and mathematically, we see that cascades eventually will happen as more people participate. And this conclusion has been proven to hold for general cases (under some technical conditions), even with multiple levels of private signals and multiple choices to adopt by each person.

It is more instructive to look at the probability of correct vs. incorrect cascades, rather than the symmetric behavior of up and down cascades. Of course that depends on the value p: how noisy the private signal is. If the signal is very noisy, $p = 0.5$, then, by symmetry, correct cascades are as likely to happen as incorrect cascades. But as p approaches 1, the probability of a correct cascade goes up pretty fast and saturates towards 1. This computation is shown in an example in Section 7.3.

Now, how long will a cascade last? Well, forever, unless there is some kind of disturbance, e.g., a release of private signals. Even a little bit of that often suffices, because, despite the number of people in the cascade, they all know they are just following a very small sample out of the desire to maximize their chance of making the correct guess. This is the counterpart of information cascade: the **Emperor's New Clothes effect**. Sometimes it only takes one kid's shouting out a private signal to stop the cascade.

So how do we break a cascade? Suppose a cascade of 1s has started. Now when one person gets a private signal of 0, she shouts *that* out, instead of her public action, which is still to guess that the correct number is 1. If the next person gets a private signal of 0, she would think that, on the one hand, now there are two private signals of 0s. On the other hand, there may have been only two private signals of 1s since that is enough to kick off a cascade of 1s (or even just one private signal of 1, since the following user receiving a private signal of 0 may break the tie by randomly choosing 1 as the public action). So she will decide to guess 0. That break the cascade. A cascade represents only what happened with a few people right around the time it started. If everyone knows that, another block of a few people may be able to break it.

Sometimes it takes the apparent conformity of more users, e.g., ten people, instead of just two, standing on a street corner looking up at the sky. For something with a low probability (e.g., something is wrong in the sky, wrong enough that I want to stop walking and take a look), you need more public actions to override your private signal. In a nosebleeding-created crowd on a street corner, a passer-by often needs more convincing than just guessing the right number with higher than 50% probability. That is why she needs to see a bigger crowd all tilting their heads before she is willing to stop and do the same. A more curious and less busy person may require a lower threshold of the crowd size for her to stop and do the same.

We have assumed that everyone's private signal has the same precision p, but each person i may have a different p_i. Suppose all people know the values of $\{p_i\}$ for everyone. Where should the high-precision person be placed in the sequence of people (if you could control that)? If we put the highest-precision person first, a cascade may start right after her. An authoritative person starts leading the pack. So if you want to start the cascade, that is the right strategy, and it partially explains why some cascades are difficult to break as private signals keep getting overwhelmed by public actions. But if you want a higher chance of reversing a cascade, you want to put higher-precision persons later in the sequence.

More importantly, we have assumed that everyone acts rationally. What each person *should* do can be quite different from what she *actually* does in real life. Researchers have observed that the number-guessing experiment does not play out in reality as the theory predicts, perhaps because people are not all rational, Bayesian agents.

There are many implications of this information-cascade model, since (1) various social phenomena exhibit the features of a crowd making a decision that ignores each individual's own private information, and yet (2) the chain reaction is fragile. Information cascade is part of the reason why US presidential primary elections use Super Tuesday to avoid sequential voting, why teenagers tend to obtain information from the experiences of peers, and why, once people suspect the underlying true signal has changed (whether it has actually changed or not), the ongoing cascade can quickly reverse.

So far we have focused on a model with essentially a *linear topology* of nodes (each node is a person) and each node's local decision is based on *strategic thinking*. In the Advanced Material, we move on to a different model, with a macroscopic view of the effect of peers on each person without the detailed strategic

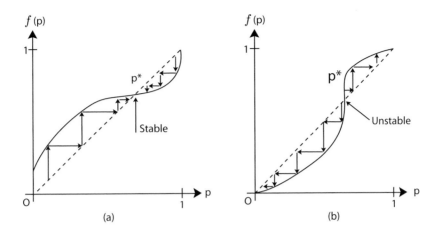

Figure 7.3 The influence function $f(p)$. Starting from a given p value, we can move vertically to hit $f(p)$, and then move horizontally to hit the 45-degree line. That represents one iteration of $p[t+1] = f(p[t])$. Graph (a) illustrates a stable equilibrium, where $p = 0$ eventually climbs to the equilibrium $p = p^*$ despite perturbation around it. Graph (b) illustrates an unstable equilibrium, where perturbation around p^* moves away from it.

thinking. In the next chapter, we will discuss models where an individual's decision depends on her local topology.

7.2.2   Tipping

But first, we will examine a different model that illustrates, in a simple way, the notions of tipping, and of the stability of equilibrium. This will be the first time we encounter *positive* externality. It also has to do with the fundamental idea of **positive feedback**, a phenomenon that cuts across many domains, from a microphone generating excessively high volume during a tune-up to a high unemployment rate lasting longer as unemployment benefits extend. In contrast, in Chapter 14 we will look at *negative* feedback in the Internet to help provide stability.

Suppose there are two possible states in each person's mind. Each person decides to flip from one state (e.g., not buying an iPad) to the other (e.g., buying an iPad) depending on whether her utility function is sufficiently high. That utility function depends on p, the product **adoption percentage** in the rest of the population (or just among her friends). In the next chapter, we will go into the details of the best response strategy of a person who learns that $p\%$ of her friends have adopted the product. Right now it suffices to say the probability of a person adopting the product is a function of p. Then taking the average across all users, we have an **influence function** f that maps p at timeslot t to p at the next timeslot $t+1$.

We can readily visualize the dynamics of p by tracing two curves on the $(p, f(p))$ plane, as shown in Figure 7.3: the $f(p)$ curve in a solid line, and the

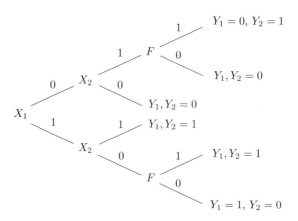

Figure 7.4 An example of information cascade, visualized as a tree. Each branching level is a new person. F denotes flipping a coin to randomly choose what number to guess.

straight line of 45 degrees in a dotted line from the origin. Given a p value at time t, the future evolution is visualized as the zig-zag trajectory bouncing vertically from the 45-degree line to the $f(p)$ curve (carrying out the mapping step $f(p[t])$), and then horizontally back to the 45-degree line (carrying out the equality step $p[t+1] = f(p)$). This graphically shows one iteration of $p[t+1] = f(p[t])$.

When the two curves intersect at a point $(p^*, f(p^*))$, we have an *equilibrium*, since if the current $p = p^*$, the next p remains the same. There can be zero, one or many equilibria. The concept of equilibrium here is not the same as strategic equilibrium in a game, or the convergence of an iterative algorithm, but instead means reaching a fixed point of an update equation.

Some equilibria are **stable**; others are **unstable**. A stable equilibrium here is one where a perturbation around it will converge back to it. And an equilibrium that is not stable is called unstable. For example, in Figure 7.3, the equilibrium in (a) is stable, as the trajectory around it converges to it, whereas the equilibrium in (b) is unstable. The shape of the influence function f matters. For $f(p)$ that cuts across the 45-degree line once, if it is concave before the cross-over point and convex afterwards, that cross-over point is stable. If it is convex before and concave afterwards, the cross-over point is unstable.

Now we can view tipping as when the adoption percentage p reaches past an unstable equilibrium $(p_1, f(p_1))$, and falls into the **attraction region** of the next stable equilibrium $(p_2, f(p_2))$: as more people adopt it, even more people will adopt it, and this process continues until the adoption percentage is p_2. This is similar to the tipping behavior of popular videos on YouTube. As more people watch a video, its viewer counter increases, and more people think it is worthy of watching. If you would like to make that happen to your video clip, you need to estimate $f(p)$ and then get into the attraction region of an equilibrium with a high p.

7.3   Examples

7.3.1   Information cascade

First recall what we discussed in Section 7.2.1 about information cascades through a simple binary-number-guessing experiment, and the symmetric behavior of up and down cascades. The example now further derives the probability of correct vs. incorrect cascades. Here we assume the correct number is 1. So an up cascade is a correct one. In contrast, in the last section we did not pre-set the correct number.

Consider the first two people. The different cases of X_1 and X_2 are shown in Figure 7.4, where F denotes a coin flip by the second user when in doubt. Then

$$\text{Prob}_{no} = P(X_1 = 0, X_2 = 1, F = 1) + P(X_1 = 1, X_2 = 0, F = 0)$$
$$= (1 - p) \cdot p \cdot \frac{1}{2} + p \cdot (1 - p) \cdot \frac{1}{2}$$
$$= p(1 - p),$$
$$\text{Prob}_{up} = P(X_1 = 1, X_2 = 1) + P(X_1 = 1, X_2 = 0, F = 1)$$
$$= p^2 + p(1 - p)\frac{1}{2}$$
$$= \frac{p(1 + p)}{2},$$
$$\text{Prob}_{down} = P(X_1 = 0, X_2 = 0) + P(X_1 = 0, X_2 = 1, F = 0)$$
$$= (1 - p)^2 + \frac{(1 - p)p}{2}$$
$$= \frac{(1 - p)(2 - p)}{2}.$$

Since the correct number is 1, $\text{Prob}_{correct} = \text{Prob}_{up}$ and $\text{Prob}_{incorrect} = \text{Prob}_{down}$. Extending to the general case of $2n$ people, it is obvious that

$$\text{Prob}_{no} = (p(1 - p))^n.$$

Furthermore, a little calculation shows that

$$\text{Prob}_{correct} = \sum_{i=1}^{n} P(\text{no cascade before } i\text{th pair}, Y_{2i-1} = 1, Y_{2i} = 1)$$
$$= \sum_{i=1}^{n} (p(1 - p))^{i-1} \frac{p(p + 1)}{2}$$
$$= \frac{p(p + 1)}{2} \sum_{i=0}^{n-1} (p(1 - p))^{i}$$
$$= \frac{p(p + 1)}{2} \frac{1 - (p(1 - p))^n}{1 - p(1 - p)}$$
$$= \frac{p(p + 1)[1 - (p - p^2)^n]}{2(1 - p + p^2)},$$

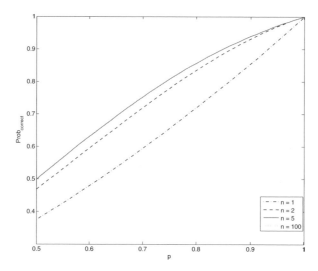

Figure 7.5 Probabilities of information cascade. As the probability p (of each person getting the correct private signal) increases, the probability of forming a correct cascade increases roughly linearly. Once the number of people n reaches 5, the curves are almost identical even as n increases further. That is why you can only see three curves in the graph. The impact of the size of the crowd saturates.

and

$$\text{Prob}_{incorrect} = \sum_{i=1}^{n} P(\text{no cascade before } i\text{th pair}, Y_{2i-1} = 0, Y_{2i} = 0)$$

$$= \sum_{i=1}^{n} (p(1-p))^{i-1} \frac{(1-p)(2-p)}{2}$$

$$= \frac{(1-p)(2-p)[1 - (p - p^2)^n]}{2(1 - p + p^2)}.$$

Figure 7.5 shows the plots of the above probabilities as functions of p, for different n. Key observations are as follows.

1. Obviously, a larger p (the probability of a private signal being correct) increases $\text{Prob}_{correct}$.
2. Increasing n reduces Prob_{no}. When $n = 1$, Prob_{no} is significant. When n is small, increasing it helps increase $\text{Prob}_{correct}$. But the effect of increasing n quickly saturates. The plots of $n = 5$ and $n = 100$ are almost indistinguishable.
3. Even for large n, $\text{Prob}_{correct}$ is significantly less than 1 for a large range of p. This is somewhat counter-intuitive because when n is large, we expect a large amount of information to be available and that a correct cascade should happen with high probability even for small p. But what happens is that cascades block the aggregation of independent information.

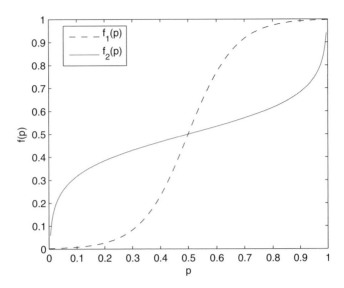

Figure 7.6 A tipping example, with two different influence functions $f(p)$. Function f_2 leads to a stable equilibrium at $(0.5, 0.5)$, and function f_1 leads to an unstable equilibrium at the same point.

7.3.2 Tipping

Now recall the influence function f that evolves the product adoption percentage p in Section 7.2.2. The following example illustrates the notion of tipping. Consider two influence functions:

$$f_1(p) = \frac{1}{1 + e^{-12(p-0.5)}}, \tag{7.2}$$

$$f_2(p) = 0.5 + \frac{1}{12} \log\left(\frac{p}{1-p}\right). \tag{7.3}$$

Figure 7.6 shows the plots of both functions, each with three equilibrium points $p^* = 0.00255, 0.5$, and 0.9975, where $f(p^*) = p^*$. For $f_1(p)$ the equilibria 0.00255 and 0.9975 are stable, while for $f_2(p)$ the equilibrium 0.5 is stable.

Now consider what happens when we try to tip from an unstable equilibrium. For $f_1(p)$, we start at $p[0] = 0.5 - 0.001 = 0.499$ and iteratively update p_i as

$$p[i+1] = f(p[i]).$$

Then $p[i]$ updates as 0.499, 0.497, 0.491, 0.473, 0.419, 0.276, 0.0639, 0.00531, 0.00264, 0.00255, 0.00255, ..., $p[i]$ stops changing when it reaches the stable equilibrium near zero: $p^* = 0.00255$. A little perturbation tips p from 0.5 all the way down to 0.00255.

For $f_2(p)$, we iterate from $p[0] = 0.00255 + 0.00001 = 0.00256$ (we perturb much less so as to achieve convergence in about 10 iterations as in the case of f_1), and $p[i]$ updates as 0.00256, 0.00291, 0.0134, 0.141, 0.350, 0.448, 0.483, 0.494,

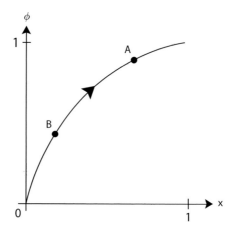

Figure 7.7 The trajectory of two pulse-coupled oscillators (A and B) shown on the state-phase plane. As time passes by, both A and B follow the trajectory. When either reaches $(1, 1)$, it fires, returns to the origin, and kicks the other oscillator vertically by ϵ. Once they fire at the same time, they are synchronized as they will travel together on this plane from that point onwards.

$0.498, 0.499, 0.500, 0.500 \ldots$ Again $p[i]$ stops changing when it reaches the stable equilibrium $p^* = 0.5$. A little perturbation from 0.00255 tips p all the way to 0.5.

7.4   Advanced Material

7.4.1   Synchronization

Sometimes herding or tipping manifests itself on the *time* dimension too, when the behaviors of many people are synchronized. This is one step beyond just converging on the same public action (in information cascade) or on adoption percentage (in tipping). The study of synchronization started with clocks. Two clocks hang on the wall next to each other. Vibration propagating through the wall produces a coupling between the two clocks. Soon their ticking will be synchronized. Synchronization happens a lot in nature: from fireflies glowing to crickets chirping, and from people walking to pacemakers adjusting.

The standard model used to study synchronization is **coupled oscillators**. We will look at a weak form of coupling called **pulse coupling**. Later we will also study *decoupling* over a network in Chapter 14, and how to use very little and implicit communication, plus a little randomization, to avoid synchronization in Chapter 18.

We will focus on the case of two weakly pulse-coupled oscillators; A and B, and visualize the dynamics in a "state–phase diagram" in the $[0, 1] \times [0, 1]$ grid, where the y-axis is the state and x-axis is the phase. Each oscillator's state x evolves over time, with magnitude normalized with repsect to $[0, 1]$. We parameterize this movement not directly as a function of time t, but instead by a phase parameter that keeps track of the asynchronism between A and B. The state x depends on the phase ϕ through a function f:

$$x(t) = f(\phi(t)),$$

where $\phi \in [0, 1]$ is the phase variable. The phase variable moves over time steadily as $d\phi/dt = 1/T$ for cycle period T, thus driving x over time as well. We can think of $\phi(t)$ as the degree marked on the clock by an arm, and $x(t)$ the state of the spring behind the arm. The boundary conditions are $f(0) = 0$ and $f(1) = 1$, as shown in Figure 7.7.

A simple and important model of $f(\phi)$ is the "pacemaker trajectory:"

$$f_\gamma(\phi) = (1 - e^{-\gamma})(1 - e^{-\gamma\phi}),$$

where γ is a constant parameter in the pacemaker model. The upper bound on the state x is now $(1 - e^{-\gamma})^2$ (close to 1 for large γ but not exactly 1), which is achieved when phase ϕ is 1.

When x reaches the boundary point (1, or $(1 - e^{-\gamma})^2$) for, say, oscillator A, three things happen.

- Oscillator A "fires."
- x_A goes back to 0.
- The other oscillator, B, gets a "kick:" its own x_B goes up to $x_B(t) + \epsilon$, unless that value exceeds 1, in which case B fires too, and the two oscillators are synchronized from that point onwards.

So if we can show that, no matter what the initial condition, the two oscillators A and B will eventually fire together, then we will have shown that synchronization must happen.

As shown in Figure 7.7, A and B move along curve f with a constant horizontal distance between them until one of them reaches the firing threshold, fires, drops back to the origin, and causes the other one to jump vertically by ϵ.

Before walking through a proof that synchronization will always happen under some mild assumptions, let us first look at a numerical example. Oscillators A and B have initial phases 0.4 and 0.6, respectively. Consider the pacemaker trajectory with $\gamma = 2$, thus the trigger threshold is $(1 - \exp(-\gamma))^2 = 0.7476$. The coupling strength is $\epsilon = 0.01$. The two oscillators start out on different trajectories, and, as they oscillate, whenever one "fires", the other's trajectory is "kicked" upwards by ϵ. As shown in Figure 7.8, the time of synchronization, $t = 737$ (eight firings), is the first instance when both oscillators achieve threshold in the same time interval.

Now we give our reasoning through the general case. Suppose A just fired, and B has phase ϕ. (This is actually an abuse of notation, as we now use ϕ as a specific phase rather than a generic notation of the phase axis.) What will be the phase of B after the *next* firing of A? We hope A and B will be closer to being

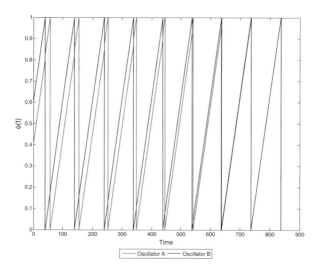

Figure 7.8 An illustration of the synchronization of two weakly coupled oscillators. The phases get closer and closer over time and become synchronized at $t = 737$ after eight firings.

synchronized. Call the answer to this question the image of a **return function** $R(\phi)$.

First, it is clear that after A's firing, B is leading A now. After a phase shift of $1 - \phi$, B fires, when A is at phase $1 - \phi$, and with state $x_A = f(1 - \phi)$. Right after B fires, A's state jumps to

$$x_A = f(1 - \phi) + \epsilon, \qquad (7.4)$$

if it is less than 1. Of course, if it is not less than 1, it must be 1, thus achieving synchronization between A and B's firing. So let us say it is less than 1. Obviously, for that to happen, ϵ needs to be less than 1 (thus the assumption of *weak coupling* here), and ϕ needs to be sufficiently far away from 0. Since $f(1-\phi)+\epsilon < 1$, we know "sufficiently away from 0" means that ϕ must be bigger than a threshold ϕ_{min}:

$$\phi > \phi_{min} = 1 - f^{-1}(1 - \epsilon). \qquad (7.5)$$

Now, going back to (7.4), the corresponding phase is

$$\phi_A = f^{-1}(f(1 - \phi) + \epsilon) = h(\phi), \qquad (7.6)$$

where h is a shorthand notation, standing for "half" of a complete round.

In summary, after one firing, i.e., half of a return round, we go on the state–phase plane from

$$(\phi_A, \phi_B) = (0, \phi)$$

to

$$(\phi_A, \phi_B) = (h(\phi), 0).$$

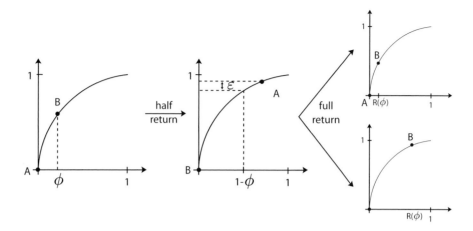

Figure 7.9 A pictorial illustration of the proof of synchronization. First is one-half of a return (from A firing to B firing). Then is a full return (from A firing to A firing again). We want to show that if ϕ, the phase of B at A's first firing, is smaller than a threshold, B will be closer to phase 0 after a full return. And if ϕ is larger than that threshold, B will be closer to phase 1 after a full return. Either way, A and B are driven closer to synchronization after each full return.

Applying h one more time, we will get to

$$(\phi_A, \phi_B) = (0, h(h(\phi))).$$

So the return function is simply

$$R(\phi) = h(h(\phi)),$$

where h is defined as in (7.6). This is illustrated in Figure 7.9.

Now we want to show that, as each of A and B gets to fire, they will be closer on the (ϕ, x) plane, either towards $\phi = 0$ or towards $\phi = 1$, so eventually they must meet. This is a **contraction-mapping** argument.

So, we want to show that there is a unique ϕ^*, such that $R(\phi) > \phi$ if $\phi > \phi^*$ and $R(\phi) < \phi$ if $\phi < \phi^*$. This will prove that synchronization must eventually happen.

So, we want to first show that there is a unique fixed point ϕ^* of R. If h has a unique fixed point, obviously R has too. Does h have a unique fixed point? Or, equivalently, is there one, and only one point ϕ^*, such that

$$F(\phi^*) = 0,$$

where $F(\phi) = \phi - h(\phi)$? Well, we can easily check, by (7.5) and (7.6), that $\phi_{min} < h(\phi_{min})$, and consequently, flipping "the arrow of time," $h^{-1}(\phi_{min}) > h(h^{-1}(\phi_{min})) = \phi_{min}$. That means

$$F(\phi_{min}) < 0 \text{ and } F(h^{-1}(\phi_{min})) > 0.$$

There must be some $\phi^* \in [\phi_{min}, h^{-1}(\phi_{min})]$ such that $F(\phi^*) = 0$, i.e., a fixed point. We have thus proved the existence of fixed points.

But there might be *many* fixed points in the above interval. Let us check F's first derivative $1 - h'(\phi)$. Taking the derivative by following the chain rule, we have

$$h'(\phi) = -\frac{f^{-1'}(f(1-\phi) + \epsilon)}{f^{-1'}(f(1-\phi))}.$$

Since f is increasing and concave, we know $f^{-1'}$ is increasing and convex, so the above ratio must be less than -1. That means F' must be greater than 2, thus F is monotonically increasing. An increasing function can only cross zero once. This proves that not only are there fixed points, but also there is a *unique* fixed point of h.

Therefore, R also has a unique fixed point ϕ^*. By the chain rule, we know $h' < -1$ implies $R' > 1$. So, if $\phi > \phi^*$, after two rounds of firing, one by A and another by B, the system is driven towards $\phi = 1$. If $\phi < \phi^*$, it is driven towards $\phi = 0$. Either way, it is driven closer to synchronization. More rounds will eventually lead to synchronization.

Over the past two decades, there have also been other generalizations: to different types and representations of coupling, to network topology-dependent coupling, and to synchronization even with delay in coupling.

The above derivation does not use a differential-equation approach as we will see next in the diffusion model. That would have been a non-scalable proof technique as we generalize from two oscillators to N of them. Instead, this *fixed-point* approach readily generalizes itself to all N, and for all increasing and concave trajectories too, without having to write down the differential equation. It goes straight to the root cause that leads to synchronization.

7.4.2   Diffusion

The tipping model is based on population dynamics. It can be further elaborated in various ways. Consider a fixed market of M people that the seller of a new product (or a song, or an app) wants it to diffuse into. Each user chooses between adopting or not, again a binary state of mind, depending on two factors.

- Something intrinsic about the product itself, which is independent of how many other people have adopted it. Some call this the innovation or external component. We will refer to it as the **intrinsic factor**.
- The network effect: either because more adopters change the value of the product itself, or because of the information-cascade effect. Some call this the internal component. We will refer to this as the **imitation factor**.

At time t, the number of adopters is $n(t)$, and the cumulative number of adopters is $N(t)$. The following equation is the starting point of the **Bass model**, which was developed in 1969:

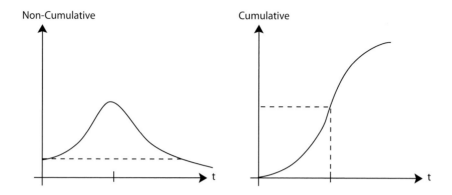

Figure 7.10 Two typical graphs illustrating the Bass diffusion model. The (non-cumulative) adoption rate is shown on the left, rising initially and then falling as the market saturates. The cumulative adoption population is shown on the right as a sigmoidal curve.

$$n(t) = \frac{dN(t)}{dt} = \left[p + q\frac{N(t)}{M}\right](M - N(t)),\tag{7.7}$$

where p is the intrinsic-factor coefficient, and q the imitation-factor coefficient that multiplies the fraction of adopters $N(t)/M$ in the market. The sum of the two effects is the growth factor, which, when multiplied by the number of non-adopters left in the population $M - N(t)$, gives the number of new adopters at timeslot t. Let us assume that $N(0) = 0$: the product starts from zero market share.

This population-based model, captured through a simple differential equation, has been extensively studied and generalized in the marketing literature. The basic version above has a closed-form solution. Solving the differential equation (7.7) with zero initial condition, we see that the number of new adopters is

$$n(t) = M\frac{p(p + q)^2 \exp(-(p + q)t)}{(p + q\exp(-(p + q)t))^2}\tag{7.8}$$

at each time t. Further integrating the resulting $n(t)$ over time, the number of total adopters at that point is

$$N(t) = M\frac{1 - \exp(-(p + q)t)}{1 + \frac{q}{p}\exp(-(p + q)t)}.\tag{7.9}$$

Figure 7.10 shows the typical evolution of diffusion over time: the cumulative adoption $N(t)$ climbs up to M through a sigmoidal function: convex first and then concave past the inflection point. The reason is clear: the adoption-rate curve $n(t)$ rises initially because there are still many non-adopters, but eventually the market saturates and there are not enough non-adopters left.

What is interesting is to see the breakdown of $N(t)$ into the number of adoptions due to the intrinsic factor, which monotonically decreases over time because

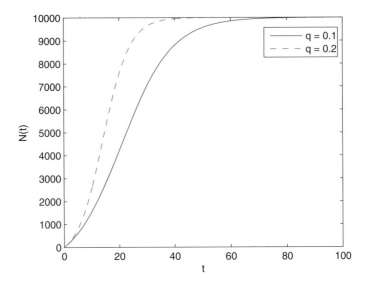

Figure 7.11 A numerical example of the trajectory of the Bass model for diffusion. The cumulative adoption is shown over time, for two different values of the imitation-factor coefficient q. Both curves are sigmoidal: convex first and then concave. A larger q leads to a faster adoption curve.

of market saturation, and the number due to the imitation factor, which rises for a while, as the imitation effect overcomes the market-saturation effect. The peak of this rise, and of the rise of $n(t)$, happens at time

$$T = \frac{1}{p+q} \log \left(\frac{q}{p} \right).$$

This makes sense since it is the ratio of the two coefficients that determines the relative weight of the two curves. The larger the overall adoption coefficient, the earlier the market reaches its peak rate of adoption.

As a numerical example, we take $p = 0.01$ and $q = 0.1$ or 0.2. The diffusion behavior is shown in Figures 7.11 and 7.12. From both graphs, we see that when q is larger, the adoption happens faster.

But how do we know p and q values? That is often done through a parameter-training phase, like the baseline predictor in the Netflix Prize in Chapter 4, using known marketing data or a new product trial. Once p and q have been estimated, we can run the model for prediction and planning purposes.

There have been many extensions in the past four decades of the above basic model.

- Incorporate individual decision-making processes.
- Divide the population into groups with different behaviors.
- Allow the market size M to vary with the effectiveness of adoption rather than being fixed a priori.

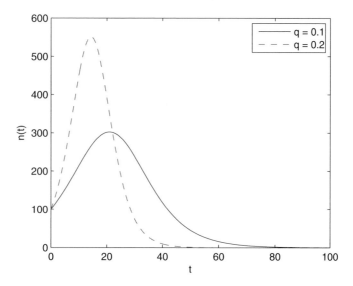

Figure 7.12 A numerical example of the trajectory of the Bass model for diffusion. The rate of adoption is shown over time, for two different values of the imitation-factor coefficient q. Both curves rise up to a peak value and then decay to zero after everyone in a market with fixed size has already adopted.

- Incorporate factors like time-varying nature of innovation, limitation of supply, and non-binary choices by consumers.

There have also been many refinements of the coefficient models, especially the imitation-factor coefficient q. We highlight a particularly interesting one that has the element of time explicitly modeled: the **flexible logistic growth** model. We have assumed q as a constant. Why not allow it to vary over time? For example,

$$q(t) = q(1 + kt)^{\frac{\mu - k}{k}},$$

where k and μ are two parameters modeling the increasing, decreasing, or constant power of imitation over time. Their values, when properly chosen, can allow the model to put the inflection point of the sigmoidal curve anywhere between 0 and 100% market diffusion.

The resulting cumulative adoption population has a somewhat complicated form. In the simpler case when $p = 0$, it is

$$N(t) = \frac{M}{1 + \exp(-(c + qt(\mu, k)))},$$

where c is some constant, and the function $t(\mu, k)$, when $\mu \neq 0$ and $k \neq 0$, is

$$\frac{(1 + kt)^{\mu/k} - 1}{\mu},$$

and, when $\mu = 0$ and $k \neq 0$, is

$$\frac{1}{k} \log(1 + kt).$$

One can also think of a model where the market size is essentially infinite, rather than at a fixed, finite size. So there is no market-saturation effect. For example, the change in $N(t)$ is the product of the imitation effect $qN(t)$ and a penalty term $D(t)$ that decreases the attractiveness of the product as time goes on. This then leads to the following differential equation:

$$n(t) = \frac{dN(t)}{dt} = qN(t)D(t) - N(t).$$

So, when time t is such that $D(t) < 1/q$, the market adoption rate will decrease even without assuming a fixed market size.

In summary, no matter which model we use for diffusion, there are two essential ingredients: an imitation effect q and a shrinkage effect (either because of time passing $D(t)$, or because of a fixed market size M), as well as an optional ingredient of an intrinsic effect p.

Summary

Box 7 Influence models in a crowd

Dependence of one's own action on peers' actions can lead to information cascades, where individuals in a crowd ignores their own private signals and follows the fallacy of crowds. Different shapes of the influence function can lead to different tipping behaviors. Oscillators' coupling can lead to synchronization as each cycle of interactions brings the two oscillators' phases closer.

Further Reading

The material in this chapter comes from a diverse set of disciplines: political economy, marketing, physics, and sociology.

1. On the information-cascade model, we follow the classic work below:
S. Bikhchandani, D. Hirshleifer, and I. Welch, "A theory of fads, fashion, custom, and cultural change as information cascades," *Journal of Political Economy*, vol. 100, no. 5, pp. 992–1026, October 1992.

2. On the synchronization model, we follow this seminal paper using fixed-point techniques to analyze synchronization:

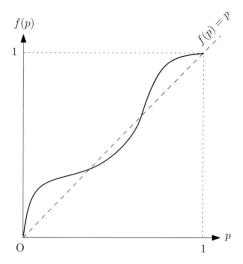

$f(p)$

$f(p) = p$

1

O 1 p

Figure 7.13 An original curve to model flipping.

R. E. Mirollo and S. H. Strogatz, "Synchronization of pulse-coupled biological oscillators," *SIAM Journal of Applied Mathematics*, vol. 50, no. 6, pp. 1645–1662, December 1990.

3. On the diffusion model, an extensive survey of the research area since 1968 can be found in

V. Mahajan, E. Muller, and F. M. Bass, "New product diffusion models in marketing: A review and directions for research," *Journal of Marketing*, vol. 54, no. 1, pp. 1–26, January 1990.

4. The following is a classic work on threshold models in sociology:

M. Granovetter, "Threshold models of collective behavior," *American Journal of Sociology*, vol. 83, no. 6, pp. 1420–1443, May 1978.

5. The music market's trends under social influence were studied in the following experiment, which we summarized in this chapter:

M. J. Salganik, P. S. Dodds, and D. J. Watts, "Experimental study of inequality and unpredictability in an artificial cultural market," *Science*, vol. 311, pp. 854–856, February 2006.

Problems

7.1   *Perturbing flipping behaviors* ★★

(a) Consider the influence function $f(p)$ in Figure 7.13, which has four equilibria $p^* = 0$, $\frac{1}{3}$, $\frac{2}{3}$ and 1. Suppose we start with $p(0) = 0.01$ (or some number slightly greater than 0). Find the equilibrium fraction, denoted as p_∞.

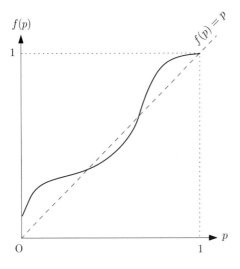

Figure 7.14 The curve is modified. It does not cross the origin now.

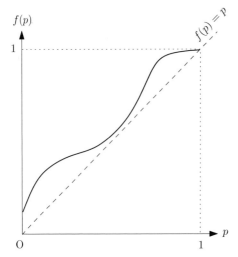

Figure 7.15 The curve is modified again. It does not touch the 45-degree line now.

(b) Suppose $f(p)$ is slightly modified as in Figure 7.14, such that the point $p = 0$ is no longer an equilibrium. Again use a graphical argument to find $p(\infty)$, starting at $p(0) = 0$.

(c) Suppose $f(p)$ is further slightly modified as in Figure 7.15, such that $f(p) > p$ for $0 \le p < 1$. Find $p(\infty)$ starting at $p(0) = 0$.

7.2   *Flocking birds* ★ ★ ★

In our study of the dynamics of collective behavior, we are often intrigued by the process that turns individual actions and local interactions into large-scale, global patterns. This happens in many ways in human crowds, bird flocks, fish schools, bacteria swarms, etc. A simple and powerful illustration is Conway's game of life.

Here is a very simple model for bird flocks that assumes away many features but suffices to illustrate the point for a homework problem. Suppose there is a two-dimensional plane with N points moving in it. Each point has neighbors, which are the points within a circle of radius r meters. All the points move with the same constant speed, say, 1 unit, but along different directions. At each timeslot, each point's direction is updated to be the average of its current direction and all the directions of its neighbors. We can think of a graph in which each node is a point, and each link is a neighbor relationship. But this graph evolves over time as the points' positions change.

Randomly place 100 points in a 10×10 units square, and initialize their directions randomly. You should try different values of r. Simulate the above model over time, and describe what happens to the directions of the points.

(For more detail of this model, see T. Vicsek, A. Czirok, E. Ben Jacob, I. Cohen, and O. Schochet, "Novel type of phase transitions in a system of self-driven particles," *Physics Review Letters*, vol. 75, pp. 1226-1229, 1995. A comprehensive survey of animal behavior can be found in I. D. Couzin and J. Krause, "Self-organization and collective behavior in vertebrates," *Advances in the Study of Behavior*, vol. 32, pp. 1–75, 2003.)

7.3   *A citation network and matrix multiplication* ★★

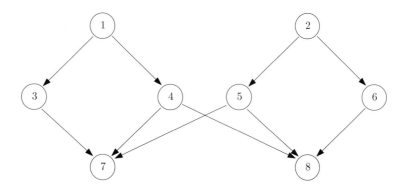

Figure 7.16 A directed graph to represent a citation network's topology.

Consider a set of eight papers with their citation relationships represented by the graph in Figure 7.16. Each paper is a node, and a directed edge from node i to node j means paper i cites paper j.

(a) Write down the adjacency matrix $\mathbf{A}$ (which we will talk much more about in the next chapter), where the (i, j) entry is 1 if node i points to node j, and 0 otherwise.

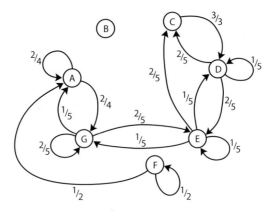

Figure 7.17 A network showing the transition diagram corresponding to the chorus of *Oh Susanna*. Each node is a diatonic pitch. Each link is a possible transition, with the link weight representing the probability of transition.

(b) Compute the matrix $\mathbf{C}$ defined as

$$\mathbf{C} = \mathbf{A}^T \mathbf{A},$$

and compare the values C_{78} and C_{75}. In general, what is the physical interpretation of the entries C_{ij}?

(c) Now compute

$$\mathbf{A}^2 = \mathbf{A}\mathbf{A},$$
$$\mathbf{A}^3 = \mathbf{A}^2\mathbf{A}.$$

Is there anything special about $\mathbf{A}^3$? In general, what do the entries in $\mathbf{A}^m$ (where $m = 1, 2, \ldots$) represent?

7.4 *Music in a graph* ★★

Melodies have been studied as mathematical objects. One way to depict a melody is through a directed graph that visualizes the conditional probabilities of transitions. Following the example in the book *Musimathics* by G. Loy, we draw the directed graph corresponding to the *Oh Susanna* chorus, where each node is a diatonic pitch, and each weighted link shows the probability of transition.

Generate a few examples of chorus synthesized by following the transitions shown in the graph in Figure 7.17. See if you can generate something very similar to the original chorus.

You may start to remember random walk on a graph or Markov chain mentioned during our discussion of PageRank in Chapter 3. We can go to a higher order Markov chain by examining the probability of transition into the next note conditioned on the previous and the current notes. It turns out such a second-order Markov chain can quite readily generate a chorus similar to the original one.

7.5   *Networked sampling* ★★★

In sociology, estimating the percentage of a hidden population, e.g., the AIDS-infected population, is difficult. One approach is to start with a few sampled "seeds," and then ask current sample members to recruit future sample members. The question is how to produce unbiased estimates.

Respondent-driven sampling is a method to address this question, and it is used by many institutions including the US Center for Disease Control and UNAIDS. The basic methodology is random walk on graphs, similar to what we saw in this chapter and in Chapter 2.

(a) First, let us characterize the sampling distribution at equilibrium. Let π_i be the stationary distribution of reaching person i.

Let K_{ij} be the probability of person i referring to person j. If each bidirectional link (i, j) has a weight $w_{(i,j)}$, and the probability of that person i recruiting person j is directly proportional to $w_{(i,j)}$, we have a recruiting mechanism similar to PageRank's spread of importance scores:

$$P[i \rightarrow j] = \frac{w_{(i,j)}}{\sum_k w_{(i,k)}}.$$

At equilibrium, what is the probability π_i that person i has been sampled?

(b) We follow a trajectory of sampling, starting with, say, one person, and running through n people sampled. If a person i on the trail of sampling has AIDS, we increment the counter of the infected population by 1. If we simply add these up and divide by n, it gives a biased estimate. The importance-sampling method weights each counter by $1/(N\pi_i)$. But we often do not know the value of N. So, in respondent-driven sampling, the estimate becomes

$$\text{(Harmonic mean of } \pi_i) \sum_{\text{infected } i} \frac{1}{\pi_i}.$$

Suppose we have two social groups of equal size, A and B, forming a network, and that the infection rates are p_A and p_B, respectively. Between groups, links have weights c, where $c \in (0, 0.5)$. Within each group, links have weights $1 - c$.

If we follow respondent-driven sampling, what do you think will happen to the sampling result? Confirm this with a simulation.

(For more details, see S. Goel and M. Salganik, "Respondent-driven sampling as Markov chain Monte Carlo," *Statistics in Medicine*, vol. 28, pp. 2202–2229, 2009.)

8   How do I influence people on Facebook and Twitter?

To study a network, we have to study both its *topology* (the graph) and its *functionalities* (tasks carried out on top of the graph). This chapter on topology-dependent influence models does indeed pursue both, as do the next two chapters.

8.1   A Short Answer

Started in October 2003 and formally founded in February 2004, Facebook has become the largest social network website, with 900 million users worldwide as of spring 2012 at the time of its IPO. Many links have been formed among these nodes, although it is not straightforward to define how many mutual activities on each other's wall constitute a "link."

Founded in July 2006, Twitter attracted more than 500 million users in six years. At the end of 2011, over 250 million tweets were handled by Twitter each day. Twitter combines several functionalities into one platform: microblogging (with no more than 140 characters), group texting, and social networking (with one-way following relationships, i.e., directional links).

Facebook and Twitter are two of the most influential communication modes, especially among young people. For example, in summer 2011's east-coast earthquake in the USA, tweets traveled faster than the earthquake itself from Virginia to New York. They have also become a major mechanism in social organization. In summer 2009, Twitter was a significant force in how the Iranians organized themselves against the totalitarian regime.

There have been all kinds of attempts at figuring out

- (1) how to quantify the statistical properties of opinions on Facebook or Twitter;
- (2) how to measure the influential power of individuals on Facebook or Twitter; and
- (3) how to leverage the knowledge of influential power's distribution to actually influence people online.

For example, on question (1), our recent study of all the tweets about Oscar-nominated movies during the month of February 2012 shows a substantial skew towards positive opinion relative to other online forums, and the need to couple tweet analysis with data from other venues like the IMDb or Rotten Tomato to get a more accurate reading of viewer reception and prediction of box office success.

Question (2) is an analysis problem and question (3) a synthesis problem. Neither is easy to answer; and there is a significant gap between theory and practice, perhaps the biggest such gap you can find in this book. Later in this chapter, we will visit some of the fundamental models that have yet to make a significant impact on characterizing and optimizing influence over these networks.

But the difficulty did not prevent people from trying out heuristics. Regarding (2), for example, there are many companies charting the influential power of individuals on Twitter, and there are several ways to approximate that influential power: by the number of followers, by the number of retweets (with "RT" or "via" in the tweet), or by the number of repostings of URLs. There are also many companies data-mining the friendship network topology of Facebook.

As to (3), Facebook uses simple methods to recommend friends, which are often based on email contact lists or common backgrounds. Marketing firms also use Facebook and Twitter to stage marketing campaigns. Some "buy off" a few influential individuals on these networks, while others buy off a large number of randomly chosen, reasonably influential individuals.

It is important to figure out who the influential people are. An often-quoted historical anecdote concerns the night rides by Paul Revere and by William Dawes on 18-19 April in 1775. Dawes left Boston earlier in the evening than did Revere. They took different paths towards Lexington, before riding together from Lexington to Concord. Revere alerted influential militia leaders along his route to Lexington, and was therefore much more effective in spreading the word of the imminent British military action. This in turn lead to the American forces winning on the next day the first battle that started the American Revolutionary War.

How do we quantify which nodes are more important? The question dates back thousands of years, and one particularly interesting example occurred during the Renaissance in Italy. The Medici family was often viewed as the most influential among the fifteen prominent families in Florence during the fifteenth and sixteenth centuries. As shown in Figure 8.1, it sat in the "center" of the family social network through strategic marriages. We will see several ideas quantifying the notion of centrality.

How do we quantify which links (and paths) are more important? We will later define strong vs. weak ties. Their effects can be somewhat unexpected. For example, Granovetter's 1973 survey in Amherst, Massachusetts showed the strength of weak ties in spreading information. We will see another surprise of weak ties' roles in social networks, on six-degree separation in Chapter 9.

Furthermore, how do we quantify which subset of nodes (and the associated links) are connected enough among themselves, and yet disconnected enough from the rest of the network, that we can call them a "group"? We save this question for the Advanced Material.

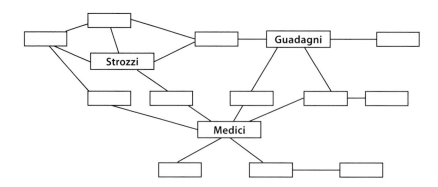

Figure 8.1 Padgett's Florentine-family graph shows the central position of the Medici family in Renaissance Florence. Each node is a family, three of them shown with their names. Each link is a marriage or kinship relationship. The Medici family clearly had the largest degree, but its influential power relative to the other families was much more than the degree distribution would indicate. Other measures of centrality, especially betweenness centrality, reveal just how influential the Medici family was.

8.1.1   Graphs and matrices

Before proceeding further, we first formally introduce two commonly used matrices to describe graphs. We have seen a few different types of graphs in the previous chapters. In general, a graph G is a collection of two sets: V is the set of vertices (nodes) and E is the set of edges (links). Each link is in turn a directed two-tuple: the starting and the ending nodes of that link.

We will construct a few other matrices later as concise and useful representations of graphs. We will see that properties of a graph can often be summarized by linear-algebraic quantities about the corresponding matrices.

The first is the **adjacency matrix A**, of dimension $N \times N$, of a given graph $G = (V, E)$ with N nodes connected through links. For the graphs we deal with in this book, A_{ij} is 1 if there is a link from node i to j, and 0 otherwise. We mostly focus on **undirected graphs** in this chapter, where each link is **bidirectional**. Given an undirected graph, **A** is symmetric: $A_{ij} = A_{ji}$, $\forall i, j$. If a link can be **unidirectional**, we have a **directed graph**, like the Twitter following relationship graph.

The second, which is less used in this book, is the **incidence matrix Â**, of dimension $N \times L$, where N is again the number of nodes and L the number of links. For an undirected graph, $\hat{A}_{ij} = 1$ if node i is on link j, and 0 otherwise. For a directed graph, $\hat{A}_{ij} = 1$ if node i starts link j, $\hat{A}_{ij} = -1$ if node i ends link j, and $\hat{A}_{ij} = 0$ otherwise.

Straightforward as the above definitions may be, it is often tricky to define what exactly constitutes a link between two persons: being known to each other by first-name as in Milgram's small-world experiment? Or "friends" on Facebook who have never met or communicated directly? Or only those to whom you text at least one message a day? Some links are also directional: I may have commented on your wall postings on Facebook but you never bothered reading my wall at all. Or I may be following your tweets, but you do not follow mine.

Even more tricky is to go beyond the simple static graph metrics and into the functionalities and dynamics on a graph. That is a much tougher subject. So we start with some simple static graph metrics first.

8.2   A Long Answer

8.2.1   Measuring node importance

You may be in many social networks, online as well as offline. How important are you in each of those networks? Well, that depends on how you define the "importance" of a node. It depends on the specific functionalities we are looking at, and it evolves over time. But we shall restrict ourselves to just static graph metrics for now. Neither is it easy to discover the actual topology of the network. But let us say for now that we are given a network of nodes and links.

There are at least four different approaches to measuring the importance, or **centrality**, of a node, say node 1.

The first obvious choice is **degree**: the number of nodes connected to node 1. If it is a directed graph, we can count two degrees: the in-degree: the number of nodes pointing towards node 1, and the out-degree: the number of nodes that node 1 points to. **Dunbar's number**, usually around 150, is often viewed as the number of friends a typical person may have, but the exact number of course depends on the definition of "friends." The communication modes of texting, tweeting, and blogging may have created new shades of definition of "friends." In Google+, you can also create your own customized notions of friends by creating new circles.

We will see there are many issues with using the degree of a node as its centrality measure. One issue is that if you are connected to more-important nodes, you will be more important than you would be if you were connected to less-important nodes. This may remind you of PageRank in Chapter 3. Indeed, we can take PageRank's importance scores as a centrality measure.

A slightly simpler but still useful view of centrality is to just look at the successive multiplication of the centrality vector $\mathbf{x}$ by the adjacency matrix $\mathbf{A}$ that describes the network topology, starting with an initialization vector $\mathbf{x}[0]$:

$$\mathbf{x}[t] = \mathbf{A}^t \mathbf{x}[0].$$

In a homework problem, you will discover a motivation for this successive multiplication.

We can always write a vector as a linear combination of the eigenvectors $\{\mathbf{v}_i\}$ of $\mathbf{A}$, arranged in descending order of the corresponding eigenvalues and indexed by i, for some weight constants $\{c_i\}$. For example, we can write the vector $\mathbf{x}(0)$ as follows:

$$\mathbf{x}[0] = \sum_i c_i \mathbf{v}_i.$$

Now we can write $\mathbf{x}[t]$ at any iteration t as a weighted sum of $\{\mathbf{v}_i\}$:

$$\mathbf{x}[t] = \mathbf{A}^t \sum_i c_i \mathbf{v}_i = \sum_i c_i \mathbf{A}^t \mathbf{v}_i = \sum_i c_i \lambda_i^t \mathbf{v}_i, \qquad (8.1)$$

where $\{\lambda_i\}$ are the eigenvalues of $\mathbf{A}$.

As $t \to \infty$, the effect of the largest eigenvalue λ_1 will dominate, so we approximate by looking only at the effect of λ_1. Now the **eigenvector centrality** measures $\{x_i\}$ constitute a vector that solves

$$\mathbf{A}\mathbf{x} = \lambda_1 \mathbf{x},$$

which means

$$x_i = \frac{1}{\lambda_1} \sum_j A_{ij} x_j, \quad \forall i. \qquad (8.2)$$

We can also normalize the eigenvector centrality $\mathbf{x}$.

The third notion, **closeness centrality**, takes a "distance" point of view. Take a pair of nodes (i, j) and find the shortest path between them. It is not always easy to compute the shortest path, as we will see in Chapters 9 and 13. But, for now, say we have found the shortest paths between each pair of nodes, and denote their lengths as $\{d_{ij}\}$. The largest d_{ij} across all (i, j) pairs is called the **diameter** of the network. The average of d_{ij} for a given node i across all other $n - 1$ nodes is an average distance $\sum_j d_{ij}/(n - 1)$. The closeness centrality is the reciprocal of this average:

$$C_i = \frac{n - 1}{\sum_j d_{ij}}. \qquad (8.3)$$

We have used the arithmetic mean of $\{d_{ij}\}$, but could also have used other "averages," such as the harmonic mean.

Closeness centrality is quite intuitive: the more nodes you know or the closer you are to other nodes, the more central you are in the network. But there is another notion that is just as useful, especially when modeling influence and information exchange: **betweenness centrality**. If you are on the (shortest) paths of many *other* pairs of nodes, then you are important. (We can also extend this definition to incorporate more than just the shortest paths, as in the context of Internet routing in Chapter 13.)

Let g_{st} be the total number of shortest paths between two diffferent nodes, source s and destination t (neither of which is node i itself), and let n_{st}^i be the number of such paths that node i sits on. Then the betweenness centrality of node i is defined as

$$B_i = \sum_s \sum_{t < s} \frac{n_{st}^i}{g_{st}}, \qquad (8.4)$$

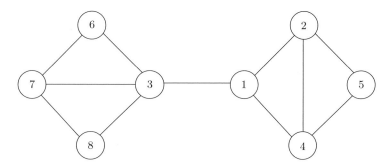

Figure 8.2 An example network to illustrate node-importance metrics, link importance metrics, and group connectedness metrics. For example, how much more important node 1 is than node 2 depends on which centrality metric we use.

where the double summation is indexed such that double counting is avoided.

A node with a large closeness centrality need not have a large betweenness centrality, and vice versa. In fact, many social networks exhibit the small-world phenomenon as explained in Chapter 9, and the closeness centrality values of different nodes tend to be close to each other. Betweenness centrality values tend to have a larger dynamic range.

We can also think of hybrid metrics. For example, first weight each node by eigenvector centrality, then weight each node pair by the product of their eigenvector centrality values, and then calculate the betweenness centrality by weighting each (st) term in (8.4) accordingly.

In the Renaissance Florence family-relationship graph, it is obvious that the Medici family has the largest degree, 6, but the Strozzi and Guadagni families' degrees are not too far behind: 4 for both. But if we look at the betweenness centrality values, Medici family has a value of 50.83, which is five times as central as Strozzi and twice as central as Guadagni, not just a mere factor of 1.5.

Now let us take a look at how different node-importance metrics turn out to be for two of the nodes, 1 and 2, in the small network in Figure 8.2.

For degree centrality, obviously $d_1 = d_2 = 3$. But nodes 1 and 2 cannot be the same in their importance according to most people's intuition that says nodes gluing the graph together are more important.

Let us compute the eigenvector centrality next. From the adjacency matrix

$$
\mathbf{A} =
\begin{bmatrix}
0 & 1 & 1 & 1 & 0 & 0 & 0 & 0 \\
1 & 0 & 0 & 1 & 1 & 0 & 0 & 0 \\
1 & 0 & 0 & 0 & 0 & 1 & 1 & 1 \\
1 & 1 & 0 & 0 & 1 & 0 & 0 & 0 \\
0 & 1 & 0 & 1 & 0 & 0 & 0 & 0 \\
0 & 0 & 1 & 0 & 0 & 0 & 1 & 0 \\
0 & 0 & 1 & 0 & 0 & 1 & 0 & 1 \\
0 & 0 & 1 & 0 & 0 & 0 & 1 & 0
\end{bmatrix},
$$

we can solve for

$$\mathbf{A}\mathbf{x} = \lambda_1 \mathbf{x}$$

to obtain

$$\lambda_1 = 2.8723,$$
$$\mathbf{x} = [0.4063 \ 0.3455 \ 0.4760 \ 0.3455 \ 0.2406 \ 0.2949 \ 0.3711 \ 0.2949]^T.$$

So node 1 is slightly more important than node 2: $0.4063 > 0.3415$. But is it really just *slightly* more important?

To compute the closeness centrality, we first write out the pairwise (shortest) distances from node 1 to the other nodes, and from node 2 to the other nodes:

$$d_{12} = 1, d_{13} = 1, d_{14} = 1, d_{15} = 2, d_{16} = 2, d_{17} = 2, d_{18} = 2,$$
$$d_{21} = 1, d_{23} = 2, d_{24} = 1, d_{25} = 1, d_{26} = 3, d_{27} = 3, d_{28} = 3.$$

Then, we can see that node 1 is again only slightly more important:

$$C_1 = \frac{7}{1+1+1+2+2+2+2} = 0.6364,$$
$$C_2 = \frac{7}{1+2+1+1+3+3+3} = 0.5.$$

Finally, to compute the betweenness centrality, it helps to first write out the quantities of interest. Let $\mathbf{G}$ be the matrix with $G_{ij} = g_{ij}$, $\mathbf{N}^1$ be the matrix with the (i,j) entry being n_{ij}^1 for node 1, and $\mathbf{N}^2$ be the matrix with the (i,j) entry being n_{ij}^2 for node 2. Also let X denote a matrix entry that is not involved in the calculations (a "don't care"). Then, we have

$$\mathbf{G} = \begin{bmatrix} X & 1 & 1 & 1 & 2 & 1 & 1 & 1 \\ 1 & X & 1 & 1 & 1 & 1 & 1 & 1 \\ 1 & 1 & X & 1 & 2 & 1 & 1 & 1 \\ 1 & 1 & 1 & X & 1 & 1 & 1 & 1 \\ 2 & 1 & 2 & 1 & X & 2 & 2 & 2 \\ 1 & 1 & 1 & 1 & 2 & X & 1 & 2 \\ 1 & 1 & 1 & 1 & 2 & 1 & X & 1 \\ 1 & 1 & 1 & 1 & 2 & 2 & 1 & X \end{bmatrix},$$

and the following two matrices for nodes 1 and 2, respectively:

$$
\mathbf{N}^1 =
\begin{bmatrix}
X & X & X & X & X & X & X & X \\
X & X & 1 & 0 & 0 & 1 & 1 & 1 \\
X & 1 & X & 1 & 2 & 0 & 0 & 0 \\
X & 0 & 1 & X & 0 & 1 & 1 & 1 \\
X & 0 & 2 & 0 & X & 2 & 2 & 2 \\
X & 1 & 0 & 1 & 2 & X & 0 & 0 \\
X & 1 & 0 & 1 & 2 & 0 & X & 0 \\
X & 1 & 0 & 1 & 2 & 0 & 0 & X
\end{bmatrix},
$$

$$
\mathbf{N}^2 =
\begin{bmatrix}
X & X & 0 & 0 & 1 & 0 & 0 & 0 \\
X & X & X & X & X & X & X & X \\
0 & X & X & 0 & 1 & 0 & 0 & 0 \\
0 & X & 0 & X & 0 & 0 & 0 & 0 \\
1 & X & 1 & 0 & X & 1 & 1 & 1 \\
0 & X & 0 & 0 & 1 & X & 0 & 0 \\
0 & X & 0 & 0 & 1 & 0 & X & 0 \\
0 & X & 0 & 0 & 1 & 0 & 0 & X
\end{bmatrix}.
$$

Applying (8.4), we clearly see an intuitive result this time: node 1 is much more important than node 2:

$$B_1 = 12,$$
$$B_2 = 2.5.$$

8.2.2   Measuring link importance

Not all links are equally important. Sometimes there is a natural and operationally meaningful way to assign weights, whether integers or real numbers, to links. For example, the frequency of Alice retweeting Bob's tweets, or of reposting Bob's URL tweets, can be the weight of the link from Bob to Alice. Sometimes there are several categories of link strength. For example, you may have 500 friends on Facebook, but those with whom you have had either one-way or mutual communication might be only 100, and those with whom you have had mutual communication might be only 20. The links to these different types of Facebook friends belong to different strength classes. Weak links might be weak in action, but strong in information exchange.

A link can also be important because it "glues" the network together. For example, if a link's two end points A and B have no common neighbors, or more generally, do not have any other (short) paths of connection, this link is *locally important* in connecting A and B. Links that are important in connecting many node pairs, especially when these nodes are important nodes, are *globally important* in the entire network. These links can be considered *weak*, since they connect nodes that otherwise have no, or very little, overlap. (The opposite is the "triad closure" which we will see in the next chapter.)

But these **weak links** are *strong* precisely for the reason that they open up communication channels across groups that normally do not communicate with each other, as seen in Granovetter's 1973 experiment. One way to quantify this notion is **link betweenness**: this is similar to the betweenness metric defined for nodes, but now we count how many shortest paths a *link* lies on.

Going back to Figure 8.2, we can compute $B_{(i,j)}$, the betweenness of link (i,j) by

$$B_{(i,j)} = \sum_{s}\sum_{t<s} \frac{n_{st}^{(i,j)}}{g_{st}}, \tag{8.5}$$

where $n_{st}^{(i,j)}$ is the number of shortest paths between two nodes s and t that traverse link (i,j).

Let $\mathbf{N}^{(i,j)}$ be a matrix with the (s,t) entry as $n_{st}^{(i,j)}$. We can compare, for example, the betweenness values of the links $(1,3)$ and $(1,2)$ in Figure 8.2:

$$\mathbf{N}^{(1,3)} = \begin{bmatrix} X & 0 & 1 & 0 & 0 & 1 & 1 & 1 \\ 0 & X & 1 & 0 & 0 & 1 & 1 & 1 \\ 1 & 1 & X & 1 & 2 & 0 & 0 & 0 \\ 0 & 0 & 1 & X & 0 & 1 & 1 & 1 \\ 0 & 0 & 2 & 0 & X & 2 & 2 & 2 \\ 1 & 1 & 0 & 1 & 2 & X & 0 & 0 \\ 1 & 1 & 0 & 1 & 2 & 0 & X & 0 \\ 1 & 1 & 0 & 1 & 2 & 0 & 0 & X \end{bmatrix},$$

$$\mathbf{N}^{(1,2)} = \begin{bmatrix} X & 1 & 0 & 0 & 1 & 0 & 0 & 0 \\ 1 & X & 1 & 0 & 0 & 1 & 1 & 1 \\ 0 & 1 & X & 0 & 1 & 0 & 0 & 0 \\ 0 & 0 & 0 & X & 0 & 0 & 0 & 0 \\ 1 & 0 & 1 & 0 & X & 1 & 1 & 1 \\ 0 & 1 & 0 & 0 & 1 & X & 0 & 0 \\ 0 & 1 & 0 & 0 & 1 & 0 & X & 0 \\ 0 & 1 & 0 & 0 & 1 & 0 & 0 & X \end{bmatrix}.$$

Now we can compute link importance by (8.5). We have the intuition quantified: link (1,3) is much more important than link (1,2) because it glues together the two parts of the graph:

$$B_{(1,3)} = 16,$$
$$B_{(1,2)} = 7.5.$$

8.2.3   Contagion

Now that we have discussed the basic (static) metrics of a graph, we continue with our discussion on influence models with the help of network topology.

Remember the last chapter's section on tipping behavior under the best response strategy? In this two-state model, the initialization has a subset of the nodes adopting one state, for example, the state of 1, while the rest of the nodes

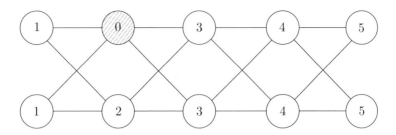

Figure 8.3 An example for the contagion model, with flipping threshold $p = 0.49$. Numbers in nodes indicate the times at which they change from state-0 to state-1, and the shaded node is the seed. At time 1 the leftmost two nodes flip because one of their two neighbors is in state-1, then this triggers a cascade of flips from left to right.

adopt the other state, the state of 0. We can consider the state-1 nodes as the early adopters of a new product, service, or trend, so it is likely they would be a small minority in the network and not necessarily aggregated together.

Now the question is when, if at all, will all the nodes flip to the new trend, i.e., flip from state-0 to state-1?

Unlike the diffusion model in the last chapter, now it is the *local* population, the set of neighbors of each node, rather than the global population, that matters. One possible local flipping rule is a memoryless, threshold model: if a fraction of p or more of your neighbors have flipped to state-1, you will flip too. For now, let us say all the nodes have the same flipping threshold: the same p for all nodes.

An example is shown in Figure 8.3 with a flipping threshold of $p = 0.49$, and the highlighted node being initialized at state-1. At time 1, the leftmost two nodes flip, and that triggers a cascade of flipping from left to right in the network.

In general, the first question we ask is will the entire network flip? It turns out there is a clear-cut answer: yes, if and only if there is no cluster of density $1 - p$ or higher, in the set of nodes with state-0 at initialization. As will be elaborated in the Advanced Material, a **cluster** of density p is a set of nodes such that each of these nodes has at least a fraction of p of its neighbors also in this set, as illustrated in Figure 8.4.

Without going through the proof of this answer, the "only if" direction of the statement is intuitively clear: a cluster of density $1 - p$ or higher, all with state-0, will never see any of its nodes flip since the inertia from within the cluster suffices to avoid flipping. *Homophily* creates a blocking cluster in the network.

The second question is, if the entire network eventually flips, how many iterations will that take? And the third question is, if only some part of the network flips in the end, how big a portion will flip (and where is it in the network)? These two questions are much harder to answer, and depend a lot on the exact network topology.

But perhaps the most useful question to answer for viral marketing is one of design: suppose each node in a given, known graph can be influenced at the initialization stage: if you pay node i $\$x_i$, it will flip. Presumably those nodes

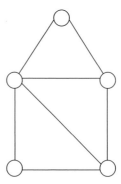

Figure 8.4 A small graph illustrating clusters with various densities. The set of nodes forming the lower left triangle is a cluster with density $1/2$, whereas the set of nodes forming the square is a cluster with density $2/3$.

that perceive themselves as more influential will charge a higher price. Under a total budget constraint, which nodes would you seed (e.g., pay to change their state and advertise that to neighbors) in order to maximize the extent of flipping at equilibrium, and furthermore, minimize the time it takes to reach that equilibrium?

While this question is hard to answer, some intuitions are clear. If you can seed just one node, it should be the most important one (by some centrality measure). But once you can seed two nodes, it is the *combined* influential power of the pair that matters. For example, you want the two nodes to be close enough to ensure some nodes will flip, but you also want them to be far apart enough from each other that more nodes can be covered. This tradeoff is further influenced by the heterogeneity of flipping thresholds: for the same cost, it is more effective to influence those easier to flip. Network topology is also important: you want to flip them in order to create a cascade so that some nodes can help flip others.

8.2.4   Infection: Population-based model

We have already seen five influence models between the last and this chapter. There is another model that is frequently used in modeling the spread of infectious disease. Unlike the other models, this one has a state transition, between two, three, or even more states that each node may find itself in. We will first describe the interaction using differential equations (over continuous time) and assuming that each node can interact with any other node (which could have been introduced in the last chapter since this model does not depend on the topology). We will then bring in network topology so that a node can directly interact only with its neighbors.

These variants of the infection model differ from each other in terms of the kind of state transitions that are allowed. We will cover only two-state and three-state models. As shown in Figure 8.5, S stands for susceptible, I stands for infected,

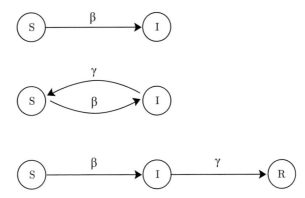

Figure 8.5 Three of the simplest state-transition models for infection: the SI model, SIS model, and SIR model. S stands for susceptible population, I for infected, and R for recovered. Each model can be mathematically described through a system of differential equations, some of which have analytic solutions while others need to be numerically solved. There is a caveat in this graphic representation of the differential equations: the β arrows indicate the transition rate that multiplies the product of the S and I populations, while the γ arrows indicate the transition rate that multiplies just the I population.

and R stands for recovered, and the symbols above the state-transition arrows represent the rates of those transitions: how many switch per unit of time. We will use $S(t), I(t)$, and $R(t)$ to represent the *proportions* of the population in that state at time t. The initial conditions at time 0 are denoted as $S(0), I(0)$, and $R(0)$. Since time is continuous here, we use round instead of square brackets around t.

The first model is called the **SI model**, which is very similar to the Bass model for diffusion in the last chapter. It is described by the following pair of differential equations, where the transition rates are proportional to the product $S(t)I(t)$ at each time t:

$$\frac{dS(t)}{dt} = -\beta S(t)I(t), \tag{8.6}$$

$$\frac{dI(t)}{dt} = \beta S(t)I(t). \tag{8.7}$$

We could have also used just one of the two equations above and the normalization equation: all population proportions need to add up to 1: $S(t) + I(t) = 1$ at all times t. Substituting $S(t) = 1 - I(t)$ into (8.6), we have

$$\frac{dI(t)}{dt} = \beta(1 - I(t))I(t) = \beta(I(t) - I^2(t)).$$

This is a simple second-order differential equation just like what we used for the Bass model in Chapter 7. The closed-form solution is indeed a special case

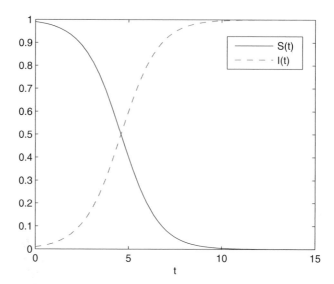

Figure 8.6 Population evolution in the SI model. Eventually everyone is infected.

of the Bass model's solution. It is a sigmoidal curve for the infected population over time, parameterized as a **logistic-growth** equation:

$$I(t) = \frac{I(0)e^{\beta t}}{S(0) + I(0)e^{\beta t}}. \tag{8.8}$$

And, of course, $S(t) = 1 - I(t)$.

We do not go into differential-equation solvers here, but it is easy to *verify*, through differentiation, that the above equation does indeed match the differential equations of the SI model.

When t is small, $I(t)$'s growth is similar to exponential growth. When t becomes large, the ratio in (8.8) approaches 1. An example of the whole curve is shown in Figure 8.6.

The SI model assumes that, once infected, a person stays infected forever. In some diseases, a person can become non-infected but still remain susceptible to further infections. As in Figure 8.5, this **SIS model** is described by the following equations:

$$\frac{dS(t)}{dt} = \gamma I(t) - \beta S(t)I(t),$$

$$\frac{dI(t)}{dt} = \beta S(t)I(t) - \gamma I(t).$$

Without even solving the equations, we can guess that, if $\beta < \gamma$, the infected proportion depletes exponentially. If $\beta > \gamma$, we will see a sigmoidal curve of $I(t)$ going up, but not to 100% since some of the infected will be going back to the susceptible state. The exact saturation percentage of $I(t)$ depends on β/γ.

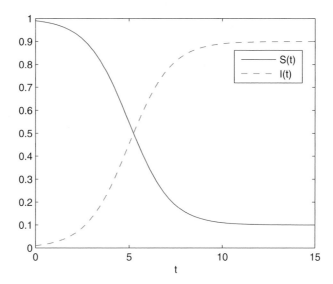

Figure 8.7 Population evolution in the SIS model. At equilibrium, some people are not infected.

These intuitions are indeed confirmed in the closed-form solution. Again using $S(t) = 1 - I(t)$ and solving the resulting differential equation in $I(t)$, we have

$$I(t) = (1 - \gamma/\beta)\frac{ce^{(\beta-\gamma)t}}{1 + ce^{(\beta-\gamma)t}}, \tag{8.9}$$

for some constant c that depends on the initial condition. A sample trajectory, where $\beta > \gamma$, is shown in Figure 8.7.

Indeed, the growth pattern depends on whether $\beta > \gamma$ or not, and the $I(t)$ saturation level (as $t \to \infty$) depends on β/γ too. This important constant,

$$\sigma = \beta/\gamma,$$

is called the **basic reproduction number**.

Both the SI and SIS models miss a common feature in many diseases: once infected and then recovered, a person becomes immunized. This is the R state. In the **SIR model** (not to be confused with the Signal-to-Interference Ratio in wireless networks in Chapter 1), one of the most commonly used, simple models for infection, the infected population eventually goes down to 0. As shown in Figure 8.5, the dynamics are described by the following equations:

$$\frac{dS(t)}{dt} = -\beta S(t)I(t),$$

$$\frac{dI(t)}{dt} = \beta S(t)I(t) - \gamma I(t),$$

$$\frac{dR(t)}{dt} = \gamma I(t).$$

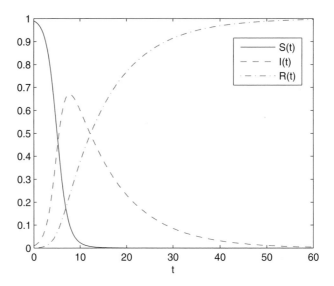

Figure 8.8 Population evolution in the SIR model, for $\sigma S(0) > 1$. Eventually everyone recovers.

Here, $\sigma = \beta/\gamma$ is the contact rate β (per unit time) times the average infection period $1/\gamma$. We can run substitution twice to eliminate two of the three equations above, but there is no closed-form solution to the resulting differential equation. Still, we can show that $\sigma S(0)$ plays the role of the threshold level that determines whether $I(t)$ will go up first before coming down. The trajectory of this SIR model has the following properties over a period of time $[0, T]$.

- If $\sigma S(0) \leq 1$, then $I(t)$ decreases to 0 as $t \to \infty$. The initial value $S(0)$ is not large enough to create an epidemic.
- If $\sigma S(0) > 1$, then $I(t)$ increases to a maximum of

$$I_{max} = I(0) + S(0) - 1/\sigma - \log(\sigma S(0))/\sigma,$$

  then decreases to 0 as $t \to \infty$. This is the typical curve of an **epidemic outbreak**.
- $S(t)$ is always a decreasing function. The limit $S(\infty)$ as $t \to \infty$ is the unique root in the range $(0, 1/\sigma)$ of the following equation:

$$I(0) + S(0) - S(\infty) + \frac{1}{\sigma} \log \left(\frac{S(\infty)}{S(0)} \right) = 0.$$

A typical picture of the evolution is shown in Figure 8.8. Eventually, everyone is recovered.

There are many other variants beyond the simplest three above, including the SIRS model where the recovered may become susceptible again, models where new states are introduced, and models where births and deaths (due to infection) are introduced.

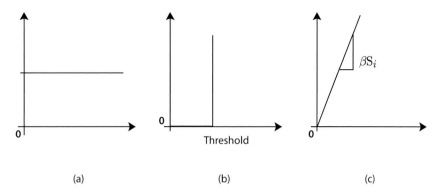

Figure 8.9 Each local node's processing models, the rate of change of node i's state (y-axis) vs. its neighbors' influence (x-axis), for (a) random walk, (b) contagion, and (c) infection. Contagion exhibits a flipping threshold, while infection has a linear dependence of the rate of change on the influence of a neighbor.

8.2.5   Infection: Topology-based model

Up to this point, we have assumed that only the global population matters: each node feels the averaged influence from the entire network. This is sometimes called the *mean-field approximation* approach to enable tractable mathematical analysis.

In reality, of course, infectious diseases spread only between two neighbors in a graph (however you may define the "link" concept for each disease). The difference between the contagion model and the infection model, as well as the random-walk model, now boils down to how we model local processing.

As illustrated in Figure 8.9, in contagion each node makes a deterministic decision to flip or not (depending on whether the local influence is strong enough), whereas in infection each node makes a probabilistic "decision", namely the likelihood of catching the disease, with the rate of change of that likelihood dependent on the amount of local influence. So the discrete state actually turns into a continuous state representing the probability of finding a node in that state.

Intuitively, if the topology is such that there is a bottleneck subset of nodes, it will be harder to spread the disease to the entire network. To model it more precisely, we need to include the adjacency matrix in the update equation.

For example, take the simplest case of the SI model. Now we have for each node i the following differential equations:

$$\frac{dS_i(t)}{dt} = -\beta \sum_j A_{ij}[S_i(t)I_j(t)], \tag{8.10}$$

$$\frac{dI_i(t)}{dt} = \beta \sum_j A_{ij}[S_i(t)I_j(t)]. \tag{8.11}$$

There are two tricky points in this seemingly simple translation from the original population-based model to the topology-based model. First, quantities S_i and I_i should be read as the *probabilities* of node i being in state S or state I, respectively. Second, it is tempting to pull S_i out of the summation over j since it does not depend on j, but actually we need to read $S_i I_j$ as one quantity: the *joint* probability that node i is in state S and its neighbor node j in state I. So the above notation is actually wrong. But, to estimate this joint probability, we need to know the probability that some neighbor of node j (other than node i) was in state I while node j itself was in state S, for that is the only way we can get to the current state of i in S and j in I. Following this line of reasoning, we have to enumerate all the possible paths of evolution of global states across the *whole network* over time. That is too much computation, and we have to stop at some level and approximate.

The first order of approximation is actually to break the joint probability exactly as in (8.10): the *joint* probability of node i being in state S and node j being in state I is approximated as the *product* of the individual probabilities of node i being in state S and node j being in state I.

For many other network computation problems, from decoding over wireless channels to identifying people by voice recognition, it is common to reduce the computational load by breaking down *global* interactions to *local* interactions. For certain topologies like trees, a low-order approximation can even be exact, or at least accurate enough for practical purposes.

With this first-order approximation, we have the following differential equation for the SI model (which can also be readily extended to the Bass model) with topology taken into account:

$$\frac{dI_i(t)}{dt} = \beta S_i(t) \sum_j A_{ij} I_j(t) \tag{8.12}$$

$$= \beta(1 - I_i(t)) \sum_j A_{ij} I_j(t). \tag{8.13}$$

The presence of the quadratic term and of the adjacency matrix makes it difficult to solve the above equation in closed form. But, during the early times of the infection, $I_i(t)$ is very small, and we can approximate the equation as a linear one by approximating $1 - I_i(t)$ with 1. In vector form, it becomes

$$\frac{d\mathbf{I}(t)}{dt} = \beta \mathbf{A} \mathbf{I}(t). \tag{8.14}$$

Here, $\mathbf{I}(t)$ is not an identity matrix, but a vector of the probabilities of the nodes being in state I at time t. We can, as in (8.1), decompose $\mathbf{I}$ as a weighted sum of eigenvectors $\{\mathbf{v}_k\}$ of $\mathbf{A}$:

$$\mathbf{I}(t) = \sum_k w_k(t) \mathbf{v}_k.$$

The eigenvectors $\{\mathbf{v}_k\}$ are determined by the topology of the graph. The weights $\{w_k(t)\}$ vary over time as the solution to the following linear, *scalar*, differential equation for each eigenvector k:

$$\frac{dw_k(t)}{dt} = \beta \lambda_k w_k(t),$$

giving rise to the solution

$$w_k(t) = w_k(0)e^{\beta \lambda_k t}.$$

Since $\mathbf{I}(t) = \sum_k w_k(t)\mathbf{v}_k$, the solution to (8.14) is

$$\mathbf{I}(t) = \sum_k w_k(0)e^{\beta \lambda_k t}\mathbf{v}_k.$$

For example, the first two terms of the above sum are

$$w_1(0)e^{\beta \lambda_1 t}\mathbf{v}_1 + w_2(0)e^{\beta \lambda_2 t}\mathbf{v}_2.$$

So the growth is still exponential at the beginning, but the growth exponent is weighted by the eigenvalues $\{\lambda_k\}$ of the adjacency matrix $\mathbf{A}$ now.

There are several other approaches to study infection with topological impact.

- A different approximation is to assume that all nodes of the same degree at the same time have the same S or I value. Like the order-based approximation above, it is clearly incorrect, but useful for generating another tractable way of solving the problem.
- So far we have assumed a detailed topology with an adjacency matrix given. An alternative is to take a generative model of topology that gives rise to features like small-world connectivity, and run infection models on those topologies. We will be studying generative models in the next two chapters.
- We can also randomly pick a link in the given topology to be in an "open" state, with probability p, that a disease will be transmitted from one node to another, and in a "closed" state with $1 - p$. Then, from any given initial condition, say, an infected node, there is a set of nodes connected to the original infected node through this maze of open links, and another set of nodes not reachable since they do not lie on the paths consisting of open links. This turns the infection model into the study of **percolation**.

8.3 Examples

8.3.1 Seeding a contagion

Contagion depends on the topology. The graph in Figure 8.10 is obtained by repositioning three links in the graph in Figure 8.3. Even with the same node is initialized as state-1, the number of eventual flips decreases sharply from ten to three. This shows how sensitive contagion outcome is with respect to network topology. We can also check that the density of the set of state-0 nodes is 2/3 after

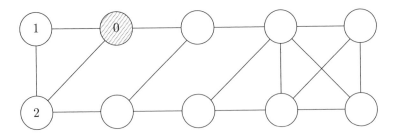

Figure 8.10 An example for contagion model with $p = 0.49$. The shaded node with 0 written in it represents the initial seed at iteration 0. Nodes with numbers written in them are flipped, and the numbers indicate the iteration at which each is flipped. One node is initialized to be at state-1 (flipped), and the final number of flipped nodes is three.

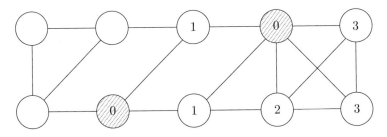

Figure 8.11 An example for contagion model with $p = 0.49$. Two nodes are initialized to be at state-1, and the final number of flipped nodes becomes seven.

the leftmost two nodes flip, which is higher than 1 minus the flipping threshold $p = 0.49$, thus preventing a complete flipping of the network.

Suppose now you want to stage a social media campaign by hiring Twitter, Facebook, and blog writers to spread their influence. Consider the problem of buying off, or seeding, nodes, assuming each node can be seeded at the same price. The aim is to maximize the number of eventual flips. If we can buy off only one node, choosing the node highlighted in Figure 8.10 is actually the optimal strategy. If we can seed two nodes, Figure 8.11 shows the optimal strategy and the final number of flips is seven, a significant improvement.

We also see that the nodes to be picked for seeding in the two-seed example do not include the node we picked to seed in the one-seed example. Optimal seeding strategies cannot be successively refined.

8.3.2   Infection: A case study

Just like the Netflix recommendation algorithm in Chapter 4 and the Bass diffusion model in Chapter 7, we actually do not know the model parameter values until we have some trial or historical data to train the model first.

In the SIR model, we can observe through historical data for each disease the initial population $S(0)$ and final population $S(\infty)$ of susceptible people. Let us assume the initial infected population $I(0)$ is negligible. This means we can approximate the key parameter σ as

$$\sigma = \frac{\log(S(0)/S(\infty))}{S(0) - S(\infty)}. \tag{8.15}$$

In making public-health decisions, a crucial one is the target vaccination rate for **herd immunity**: we want the immunized population to be large enough that infection does not go up at all, i.e., $S(0) < 1/\sigma$, or,

$$R(0) > 1 - 1/\sigma. \tag{8.16}$$

That means the fraction of the population with immunity, either through catching and recovering from the disease, or through vaccination (the more likely case), must be large enough. Using (8.15) to estimate σ, we can then estimate the vaccination rate $R(0)$ needed. These estimates are not exact, but at least they provide a sense of relative difficulty in controlling different infectious diseases.

Let us take a look at the case of measles. It causes about 1 million deaths worldwide each year, but in developed countries populations are sufficiently vaccinated that it affects very few. Its σ is quite large, having been estimated to be 16.67. By (8.16), this translates into a vaccination rate of 94% needed for herd immunity. But the vaccine's efficacy is not 100%, but more like 95% for measles. So the vaccination rate needs to be 99% to achieve herd immunity. This is a very high target number, and can be achieved only through a two-dose program, which is more expensive than the standard single-dose program.

When measles vaccination was first introduced in 1963 in the USA, the measles infection population dropped but did not disappear: it stayed at around 50000 a year. In 1978, the US government increased coverage of immunization in an attempt to eliminate measles, and the infection population further dropped to 5000, but was still not near 0. In fact the number went back up to above 15000 in 1989–1991. Just increasing the coverage of immunization did not make the immunization rate high enough. In 1989, the US government started using the two-dose program for measles: one vaccination at around 1 year old and another in around 5 years' time. This time the immunization rate went up past the herd-immunity threshold of 99% before children reach school age. Consequently, the number of reported cases of measles dropped to just 86 ten years later.

In a 2011 US movie "Contagion" (we use the term "infection" for spreading of disease), the interactions among three types of networks: social networks, information networks, and disease-spreading networks were depicted. Kate Winslet explained the basic reproduction number too (using the notation R_0, which is equivalent to σ for the cases we mentioned here). Some of the scenes in this drama actually occurred in real life during the last major epidemic, SARS in 2003, e.g., the Chinese government suppressing the news of the disease, some healthcare workers staying on their jobs despite there being a very high basic reproduction number and mortality rate, and the speed of information transmission exceeding that of disease transmission.

8.4   Advanced Material

8.4.1   Random walk

One influence model with network topology has already been introduced in Chapter 3: the PageRank algorithm. In that chapter, we wanted to see what set of numbers, one per node, is *self-consistent* on a directed graph: if each node spreads its number evenly across all its outgoing neighbors, will the resulting numbers be the same? It is a state of equilibrium in the influence model, where the influence is spread across the outgoing neighbors.

In sociology, the **DeGroot model** is similar, except that it starts with a *given* set of numbers $\mathbf{v}$, one per node (so the state of each node is a real number rather than a discrete one), and you want to determine what happens over time under the above influence mechanism.

The evolution of $\mathbf{x}[t]$, over discrete timeslots indexed by t, can be expressed as follows:

$$\mathbf{x}[t] = \mathbf{A}^t \mathbf{v}. \tag{8.17}$$

Here $\mathbf{A}$ is an influence-relationship adjacency matrix. If $A_{ii} = 1$ and $A_{ij} = 0$ for all j, that means node i is an "opinion seed" that is not influenced by any other nodes.

We have seen this linear equation many times by now: power control, PageRank, centrality measures. It is also called random walk on graphs. When does it converge? Will it converge only on a subset of nodes (and the associated links)?

Following standard results in linear algebra, we can show that, for any subset of nodes that is **closed** (there is no link pointing from a node in the subset to a node outside), the necessary and sufficient conditions on matrix $\mathbf{A}$ for convergence are that it is

- *irreducible*: an adjacency matrix being irreducible means that the corresponding graph is connected: there is a directed path from any node to any other node; and
- *aperiodic*: an adjacency matrix being aperiodic means that the lengths of all the cycles in the directed graph have the greatest common denominator of 1.

What about the *rate of convergence*? That is much harder to analytically characterize. But, to first order, an approximation is that the convergence rate is governed by the *ratio* of the second-largest eigenvalue λ_2 and the largest eigenvalue λ_1. An easy way to see this is to continue the development of eigenvector centrality (8.2). The solution to (8.17) can be expressed through the eigenvector decomposition $\mathbf{v} = \sum_i c_i \mathbf{v}_i$:

$$\mathbf{x}[t] = \mathbf{A}^t \sum_i c_i \mathbf{v}_i = \sum_i c_i \lambda_1^t \left(\frac{\lambda_i}{\lambda_1} \right)^t \mathbf{v}_i.$$

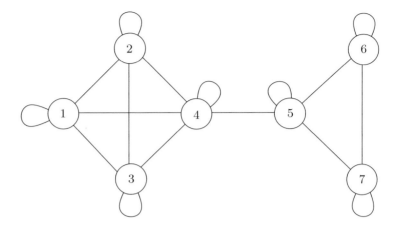

Figure 8.12 An example of the DeGroot model and the contagion model. The initial scores on nodes $1-4$ eventually spread evenly to all nodes. The rate of convergence to this equilibrium is governed by the second-largest eigenvalue of the adjacency matrix.

Dividing both sides by the leading term $c_1 \lambda_1^t$, since we want to see how accurate the first-order approximation is, and rearranging the terms, we have

$$\frac{\mathbf{x}[t]}{c_1 \lambda_1^t} = \mathbf{v}_1 + \frac{c_2}{c_1} \left(\frac{\lambda_2}{\lambda_1} \right)^t \mathbf{v}_2 + \cdots .$$

This means that the leading term of the *error* between $\mathbf{x}(t)$ and the first-order approximation $c_1 \lambda_1^t \mathbf{v}_1$ is the second-order term in the eigenvector expansion, with a magnitude that evolves over time t proportionally to

$$\left(\frac{\lambda_2}{\lambda_1} \right)^t .$$

For certain matrices like the Google matrix $\mathbf{G}$, the largest eigenvalue λ_1 is 1. Then it is the second-largest eigenvalue λ_2 that governs the rate of convergence.

As a small example, consider the network shown in Figure 8.12 consisting of two clusters with a link between them. Suppose the state of each node is a score between 0 and 100, and the initial score vector is

$$\mathbf{v} = \begin{bmatrix} 100 & 100 & 100 & 100 & 0 & 0 & 0 \end{bmatrix}^T ,$$

i.e., all nodes in the left cluster have an initial score of 100, and the right cluster has an initial score of 0.

From the network we can also write out $\mathbf{A}$, normalized per row as in Google's PageRank matrices:

$$\mathbf{A} = \begin{bmatrix} 1/4 & 1/4 & 1/4 & 1/4 & 0 & 0 & 0 \\ 1/4 & 1/4 & 1/4 & 1/4 & 0 & 0 & 0 \\ 1/4 & 1/4 & 1/4 & 1/4 & 0 & 0 & 0 \\ 1/5 & 1/5 & 1/5 & 1/5 & 1/5 & 0 & 0 \\ 0 & 0 & 0 & 1/4 & 1/4 & 1/4 & 1/4 \\ 0 & 0 & 0 & 0 & 1/3 & 1/3 & 1/3 \\ 0 & 0 & 0 & 0 & 1/3 & 1/3 & 1/3 \end{bmatrix}.$$

Then we iterate the equation

$$\mathbf{x}[t] = \mathbf{A}\mathbf{x}[t-1]$$

to obtain

$$\mathbf{x}[0] = \begin{bmatrix} 100 & 100 & 100 & 100 & 0 & 0 & 0 \end{bmatrix}^T,$$

$$\mathbf{x}[1] = \begin{bmatrix} 100 & 100 & 100 & 80 & 25 & 0 & 0 \end{bmatrix}^T,$$

$$\mathbf{x}[2] = \begin{bmatrix} 95 & 95 & 95 & 81 & 26.25 & 8.333 & 8.333 \end{bmatrix}^T,$$

$$\mathbf{x}[3] = \begin{bmatrix} 91.5 & 91.5 & 91.5 & 78.45 & 30.98 & 14.31 & 14.31 \end{bmatrix}^T,$$

$$\vdots$$

$$\mathbf{x}[\infty] = \begin{bmatrix} 62.96 & 62.96 & 62.96 & 62.96 & 62.96 & 62.96 & 62.96 \end{bmatrix}^T.$$

We see that the network is connected, i.e., $\mathbf{A}$ is irreducible. The existence of self-loops ensures the network is also aperiodic. Convergence to an equilibrium is therefore guaranteed. The scores at equilibrium are biased towards the initial scores of the left cluster because it is the larger cluster.

8.4.2   Measuring subgraph connectedness

We have finished our tour of seven influence models in two chapters. We conclude this cluster of chapters with a discussion of what constitutes a group in a network.

Intuitively, a set of nodes form a group if there are many connections (counting links or paths) among them, but not that many between them and other sets of nodes. Suppose we divide a graph into two parts. Count the number of links between the two parts; call that A. Then, for each part of the graph, count the total number of links with at least one end in that part, and call those B_1 and B_2. A different way to divide the graph may lead to a different ratio

$$\frac{A}{\min(B_1, B_2)}.$$

The smallest possible value of the ratio is called the **conductance** of this graph. It is also used to characterize convergence speed in random walk.

To dig deeper into what constitutes a group, we need some metrics that quantify the notion of connectedness in a **subgraph**: a subset of nodes, and the links that start and end with these nodes. First of all, let us focus on end-to-end

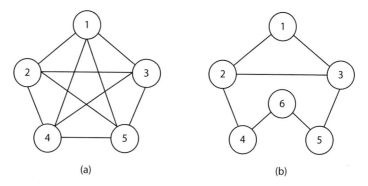

Figure 8.13 Two graphs to illustrate some of the definitions discussed. The entirety of (a) is a 4-component, as any node can reach any other using one of 4 node-disjoint paths. It is also a clique, since all nodes are neighbors. In (b), nodes 1-5 form a 2-clique, whereas either node 4 or 5 must be removed from this set to make a 2-club.

connectedness, i.e., the existence of paths. We say a subset is a **connected component**, or just a **component**, if each node in the subset can reach any other node in the subset through some path. For directed graphs, the path needs to be directed too, and it is called a **strongly connected component**. In almost all our networks, there is just one component: the entire network itself. We can also further strengthen the notion of component to k-**component**: a maximal subset of nodes in which each node can be connected to any other node through not just one path, but k different paths.

As is typical in this subsection, to make the definitions useful, we refer to the *maximal* subset of nodes: you cannot add another node to the subset and still satisfy the definition.

What if we shift our attention from end-to-end path connectedness to one-hop connectedness? We call a maximal subset of nodes a **clique** if every node in the subset is every other node's neighbor, i.e., there is a link between every pair of nodes in the subset. It is sometimes called a **full mesh**.

A clique is very dense. More likely we will encounter a cluster of density p as defined before in the analysis of the contagion model: a maximal subset of nodes in which each node has at least $p\%$ of its neighbors in this subset.

In-between a component and a clique, there is a middle ground. A k-**clique** is a maximal subset of nodes in which any node can reach any other node through no more than k links. If these k links also are all in-between nodes belonging to this subset, we call the subset a k-**club**.

The two graphs in Figure 8.13 will help illustrate these terms. In graph (a), the star in the middle adds various connections between the nodes. Consider node 1. It has a direct connection to each of the other nodes, making each its neighbor. By symmetry of the graph, all the nodes are neighbors, which means the graph is a clique. Obviously, graph (a) is a component, as each node can reach any other through some path. But we can say more. Again, consider node

1. It can reach node 2 by going directly to it, or by first visiting any of nodes 3, 4, or 5, before going to it. All of these paths are difference in the sense of being node-disjoint, as none of the intermediate nodes are the same. We can say the same for the other nodes. Hence, graph (a) is a 4-component.

Now consider graph (b). Most of the middle links from (a) have been removed, and a node has been added. We can no longer say the graph is a 4-component or a clique. But consider the set of nodes $V = \{1, 2, 3, 4, 5\}$. Any of these nodes can reach any other using no more than *two* links, and therefore the subgraph V is a 2-clique. If we had included node 6, it would have been a 3-clique, as any path between nodes 1 and 6 has at least 3 links. But notice that for node 4 to reach node 5 using two links, it has to use links $(4, 6)$ and $(6, 5)$, which are *not* in V. Hence, V is not a 2-club. But the subgraphs $V \backslash \{4\}$ and $V \backslash \{5\}$ (V without nodes 4 and 5, respectively) are both 2-clubs.

We have so far assumed that geographic proximity in a graph reflects social distance too. But that does not have to be the case. Sometimes, we have a system of labeling nodes by some characteristics, and we want a metric quantifying the notion of **associative mixing** based on this labeling: that nodes which are alike tend to associate with each other.

Consider labeling each node in a given graph as belonging to one of M given types, e.g., M social clubs, M undergraduate majors, or M dorms. We can easily count the number of links connecting nodes of the same type. From the given adjacency matrix $\mathbf{A}$, we have

$$\sum_{ij \in \text{same type}} A_{ij}. \tag{8.18}$$

But this expression is not quite right to use for our purpose. Some nodes have large degrees anyway. So we have to calibrate with respect to that. Consider an undirected graph, and pick node i with degree d_i. Each of its neighbors, indexed by j, has a degree d_j. Let us pick one of node i's links. What is the chance that on the other end of this link is node j? That would be d_j/L, where L is the total number of links in the network. Now multiply by the number of links node i has, and sum over node pairs (ij) of the same type, we have

$$\sum_{ij \in \text{same type}} \frac{d_i d_j}{L}. \tag{8.19}$$

The difference between (8.18) and (8.19), normalized by the number of links L, is the **modularity** Q of a given graph (with respect to a particular labeling of the nodes):

$$Q = \frac{1}{L} \sum_{ij \in \text{same type}} \left(A_{ij} - \frac{d_i d_j}{L} \right). \tag{8.20}$$

Q can be positive, in which case we have *associative* mixing: people of the same type connect more with each other (relative to a random drawing). It can be negative, in which case we have *dissociative* mixing: people of different types

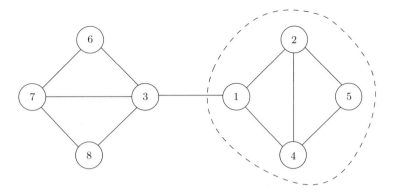

Figure 8.14 An associative labeling of a small graph. Nodes close to each other are labeled into the same type, and the modularity Q is positive.

connect more with each other. With the normalization by L in its definition, we know $Q \in [-1, 1]$.

Still using the same graph in Figure 8.2, we consider a labeling of the nodes into two types. In the following example, nodes enclosed in dashed lines belong to type 1, and the remaining nodes belong to type 2.

Obviously, the degrees are

$$\mathbf{d} = \begin{bmatrix} 3 & 3 & 4 & 3 & 2 & 2 & 3 & 2 \end{bmatrix}^{T}.$$

We also have

$$L = 11 \times 2 = 22,$$

since the links are undirected. In the computation of modularity, all pairs (ij) (such that $ij \in$ same type) are considered, which means every undirected link is counted twice; thus the normalization constant L should be counted the same way.

Now consider associative mixing with the grouping in Figure 8.14. The modularity can be expressed as

$$
\begin{aligned}
Q &= \frac{1}{L} \sum_{ij \in \text{same type}} \left(A_{ij} - \frac{d_i d_j}{L} \right) \\
&= \frac{1}{L} \sum_{ij} S_{ij} \left(A_{ij} - \frac{d_i d_j}{L} \right),
\end{aligned}
$$

where the index S_{ij} denotes whether i and j are of the same type as specified by the labeling. We can also collect these indices into a binary matrix:

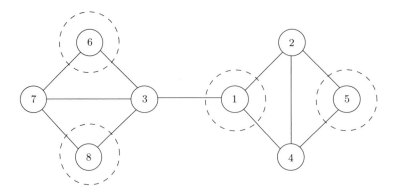

Figure 8.15 A disassociative labeling of a small graph. Nodes close to each other are labeled into different types, and the modularity Q is negative.

$$\mathbf{S} = \begin{bmatrix} 0 & 1 & 0 & 1 & 1 & 0 & 0 & 0 \\ 1 & 0 & 0 & 1 & 1 & 0 & 0 & 0 \\ 0 & 0 & 0 & 0 & 0 & 1 & 1 & 1 \\ 1 & 1 & 0 & 0 & 1 & 0 & 0 & 0 \\ 1 & 1 & 0 & 1 & 0 & 0 & 0 & 0 \\ 0 & 0 & 1 & 0 & 0 & 0 & 1 & 1 \\ 0 & 0 & 1 & 0 & 0 & 1 & 0 & 1 \\ 0 & 0 & 1 & 0 & 0 & 1 & 1 & 0 \end{bmatrix}.$$

Given $\mathbf{S}, \mathbf{A}$, and $\mathbf{d}$, we can compute Q. We see the modularity value is indeed positive for this labeling system, and reasonably high:

$$Q = 0.5413.$$

But suppose the labeling is changed to what is shown in Figure 8.15, i.e.,

$$\mathbf{S} = \begin{bmatrix} 0 & 0 & 0 & 0 & 1 & 1 & 0 & 1 \\ 0 & 0 & 1 & 1 & 0 & 0 & 1 & 0 \\ 0 & 1 & 0 & 1 & 0 & 0 & 1 & 0 \\ 0 & 1 & 1 & 0 & 0 & 0 & 1 & 0 \\ 1 & 0 & 0 & 0 & 0 & 1 & 0 & 1 \\ 1 & 0 & 0 & 0 & 1 & 0 & 0 & 1 \\ 0 & 1 & 1 & 1 & 0 & 0 & 0 & 0 \\ 1 & 0 & 0 & 0 & 1 & 1 & 0 & 0 \end{bmatrix}.$$

Then the modularity value is negative, as we would expect from dissociative mixing of putting nodes of different types next to each other in Figure 8.15.

$$Q = -0.2025.$$

8.4.3   Graph partition and community detection

It is actually not easy to infer either the network topology or the traffic pattern from limited (local, partial, and noisy) measurement. In the last part of this chapter, we assume someone has built an accurate topology already, and our job is to detect the non-overlapping communities, or subgraphs, through some centralized, off-line computation.

The easiest version of the problem statement is that we are given the number of communities and the number of nodes in each community. This is when you have pretty good prior knowledge about the communities in the first place. It is called the **graph-partition** problem. We will focus on the special case in which we partition a given graph into two subgraphs, the **graph-bisection** problem, with a fixed target size (number of nodes) in each subgraph. The input is a graph $G = (V, E)$, and the output is two sets of nodes that add up to the original node set V.

How do we even define that one graph partition is "better" than another? One metric is the number of links between the two subgraphs, called the **cut size**. Later you will see the "max flow min cut" theorem in routing. For now, we just want to find a bisection that minimizes the cut size.

An algorithm that is simple to describe although heavy in computational load is the **Kernighan–Lin algorithm**. There are two loops in the algorithm. In each step of the outer loop indexed by k, we start with a bisection: graphs $G_1[k]$ and $G_2[k]$. To initialize the first outer loop, we put some nodes in subgraph $G_1[1]$ and the rest in the other subgraph $G_2[1]$.

Now we go through an inner loop, where at each step we pick the pair of nodes (i, j), where $i \in G_1$ and $j \in G_2$, such that swapping them reduces the cut size most. If cut size can only be increased, then pick the pair such that the cut size increases by the smallest amount. After each step of the inner loop, the pair that has been swapped can no longer be considered in future swaps. When there are no more pairs to consider, we complete the inner loop and pick the configuration $(G_1^*[k], G_2^*[k])$ (i.e., which nodes belong to which subgraph) with the smallest cut size $c^*[k]$.

Then we take that configuration $(G_1^*[k], G_2^*[k])$ as the initialization of the next step $k+1$ in the outer loop. This continues until the cut size cannot be decreased further through the outer loops. The configuration with the smallest cut size throughout the outer loop,

$$\min_{k}\{c^*[k]\},$$

is the bisection returned by this algorithm.

More often, we do *not* know how many communities there are, or how many nodes are in each. That is part of the job of **community detection**. For example, you may wonder about the structure of communities in the graph of Facebook connections. And we may be more interested in the richness of connectivity within each subgraph than in the sparsity of connectivity between them.

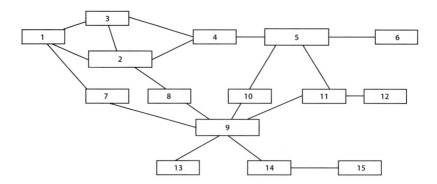

Figure 8.16 The original graph of the Florentian families, viewed as a single community.

Again, we focus on the simpler case of two subgraphs (G_1, G_2), but this time not imposing the number of nodes in each subgraph a priori.

Modularity is an obvious metric to quantify how much more connected a set of nodes is relative to the connectivity if links were randomly established among the nodes. Modularity is defined with respect to a labeling system. Now there are two labels on the nodes, those belonging to G_1 and those to G_2.

So we can simply run the Kernighan–Lin algorithm again. But instead of picking the pair of nodes to *swap* across G_1 and G_2 in order to minimize the cut size, now we select one node to *move* from G_1 to G_2 (or the other way around), in order to maximize the modularity of the graph.

A very different approach gets back to the cut-size-minimization idea, and tries to disconnect the graph into many pieces by *deleting* one link after another. This is useful for detecting not just two communities, but any number of communities. Which link should we delete first? A greedy heuristic computes the betweenness metric of all the links (8.5), and then deletes the link with the highest betweenness value. If that does not break the graph into two subgraphs, then compute the betweenness values of all the remaining links, and delete the link with the highest value again. Eventually, this process will break the graph and give you 2 subgraphs (and $3, 4, \ldots, N$ graphs as you keep deleting links).

As an example, we consider the Renaissance Florentian family graph again, shown in Figure 8.16.

Now we run the following community-detection algorithm.

1. From graph $G = (V, E)$, write down the adjacency matrix $\mathbf{A}$.
2. Compute the betweenness of all links $(i, j) \in E$ from $\mathbf{A}$.
3. Find the link $(i, j)^*$ that has the highest betweenness value. If more than one such link exists, select one of these randomly.
4. Remove link $(i, j)^*$ from G. Check whether any communities have been detected. If not, return to step 1.

Link	Betweenness	Link	Betweenness
(1, 2)	5	(5, 11)	17
(1, 3)	6	(7, 9)	19
(1, 7)	13	(8, 9)	18.17
(2, 3)	4	(9, 10)	15.5
(2, 4)	9.5	(9, 11)	23
(2, 8)	13.5	(9, 13)	14
(3, 4)	8	(9, 14)	**26**
(4, 5)	18.83	(11, 12)	14
(5, 6)	14	(14, 15)	14
(5, 10)	12.5		

Table 8.1 Betweenness values of the links from the initial graph in Figure 8.16. Link (9, 14) has the largest betweenness and will be eliminated first.

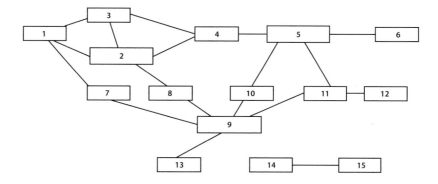

Figure 8.17 Eliminating the link with the highest betweenness, (9, 14), detects two communities within the graph: Node sets $V_1 = \{1, ..., 13\}$ and $V_2 = \{14, 15\}$.

The betweenness values of all the links in the initial graph are shown in Table 8.1. The largest betweenness is that of link (9, 14), with a value of 26. As a result, it is eliminated from the graph. The result is shown in Figure 8.17. As one can see, removing (9, 14) has created two distinct communities: the first is the node set $V_1 = \{1, ..., 13\}$, and the second is the set $V_2 = \{14, 15\}$. If we just want to detect two communities, we can stop now. If we want to detect one more community, we have to keep going.

Next, the adjacency matrix is modified according to the new graph, and the betweenness values are calculated again. The results are shown in Table 8.2.

The largest betweenness is that of link (4, 5): 17.5. As a result, it will be eliminated from the graph. However, removing this link does not detect additional communities, so the process is repeated. Computing betweenness with (4, 5) eliminated, the maximum is that of link (8, 9): 25.5. Again, this link is eliminated. Finally, after running the procedure again, link (7, 9) is found to

Link	Betweenness	Link	Betweenness
(1, 2)	5	(5, 11)	14.33
(1, 3)	5	(7, 9)	14
(1, 7)	10	(8, 9)	12.5
(2, 3)	3	(9, 10)	10.83
(2, 4)	8.83	(9, 11)	16.33
(2, 8)	9.83	(9, 13)	12
(3, 4)	8	(9, 14)	–
(4, 5)	**17.5**	(11, 12)	12
(5, 6)	12	(14, 15)	1
(5, 10)	9.83		

Table 8.2 Betweenness values of the links from the two-component graph in Figure 8.17. Link (4, 5) has the largest betweenness and will be eliminated.

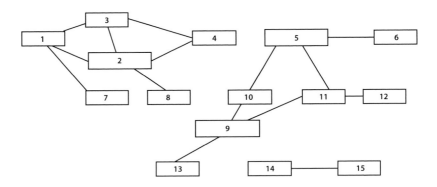

Figure 8.18 Eliminating links (4, 5), (8, 9), and (7, 9) detects three communities: Node sets $V_1 = \{1, 2, 3, 4, 7, 8\}$, $V_2 = \{5, 6, 9, 10, 11, 13\}$, and $V_3 = \{14, 15\}$.

have the highest betweenness value of 42 and is eliminated. This separates the graph into three communities: $V_1 = \{1, 2, 3, 4, 7, 8\}$, $V_2 = \{5, 6, 9, 10, 11, 13\}$, and $V_3 = \{14, 15\}$, as shown in Figure 8.18.

In addition to modularity maximization and betweenness-based link removal, there are several other algorithms for community detection, including graph Laplacian optimization, maximum-likelihood detection, and latent-space modeling. When it comes to a large-scale community-detection problem in practice, it remains unclear which of these will be most helpful in attaining the eventual goal of detecting communities, while remaining robust against measurement noise.

Instead of deleting links, how about *adding* links from a set of disconnected nodes? This way of constructing communities is called **hierarchical clustering**. Now, finding one pair of similar nodes is easy, for example, by using node-similarity metrics like the cosine coefficient in Chapter 4. The difficulty is in defining a consistent and useful notion of similarity between two *sets* of nodes.

If there are N_1 nodes in G_1 and N_2 nodes in G_2, there are then $N_1 N_2$ node pairs. Therefore, there are $N_1 N_2$ similarity-metric values. We need to scalarize this long vector. We can take the largest, the smallest, the average, or any scalar representation of this vector of values as the similarity metric between the two sets of nodes. Once a similarity metric has been fixed and a scalarization method picked, we can hierarchically run clustering by greedily adding nodes, starting from a pair of nodes until there are only two groups of nodes left.

Summary

Box 8 Influence models in a network

Contagion creates a complete flip of all the nodes if and only if no cluster of non-flipped nodes is dense enough initially. Infection models the change of states in each node through differential equations. The importance of nodes and links in a graph can be quantified through a variety of centrality measures. Communities in a graph can be detected by deleting links, clustering nodes, or computing how connected a set of nodes are among themselves relative to the other nodes.

Further Reading

Similar to the last chapter, there is a gap between the rich foundation of graph theory and algorithms on the one hand, and the actual operation of Facebook and Twitter and their third-party service providers on the other.

1. The standard reference on contagion models is the following one:
S. Morris, "Contagion," *Review of Economic Studies*, vol. 67, pp. 57–78, 2000.

2. Our discussion of infection models follows the comprehensive survey article:
H. W. Hethcote, "The mathematics of infectious diseases," *SIAM Review*, vol. 42, no. 4, pp. 599–653, October 2000.

3. A classic work on innovation diffusion, both quantitative models and qualitative discussions, can be found in the following book:
E. M. Rogers, *Diffusion of Innovation*, 5th edn., Free Press, 2003.

4. Many graph-theoretic quantities on node importance, link importance, and group connectedness can be found in the following textbook:
M. E. J. Newman, *Networks: An Introduction*, Oxford University Press, 2010.

5. Another standard reference for social network analysis is the following text-book:

S. Wasserman and K. Faust, *Social Network Analysis*, Cambridge University Press, 1994.

Problems

8.1 *Computing centrality and betweenness* ⋆

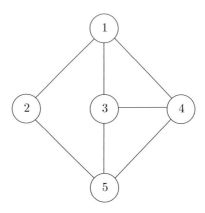

Figure 8.19 A simple graph for computing centrality measures.

(a) Compute the degree, closeness, and eigenvector centrality of each node in the graph in Figure 8.19.

(b) Compute the node betweenness centrality of nodes 2 and 3.

(c) Compute the link betweenness centrality of the links $(3, 4)$ and $(2, 5)$.

8.2 *Contagion* ⋆

Consider the contagion model in the graph in Figure 8.20 with $p = 0.3$.
(a) Run the contagion model with node 1 initialized at state-1 and the other nodes initialized at state-0.

(b) Run the contagion model with node 3 initialized at state-1 and the other nodes initialized at state-0.

(c) Contrast the results from (a) and (b) and explain in terms of the cluster densities of the sets of initially state-0 nodes.

8.3 *SIRS infection model* ⋆⋆

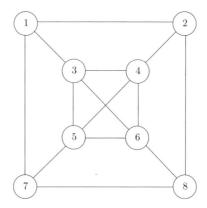

Figure 8.20 A simple graph for studying contagion.

We consider an extension to the SIR model that allows nodes in state R to go to state S. This model, known as the SIRS model, accounts for the possibility that a person loses the acquired immunity over time.

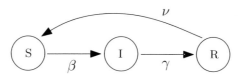

Figure 8.21 The state-transition diagram for the SIRS infection model.

Consider the state diagram in Figure 8.21. We can write out the set of differential equations as

$$\frac{dS(t)}{dt} = -\beta S(t)I(t) + \nu R(t)$$

$$\frac{dI(t)}{dt} = \beta S(t)I(t) - \gamma I(t)$$

$$\frac{dR(t)}{dt} = \gamma I(t) - \nu R(t).$$

Modify the Matlab code www.network20q.com/hw/simulate_SIR.m for the numerical solution of the SIR model. Solve for $t = 1, 2, \ldots, 200$ (set the tspan vector in code accordingly) with the following parameters and initial conditions: $\beta = 1$, $\gamma = 1/3$, $\nu = 1/50$, $I(0) = 0.1$, $S(0) = 0.9$, and $R(0) = 0$. Describe and explain your observations.

8.4 *Information centrality* ★★

Consider a weighted, undirected, and connected graph with N nodes, where the weight for link (i,j) is x_{ij}. First construct a matrix $\mathbf{A}$ where the diagonal entries $A_{ii} = 1 + \sum_j x_{ij}$, $A_{ij} = 1 - x_{ij}$ if nodes i and j are adjacent, and $A_{ij} = 1$ otherwise.

Now compute the inverse: $\mathbf{C} = \mathbf{A}^{-1}$. The following quantity is called the **information centrality** of node i:

$$C_I(i) = \frac{1}{C_{ii} + (T - 2R)/N},$$

where $T = \sum_i C_{ii}$ is the trace of matrix $\mathbf{C}$ and $R = \sum_j C_{ij}$ is (any) row sum of matrix $\mathbf{C}$.

Can you think of why this metric is called information centrality?

8.5 *Hypergraphs and bipartite graphs* ★★

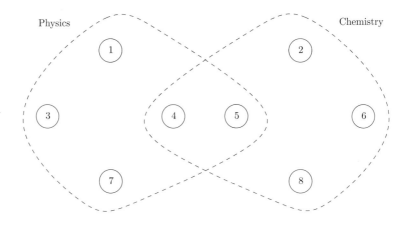

Figure 8.22 Hypergraphs allow each link to connect more than two nodes. Each dotted curve represents a hyperlink.

Why must a link be defined as the connection between just *two* nodes? Suppose eight papers are in the fields of physics or chemistry. Group membership, i.e., to which field a paper belongs, is presented as a **hypergraph** in Figure 8.22. Each dotted area is a group or a **hyperedge**, which is a generalization of an undirected edge to connect possibly more than two nodes. Papers 4 and 5 are "interdisciplinary" papers, so their nodes are contained in both hyperedges.

(a) We can transform the hypergraph in Figure 8.22 into an undirected bipartite graph by introducing two more nodes, each representing one of the hyperedges, and linking a "standard" node to a "hyperedge" node if the former is contained in the corresponding hyperedge. Draw this bipartite graph.

(b) Define an *incidence matrix* **B** of size 2×8 with

$$B_{ij} = \begin{cases} 1 & \text{node } j \text{ is contained in group } i, \\ 0 & \text{otherwise,} \end{cases}$$

where group 1 is "Physics" and group 2 is "Chemistry." Write down **B** for this graph.

(c) Compute the matrix $\mathbf{B}^T\mathbf{B}$. What is its interpretation?

(d) Compute the matrix $\mathbf{BB}^T$. What is its interpretation?

9 Can I really reach anyone in six steps?

In the last chapter, we saw the importance of topology to functionality. In this and the next chapters, we will focus on **generative models** of network topology and reverse-engineering of network functionality. These are mathematical constructions that try to *explain* widespread empirical observations about social and technological networks: the "small world" property and the "scale free" property. We will also highlight common misunderstandings and misuse of generative models.

9.1   A Short Answer

Since Milgram's 1967 experiment, the **small world** phenomenon, or the **six degrees of separation**, has become one of the most widely told stories in popular science books. Milgram asked 296 people living in Omaha, Nebraska to participate in the experiment. He gave each of them a passport-looking letter, and the destination was in a suburb of Boston, Massachusetts, with the recipient's name, address, and occupation (stockbroker) shown. Name and address sound obvious, and it turned out that it was very helpful to know the occupation. The goal was to send this letter to one of your friends, defined as someone you knew by first name. If you did not know the recipient by first name, you had to send the letter via others, starting with sending it to a friend (one hop), who then sent it to one of her friends (another hop), until the letter finally arrived at someone who knew the recipient by first name and sent it to the recipient. This is illustrated in Figure 9.1.

Of these letters, 217 were actually sent out and 64 arrived at the destination, a seemingly small arrival rate of 29.5% but actually quite impressive, considering that a later replica of the experiment via email had only a 1.5% arrival rate. The other letters might have been lost along the way, and needed to be treated carefully in the statistical analysis of this experiment's data. But, out of those 64 that arrived, the average number of hops was 5.2 and the median 6, as shown in Figure 9.2.

Researchers have long suspected that the **social distance**, the average number of hops of social relationships it takes (via a short path) to reach anyone in a population, grows very slowly as the population size grows, often logarithmically.

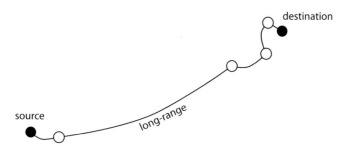

Figure 9.1 A picture illustrating the Milgram experiment in 1967. A key phenomenon is that there is often one or two long-range links in these short paths between Omaha and Boston. It turns out that the long-range links substantially reduced the shortest paths' and the searchable paths' lengths without reducing the clustering coefficient significantly in the social network.

Milgram's celebrated experiment codified the viewpoint. From the 1970s to the online social media era, much empirical evidence suggested the same: from the Erdos number among mathematicians to co-starring relationships in the IMDb.

Should we be surprised by this seemingly universal observation of social networks? There are two issues here, echoing the dichotomy between topology and functionality.

- One is *structural*: There exist short paths in social networks.
- Two is *algorithmic*: With very limited local information a node can navigate through a social network and find a short path to a given destination.

The second kind of small worlds is more surprising than the first and requires more careful modeling of the functionality of **social search**. For example, a report in November 2011 computed the degrees of separation on Facebook to be 4.74. That concerned only with the existence of short paths, not the more relevant and more surprising discoverability of short paths from local information. As we will see, it is also more difficult to create a robust explanation for the observation of an algorithmic small world.

But first, we focus on the existence of short paths. On the surface, it seems fascinating that you can likely reach anyone in six steps or fewer. Then, on second thoughts, you may reason that, if I have twenty friends, and each of them has twenty friends, then in six steps, I can reach 20^6 people. That is already 64 million people. So of course six steps often suffice.

But then, giving it further thought, you realize that social networks are filled with "triangles," or **triad closures**, of social relationships. This is illustrated in Figure 9.3: if Alice and Bob both know Chris, Alice and Bob likely know each other directly too. This is called **transitivity** in a graph (not to be confused with transitivity in voting). In other words, the catch of the 20^6 argument above is that you need your friend's friends to not overlap with your own friends. Otherwise,

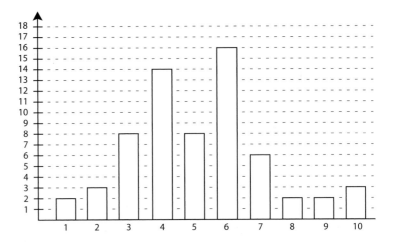

Figure 9.2 The histogram of the length of the search paths from different sources in Omaha to the common destination in Boston in Milgram's experiment. The median value is six, leading to the famous six degrees of separation.

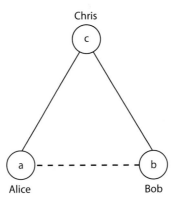

Figure 9.3 An illustration of triad closure in social networks. If Alice knows Chris and Bob knows Chris (the solid lines), it is likely that Alice and Bob also know each other (the dotted line). If so, the connected triple forms a triangle. The clustering coefficient quantifies the ratio between connected triples and triangles in a graph.

the argument fails. But of course, many of your friend's friends are your own friends too. There is a lot of overlap. The phenomenon of people who are alike tending to form social links is called **homophily**, and it can be quantified by the **clustering coefficient** as discussed later. Now, six degrees of separation is truly surprising.

Milgram-type experiments suggest something even stronger: not only are there short paths, but they can be discovered by each individual node using very limited information about the destination and its local view of the network.

Compared with routing packets through the Internet, social search is even harder since nodes do not pass messages around to help each other construct some global view of the network topology. That help is implicitly embedded in the address and occupation of the recipient, and possibly in the name that can reveal something about the destination's sex and ethnicity. Some kind of *distance metric* must have been constructed in each person's mind throughout Milgram's experiment. For example, New York is closer to Boston than Chicago is, on the geographic proximity scale measured in miles. Or, a financial advisor is perhaps closer to a stockbroker than a nurse is, on some occupation proximity scale, which is more vague but nonetheless can be grossly quantified. Suppose each person uses a simple, "greedy" algorithm to forward the letter to her friend who is closest to the destination, where "closeness" is defined by a composite of these kinds of distance metrics. Is it a coincidence that this social search strategy discovers a short path?

We will walk through several models that address the above issues.

9.2   A Long Answer

9.2.1   Structural small worlds: Short paths

There are several ways to measure how "big" a (connected) graph is. One is diameter: it is the length of the longest shortest path between any pair of nodes. "Longest shortest" may sound strange. Here, "shortest" is with respect to all the paths between a given pair of nodes, and "longest" is with respect to all possible node pairs.

When we think of a network as a small world, however, we tend to use the *median* of the shortest paths between all node pairs, and look at the growth of that metric as the number of nodes increases. If it grows at a rate on the order of the log of the number of nodes, we say the network is (structurally) a small world.

Suppose we are given a fixed set of n nodes, and for each pair of nodes decide with probability p that there is a link between them. This is the basic idea of a **Poisson random graph**, or the **Erdos–Renyi model**.

Of course, this process of network formation does not sound like most real networks. It turns out that neither does it provide the same structures as those we encounter in many real networks. For example, while in a random graph the length of the average shortest path is small, it does not have the right clustering coefficient. For a proper explanatory model of small world networks, we need the shortest path's length to be small and the clustering coefficient to be large.

What is the clustering coefficient? Not to be confused with the density of a cluster from Chapter 8, it is a metric to quantify the notion of triad closure. As in Figure 9.3, we define a set of three nodes (a, b, c) in an undirected graph as a **connected triple**, if there is a path connecting them.

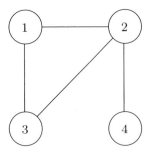

Figure 9.4 A small example for the calculation of the cluster coefficient. $C = 3/5$ in this graph.

- If there are links (a, b) and (b, c) (whether there is a link (a, c) or not), we have a connected triple (a, b, c).
- Similarly, if there are links (a, c) and (c, b), we have a connected triple (a, c, b).
- If there are links (b, a) and (a, c), we have a connected triple (b, a, c).

But if there are links (ab, bc, ca), we have not just a connected triple, but also a triangle. We call this a triad closure.

The clustering coefficient C summarizes the above countings in the graph:

$$C = \frac{\text{Number of triangles}}{\text{Number of connected triples}/3}. \tag{9.1}$$

The division by 3 normalizes the metric, so that a triangle's clustering coefficient is exactly 1, and that $C \in [0, 1]$.

For example, consider the toy example in Figure 9.4. It has one triangle $(1, 2, 3)$, and the following five connected triples: $(2, 1, 3)$, $(1, 2, 3)$, $(1, 3, 2)$, $(1, 2, 4)$, and $(3, 2, 4)$. Hence, its clustering coefficient is

$$C = \frac{1}{5/3} = \frac{3}{5}.$$

It is easy to calculate that the *expected* clustering coefficient of a random graph (we have to use expectation here because random graphs are probabilistic objects):

$$C = \frac{c}{n - 1},$$

where c is the expected degree of a node and n the total number of nodes. This is because the probability of any two nodes being connected is $c/(n - 1)$ in a random graph, including the case when the two nodes are known to be indirectly connected via a third node. So, if there are 100 million people, and each has 1000 friends, the clustering coefficient is about 10^{-3}, way too small for a realistic social network.

What about a very regular ring like in Figure 9.5 instead? This **regular graph** is parameterized by an even integer c: the number of neighbors each node has.

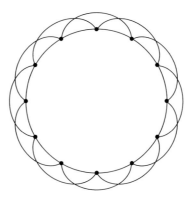

Figure 9.5 An example of a regular graph, with n nodes living on a ring and each having c links. In this example, $n = 12$ and $c = 4$. Each node has two links pointing to the right and two to the left. The clustering coefficient C is independent of n and grows as c becomes larger.

Because all the nodes "live" on a ring, each node can have $c/2$ left-pointing links to its closest $c/2$ neighbors on the left, and similarly $c/2$ on the right.

It is easy to see that the clustering coefficient is large for a regular graph.

- To form a triangle, we need to go along one direction on the ring two steps, then take one step back towards where we started. And the farthest we can go along the ring in one direction and still be back in one hop is $c/2$. So, the number of triangles starting from a given node is simply $c/2$ choose 2, the number of distinct choices of picking two nodes out of $c/2$ of them. This gives us $\frac{1}{2}\frac{c}{2}(\frac{c}{2} - 1)$ triangles per node.
- On the other hand, the number of connected triples centered on each node is just c choose 2, i.e., $\frac{1}{2}c(c - 1)$.

Putting everything together, the clustering coefficient is

$$C = \frac{\frac{1}{2}\frac{c}{2}(\frac{c}{2} - 1)}{(\frac{1}{2}c(c - 1))/3} = \frac{3(c - 2)}{4(c - 1)}.$$

It is independent of the number of nodes n. This makes intuitive sense since the graph is symmetric: every node looks like any other node in its neighborhood topology.

- When $c = 2$, the smallest possible value of c, we have a circle of links, so there are obviously no triangles. Indeed, $C = 0$.
- But as soon as $c = 4$, we have $C = 1/2$, which is again intuitively clear and shown in another way by drawing a regular ring graph with $c = 4$ in Figure 9.6.
- When c is large, the clustering coefficient approaches the largest it can get on a regular ring topology: $3/4$. This is many orders-of-magnitude larger than that of a random graph.

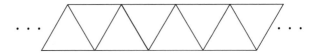

Figure 9.6 Another way to draw a regular ring graph with $c = 4$. This visualization clearly shows that half of the triad closures are there in the graph, and C should be 0.5.

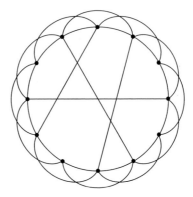

Figure 9.7 An example of the Watts–Strogatz model. It adds random links, possibly long-range ones, to a regular ring graph. When there is a short-range link (a link in the regular ring graph), there is now a probability p of establishing a long-range link randomly connecting two nodes. The resulting graph has both a large clustering coefficient C and a small expected shortest distance L.

The regular (ring) graph model stands in sharp contrast to the random graph model: it has a high clustering coefficient C, which is realistic for social networks, but has a large value of the (median or average, across all node pairs) shortest path's distance L, since there are only short-range connections. Random graph networks are exactly the opposite: small L but also small C. If only we could have a "hybrid" graph that combines both models to get a small L and a large C on the same graph.

That is exactly what the **Watts–Strogatz model** accomplished in 1998. It is the canonical model explaining small world networks with large clustering coefficients. As shown in Figure 9.7, it has two parameters.

- c is the number of nearest neighbors each node has.
- p is the probability that any pair of nodes, including those far apart, are connected by a link whenever there is a short-range link.

Actually, the Watts–Strogatz model deletes one regular link for each of the random links added. The model we just showed does not include the deletion step. But the main analysis for our purpose remains effectively the same.

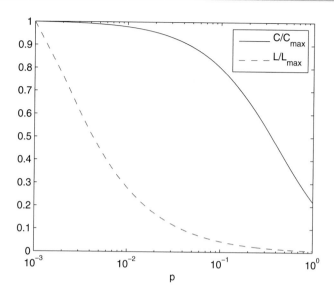

Figure 9.8 A numerical example of the impact of p on C and L in the Watts–Strogatz model. When p is small, like 0.01, C is almost the same as in a regular ring graph, but L is substantially smaller.

The key point is that with the additional links, we get to preserve the large clustering coefficient of a regular graph while achieving the small world effect. With just a little randomization p, the expected shortest path's distance can be reduced substantially.

How much randomization do we need? While we postpone the detailed calculation of C and L to the Advanced Material, it suffices to summarize at this point that, as long as p is small, e.g., 0.1, its impact on the clustering coefficient is almost negligible. Yet the average shortest-path distance behaves like a small world: it grows at a rate on the order of $\log n$. It fundamentally changes the order of growth of L with respect to the number of nodes in the network. This is illustrated in Figure 9.8. Fixing $n = 600$ and $c = 6$, we plot C/C_{max} and L/L_{max} against p, where C_{max} and L_{max} are their respective maxima over all computed values. When p is in the range of 0.1 to 0.01, we have both a large C and a small L.

Where does this asymmetry in p's impact come from? Fundamentally it has to do with the very definition of our metrics.

- The shortest path is an *extremal* quantity: we only care about the *shortest* path, there is no need to reduce all the paths' lengths. Just add a few long-range links, and even add them randomly, and then the shortest path will be much shorter.
- In contrast, the clustering coefficient is an *average* quantity: it is defined as the average number of triangles (involving a node) divided by the average number of connected triples (centered at the node). So adding a small

proportion of non-triangular, connected triples does not hurt the clustering coefficient that much.

There lies the magic of small world with a large clustering coefficient: we have triad closure relationships with most of our friends, but a very small fraction of our friends are outside our normal social circle. All Milgram needed to observe six degrees of separation was that very small fraction of long-range links.

However, this begs a deeper question: to be "fair" with respect to how we define clustering coefficients, why not define a small world as one where, for most node pairs, the *average*, not just the shortest, path length between these two nodes is small? "Average" here refers to the average over all the paths between a given pair of nodes. Well, we believe that for social search, we do not *need* that stringent a definition. The existence of some short paths suffices. This means we implicitly assume the following: each node can actually *find* a short path with a hop count not much bigger than that along the shortest path. How *that* can be accomplished with only local information at each node is a deeper mystery than just the existence of short paths.

Before we move on to look at models explaining this, we should mention that a little randomization also showed up in Google's PageRank in Chapter 3, and a locally dense but globally sparse graph will be a key idea in constructing peering graphs in P2P content distribution in Chapter 15.

9.2.2   Algorithmic small worlds: Social search

If you were one of those people participating in Milgram's experiment, or one of those along the paths initiated by these people, how would you decide the next hop by just looking at the destination's name, address, and occupation? You probably would implicitly define a metric that can measure distance, in terms of both geographic proximity (relatively easy) and occupational proximity (relatively hard). And then you would look at all your friends whom you know by first name, and pick the one closest to the destination in some combination of these two distances. This is a **greedy social search**, with an average length (i.e., hop count) denoted by l. And we wonder whether it can discover very short paths: can l be close to L? (One could also have picked the friend with the highest degree, and the letter will soon reach a highly-connected person in the network. But we will not consider this or other alternative strategies.)

As mentioned before, compared with IP routing in the Internet in Chapter 13, social search is even harder, since people do not pass messages to tell each other exactly how far they are from the destination. But here, we are not asking for an exact guarantee of discovering the shortest path either.

There have been several models for social search in the past decade beyond the original Watts–Strogatz model. The first was the **Kleinberg model**, and was defined for any dimension d. We examine the one-dimensional case now, like a ring with n nodes, which is really a line wrapped around so that we can ignore edge effects. Links are added at random as in the Watts–Strogatz model. But the random links are added with a refined detail in the Kleinberg model. The probability of having a random link of length r is proportional to $r^{-\alpha}$, where $\alpha \geq 0$ is another model parameter. The longer the link, the less likely it will show up. When $\alpha = 0$, we are back to the Watts–Strogatz model.

It turns out that only when $\alpha = 1$ will the algorithmic small world effect appear: l is upper bounded by $\log^2 n$. It is not quite $\log n$, but at least it is an upper bound that is a polynomial function of $\log n$. But when $\alpha \neq 1$, l grows as a polynomial of n, and thus is not a small world. This is even under the assumption that each node knows the exact locations of the nodes it is connected to, and therefore the distances to the destination.

In a homework problem, we will see that the exponent α needs to be exactly two for a network where nodes live on a two-dimensional rectangular grid. Generally, the exponent α must be exactly the same as the dimension of the space where the network resides. The intuition is clear: in k-dimensional space, draw spheres with radius $(r, 2r, \ldots)$ around any given node. The number of nodes living in the space between a radius-r sphere and a radius-$2r$ sphere is proportional to r^k. But according to the Kleinberg model, the probability of having a link to one of those nodes also drops as r^{-k}. These two terms cancel each other out, and the probability of having *some* connection d hops away becomes independent of d. This independence turns out to give us the desired **searchability** as a function of n: l grows no faster than a polynomial function of $\log n$.

The underlying reasoning is therefore as follows: the chance of you having a friend outside of your social/residential circle gets smaller and smaller as you go farther out, but the number of people also becomes larger and larger. If the two effects cancel each other out, you will likely be able to find a short path.

A similar argument runs through the **Watts–Dodds–Newman model**. In this hierarchical model shown in Figure 9.9, people live in different "leaf nodes" of a binary tree that depicts the geographic, occupational, or their combination's proximity. If two leaf nodes, A and B, share the first common ancestry node that is lower than that shared by A and C, then A is closer to B than to C.

For simplicity, let us assume each leaf node is a group of g people. So there are n/g leaf nodes (for notational simplicity, assume this is an integer) and $\log_2(n/g)$ (base-2 logarithm) levels in the binary tree. For example, if there are $n = 80$ people in the world, and $g = 10$ people live in a social circle where they can reach each other directly, we will have a $\log(80/10) = 3$-level tree.

In this model, each person measures the distance to another person by counting the number of tree levels one needs to go up before finding the first common-ancestry node between the two people. We also assume that the probability that two people know each other decays *exponentially* with m, the level of their first common-ancestry node, with an exponent of α:

$$p_m = K2^{-\alpha m}, \tag{9.2}$$

where K is a normalization constant so that $\sum_m p_m = 1$. You probably can recognize the similarity between this way of assigning probabilities and the random link probabilities in Kleinberg's model. Indeed, the results are similar.

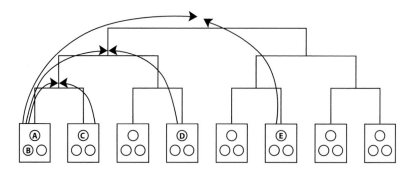

Figure 9.9 An example of the Watts–Dodds–Newman model. The three-level hierarchy is represented through a binary tree. Each leaf node of this tree has a population of g people who are directly connected to each other. The probability that two people know each other decays exponentially with m, the level of their first common-ancestry node. For examples, nodes A and B are directly connected since they live in the same leaf node. Nodes A and C have their first common-parent of the tree at level 1 (counting from below); A and D at level 2; and A and E at level 3.

The *expected* number of people that a person can be connected to through her mth ancestry, denoted as N_m, is clearly the product of two numbers: $g2^m$, the number of people connectable through the mth ancestry, and p_m, the probability that the level of first common ancestry is indeed m:

$$N_m = g2^{1\times m}p_m. \tag{9.3}$$

We highlight the fact that the number of levels in a tree grows exponentially in m and the exponent of this growth is 1. This turns out to be the mathematical root cause for the main result below.

Plugging the formula for p_m (9.2) in (9.3), we have

$$N_m = gK2^{(1-\alpha)m}. \tag{9.4}$$

In passing the message towards the destination, if the first common-ancestry level shared by a node with the destination is m, the expected number of hops needed to pass the message (before getting to someone who is on the same side of the mth hierarchy as the destination) is simply $1/N_m$. Summing over all levels m, we determine that the expected length of the path by greedy search is

$$l = \sum_m \frac{1}{N_m} = \frac{1}{Kg} \sum_{m=0}^{\log(n/g)-1} 2^{(\alpha-1)m} = \frac{1}{Kg}\frac{(n/g)^{\alpha-1}-1}{2^{\alpha-1}-1}. \tag{9.5}$$

But we cannot directly use (9.5) yet, because there is a normalization constant K that depends on n/g. We need to express it as a function of observable quantities in the tree. Here is one approach: summing N_m in (9.4) over all m levels, we get the average degree $\bar{d}$ of a person:

$$\bar{d} = \sum_m N_m = \sum_{m=0}^{\log(n/g)-1} Kg2^{(1-\alpha)m} = Kg\frac{(n/g)^{1-\alpha}-1}{2^{1-\alpha}-1}. \tag{9.6}$$

Now we express K in terms of $(\bar{d}, g, n, \alpha)$ from (9.6), and plug it back into (9.5):

$$l = \frac{1}{\bar{d}}\frac{(n/g)^{\alpha-1}-1}{2^{\alpha-1}-1}\frac{(n/g)^{1-\alpha}-1}{2^{1-\alpha}-1}. \tag{9.7}$$

Since we want to understand the behavior of l as a function of n when n grows, we take the limit of (9.7) as n becomes large.

- If $\alpha \neq 1$, by (9.7), clearly l as a function of n grows like $(n/g)^{|\alpha-1|}$, a polynomial in n. That is not an algorithmic small world.
- If $\alpha = 1$, we can go back to the formula of N_m in (9.4), which simplifies to just $N_m = gK$ irrespective of the level m. This independence is the critical reason why the small world property now follows: it becomes straightforward to see that l now grows like $\log^2(n/g)$, i.e., the square of a logarithm:

$$\frac{1}{Kg}\sum_{m=0}^{\log(n/g)-1} 1 = \frac{1}{g}\log(n/g)\frac{1}{K} = \frac{1}{\bar{d}}\log^2(n/g).$$

One interpretation of the above result is as follows: since a binary tree is really a one-dimensional graph, we need α to be exactly the same as the number of dimensions in order to get an algorithmic small world.

This condition on α makes these explanatory models of an algorithmic small world brittle, in contrast to the robust explanatory model of Watts–Strogatz where p can be over a range of values and still lead to a structural small world. In our three-dimensional world, α does not always equal 3, and yet we still observe algorithmic small worlds. In the Advanced Material, we will summarize an alternative, less brittle explanatory model of an algorithmic small world.

9.3 Examples

Since the results we saw are intrinsically asymptotic ones, we have to numerically illustrate the social search model by Watts Dodds and Newman.

Fix $g = 100$ (as in the original paper), $\bar{d} = 100$ (a number close to Dunbar's number). Figure 9.10 shows how l grows with an increasing population n. Then Figure 9.11 shows the scaling behavior of l with respect to n for different values of α. We see that l grows much more slowly when α is right around 1.

9.4 Advanced Material

9.4.1 Watts–Strogatz model: Clustering coefficient and shortest path length

Consider the Watts–Strogatz model again, with n nodes, c short-range links per node, and probability p of a long-range link. We want to first count C and then approximate L.

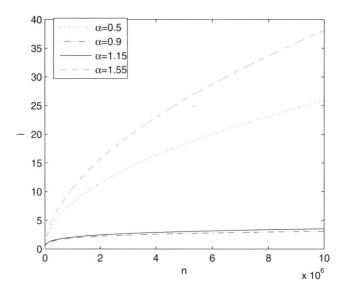

Figure 9.10 The impact of n in the Watts–Dodds–Newman model. When the population grows from 1 million to 10 million, a factor of two increase on the log scale, the length of greedy social search's path barely increases, as long as α is close to 1. But when α is a little farther from 1, l increases almost 20-fold.

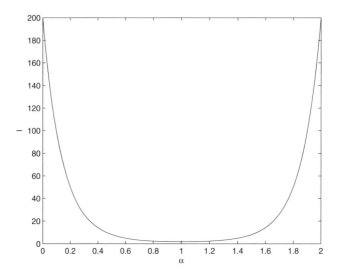

Figure 9.11 Impact of α in the Watts–Dodds–Newman model. The value of n is 1 million. When α is very close to 1, the dimension of the binary tree, l remains very small. As α moves away from 1, l quickly becomes very large.

The number of triangles remains the same irrespective of whether it is determined before or after the long-range links are added. But there are more connected triples thanks to the long-range links.

- The expected number of long-range links is $\frac{1}{2}ncp$, since for each node's short-range links there is a probability p of establishing a long-range link, and there are nc such opportunities altogether. The factor $1/2$ just avoids double counting, since each link has two end points. Each of these long-range links, together with the c short-range links per node, can form two connected triples. So we multiply $\frac{1}{2}ncp$ by $2c$, and this gives rise to nc^2p connected triples.

- The long-range links themselves can also create connected triples similar to the short-range links, with an expected total number of $\frac{1}{2}nc^2p^2$. This is because starting from each of the n nodes, we need two long-range links to create a connected triple, and the probability of that is c^2p^2.

So the clustering coefficient now becomes

$$C = \frac{\frac{n}{2}\frac{c}{2}(\frac{c}{2}-1)}{\left(\frac{n}{2}c(c-1) + nc^2p + \frac{1}{2}nc^2p^2\right)/3} = \frac{3(c-2)}{4(c-1) + 8cp + 4cp^2}.$$

Staying with the Watts–Strogatz model, the derivations of L are too technically involved for inclusion here. But the following is one of the possible approximations when ncp is large:

$$L \approx \frac{\log(ncp)}{c^2p}.$$

We can clearly see that for small p, $C(p)$ behaves like $1/p$. But $L(p)$ does not drop much as p increases, when p is small. At the same time, L grows very slowly as a function of n. This is the mathematical representation of our discussion before: C becomes large but L is small as we add just a few long-range links.

9.4.2   Generalized Watts–Strogatz model: Search path length

We have argued intuitively why the dimension of the space in which the graph lives must be equal to the rate of exponential decay in the probability of having a long-range link. But they must be *exactly* the same. Maybe if the probability-decay model is not exactly exponential, we will not need α to be exactly the same as the space dimension.

It turns out that we do not need p to decay like $r^{-\alpha}$ in order to get searchability and an algorithmic small world. Consider the Watts–Strogatz model, where α is effectively 0. Long-range links are created independently of distance, but now follow some probability distribution, e.g., binomial, geometric, Poisson, or Pareto (which we will discuss when introducing scale-free networks in the next chapter). This we call a **Generalized Watts–Strogatz model**.

For this model, we can show that l is small. In fact, we can show this through an analytic recursive formula for the entire distribution of search path lengths (not just the average), and for any general "metric space" (not just a ring or a grid).

For example, consider the special case of a ring network with n nodes and c short-range links per node. Suppose each long-range link is independently established, and the number of long-range connections, N, is drawn from a Poisson distribution with rate λ:

$$\text{Prob}(N = n) = \frac{\lambda^n e^{-\lambda}}{n!}.$$

If the source–destination distance is between kc and $(k+1)c$ for some integer k, the expected search length for such a source–destination pair has been shown to be

$$l_k = 1 + \sum_{j=1}^{k-1} \prod_{i=1}^{j} \exp(-\lambda(1 - \beta_i)),$$

where β_i is a shorthand notation:

$$\beta_i = \frac{\pi - ic/(2n)}{\pi - c/(2n)}.$$

What happens in this model, and in empirical data from experiments, is that short-range links are used, and l increases linearly but remains small, as the distance between the source and destination rises. When this distance becomes sufficiently large, long-range links start to get used, often just one or two of them and at the early part of the social search paths. Then l quickly saturates and stays at about the same level even as the source–destination distance continues to rise.

We have seen four small world models in this chapter. In the end, it is probably a combination of the Generalized Watts–Strogatz model and the Kleinberg model that matches reality the best, where we have a nested sequence of "social circles", with distance-dependent link formation within each circle and distance-independent link formation across the circles.

Summary

Box 9 Structural and algorithmic small worlds

A structural small world has a large clustering coefficient and a small shortest-path length. This can be explained through the Watts-Strogatz model with random long-range links on top of a regular graph. An algorithmic small world allows the discovery of short paths through local views, and can be explained by several models. Small worlds are surprising because short paths exist despite homophily and can even be readily found by the nodes.

Further Reading

There are many interesting popular science books on six degrees of separation: *Six Degrees*, *Linked*, *Connected*, and many more. The analytic models covered in this chapters come from the following papers.

1. The experiment by Travers and Milgram was reported in the following heavily cited paper:

J. Travers and S. Milgram, "An experimental study of the small world problem," *Sociometry*, vol. 32, no. 4, pp. 425–443, December 1969.

2. The Watts–Strogatz model is from the following seminal paper explaining the six degrees of separation:

D. J. Watts and S. H. Strogatz, "Collective dynamics of small-world networks," *Nature*, vol. 393, pp. 440–442, 1998.

3. The Kleinberg model explaining the short length of social search's path is from the following paper:

J. M. Kleinberg, "The small world phenomenon: An algorithmic perspective," in *Proceedings of ACM Symposium on Theory of Computing*, 2000.

4. The related Watts–Dodds–Newman model is from the following paper:

D. J. Watts, P. S. Dodds, and M. Newman, "Identity and search in social networks," *Science*, vol. 296, pp. 1302–1304, 2002.

5. The following recent paper presented the Generalized Watts–Strogatz model (called the Octopus Model in the paper) that explains the short length of social search's path without imposing the condition of α being equal to the dimension d of the space that the graph resides in:

H. Inaltekin, M. Chiang, and H. V. Poor, "Delay of social search on small-world random geometric graphs," *Journal of Mathematical Sociology*, vol. 36, no. 4, 2012.

Problems

9.1   *Computation of C and L* ⋆

(a) Compute the clustering coefficient C and the average shortest path L for the graph in Figure 9.12.

(b) Compute C and L for the two graphs in Figure 9.13. Contrast their values with (a).

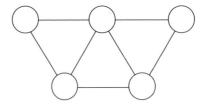

Figure 9.12 A simple graph for computing network measures.

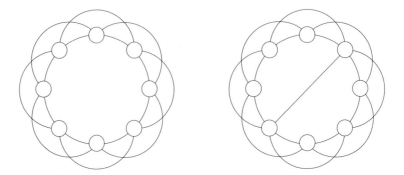

Figure 9.13 The Watts–Strogatz model with $n = 8$ and $c = 4$.

9.2 *Generalization of triadic closure* ★★

We have seen the definition of the clustering coefficient, which quantifies the amount of triadic closure. In general, "closure" refers to the intuition that if there are many pairwise connections among a set of nodes, there might be connection for any pair in the set as well. We do not have to limit closure to node-triples as in triadic closure.

Here we consider a simple extension called "quad closure." As shown in Figure 9.14; if node pairs (a, b), (a, c), (a, d), (b, c), and (b, d) are linked, then the pair (c, d) is likely to be linked too.

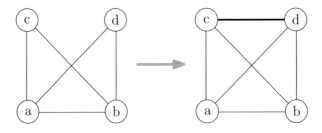

Figure 9.14 Quad closure: for nodes a, b, c, and d if currently five out of all six possible links exist (left), then the remaining link is likely to appear too (right).

To quantify the amount of quad closure, we define a "quad clustering coefficient" as

$$Q = \frac{\text{Number of cliques of size 4}}{\text{Number of connected quadruples with 5 edges}/K}$$

where K is some normalizing constant. But this definition is incomplete unless
we specify the value of K to normalize Q. Find the value of K such that the
value of Q for a clique is exactly 1.

9.3 *Metrics of class social graph* $\star\star\star$

Consider a class graph, where each student is a node, and a link between A
and B means that A and B know each other on a first-name basis before coming
to this class.

Download an anonymized class graph from
`www.network20q.com/hw/class_graph.graphml`
and, using your favorite software (e.g., NodeXL, gephi, Matlab toolboxes), or by
hand, compute C and L, compute eigenvector centrality, and partition the graph
into communities. Attach a few screenshots to show the results.

9.4 *Kleinberg model* $\star\star\star$

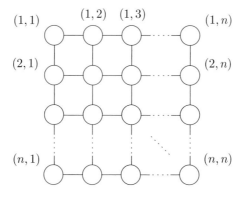

Figure 9.15 A two-dimensional lattice to illustrate the Kleinberg model.

In this question, $\log(\cdot)$ is in base 2 and $\ln(\cdot)$ is in base e. Consider the two-
dimensional lattice in Figure 9.15 with n^2 nodes labeled with their coordi-
nates: $(1,1),(1,2),\ldots,(n,n)$. The distance between two nodes $u = (x_1,y_1)$ and
$v = (x_2,y_2)$ is $d(u,v) = |x_1 - x_2| + |y_1 - y_2|$.

(a) What is the number of nodes of distance 1 from node $(1,1)$, excluding
itself? How about the numbers of nodes of distances 2 and 3, and of general
distance i, where $1 \le i \le n$?

(b) Let B_j be the set of nodes of distance at most 2^j from node $(1,1)$, excluding itself. Calculate the size of B_j for $0 \le j \le \log n$. Use the summation identity

$$\sum_{k=1}^{r} k = \frac{r(r+1)}{2}.$$

(c) Let R_j be the set of nodes contained in B_j but outside B_{j-1}, i.e., $R_j = B_j \setminus B_{j-1}$ for $j \ge 1$, and $R_0 = B_0$. Calculate a lower bound on the size of R_j. Specifically, if you obtain something of the form

$$2^s + 2^t$$

with $s > t > 0$, you should report 2^s as your answer.

(d) Let $\alpha = 2$. According to the Kleinberg model, given that two nodes u and v are separated by distance r, the probability of the nodes being connected by a random link is lower bounded as follows:

$$\Pr(u \leftrightarrow v) \ge \frac{r^{-\alpha}}{4\ln(6n)}.$$

Given that node $(1,1)$ has only one random link, use the result from (c) to lower bound the probability that node $(1,1)$ has a random link to a node in R_j, $\Pr(u \leftrightarrow R_j)$, for $0 \le j \le \log n$. Does the answer depend on the value of j?

(e) What happens to the result in (d) if $\alpha = 1$ or $\alpha = 3$?

9.5   *De Bruijn sequence, Eulerian cycle, and card magic* ★★

A *de Bruijn sequence* of order k is a binary vector of 2^k 0s or 1s, such that each sequence of k 0s or 1s appears only once in the vector (wrapping around the corner). For example, 0011 is a de Bruijn sequence of order $k = 2$ because each of the sequences $00, 01, 10, 11$ appears only once, as shown in Figure 9.16.

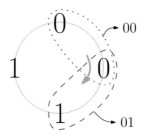

Figure 9.16 Illustration of a de Bruijn sequence. The area with the dotted/dashed line is a window which shifts in the clockwise direction, covering all possible combinations of 0s and 1s of length 2.

There are two basic questions regarding de Bruijn sequences: for any order k, (1) do they exist? and (2) how can we find one?

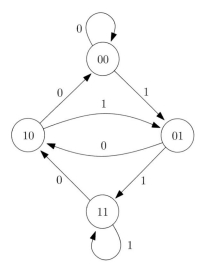

Figure 9.17 A de Bruijn graph that can help visualize and construct de Bruijn sequences of desired lengths.

The answers can be obtained by studying a *de Bruijn graph*. Consider the case of $k = 3$ with the corresponding two-dimensional de Bruijn graph shown in Figure 9.17. We traverse the graph (starting at any node) while writing down the labels (0 or 1) on the edges. It is not difficult to see that for any **Eulerian cycle** on the graph, i.e., a traversal of the nodes through a cycle using every edge once and only once, the corresponding sequence of edge labels is a de Bruijn sequence. Hence the problem of finding a de Bruijn sequence reduces to finding an Eulerian cycle in the corresponding de Bruijn graph, and this answers question (2). For question (1), the answer is affirmative if every de Bruijn graph has an Eulerian cycle, which indeed is true because each node's in-degree and out-degree are equal (a basic result in graph theory due to Euler).

So what are de Bruijn sequences good for? Among their important applications is a famous card trick, where the magician can name the card held by k people in the audience even after random cuts have been made to the deck of cards. (The details can be found in a unique book by Persi Diaconis and Ron Graham: *Magical Mathematics: The Mathematical Ideas that Animate Great Magic Tricks*, Princeton University Press, 2011.)

Now, your task for this homework problem is simple: For $k = 3$ there are two distinct de Bruijn sequences. What do we mean by "distinst sequences"? Sequences 01011 and 00111 are distinct, but sequences 01011 and 10101 are not (try to write out the sequences as in Figure 9.16). Draw two distinct Eulerian cycles on the graph in Figure 9.17, and report the two distinct de Bruijn sequences found.

10 Does the Internet have an Achilles' heel?

10.1 A Short Answer

It does not.

10.2 A Long Answer

10.2.1 Power-law distribution and scale-free networks

Sure, the Internet has many security loopholes, from cyber-attack vulnerability to privacy-intrusion threats. But it does not have a few highly-connected routers in the center of the Internet that an attacker can destroy to disconnect the Internet, which would have fit the description of an "Achilles' heel". So why would there be rumors that the Internet has an Achilles' heel?

The story started in the late 1990s with an inference result: the Internet topology exhibits a **power-law distribution** of node degrees. Here, the "topology" of the Internet may mean any of the following:

- the graph of webpages connected by hyperlinks (like the one we mentioned in Chapter 3),
- the graph of Autonomous Systems (ASs) connected by the physical and business relationships of peering (we will talk more about that in Chapter 13), and
- the graph of routers connected by physical links (the focus of this chapter).

For the AS graph and the router graph, the actual distribution of the node degrees (think of the histogram of the degrees of all the nodes) is not clear due to measurement noise. For example, the AS graph data behind the power-law distribution had more than 50% of links missing. Internet exchange points further lead to many peering links among ASs. These are "shortcuts" that enable settlement-free exchange of Internet traffic, and cannot be readily measured using standard network-layer measurement probes.

To talk about the Achilles' heel of the Internet, we have to focus on the graph of routers as nodes, with physical links connecting the nodes. No one knows for sure what that graph looks like either, so people use proxies to estimate it through measurements like trace-route. Studies have shown that such estimates

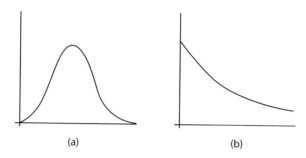

Figure 10.1 (a) Gaussian vs. (b) long-tail distribution. A Gaussian distribution has a characteristic scale, e.g., the standard deviation from the mean, whereas a long-tail distribution does not.

lead to biased sampling in the first place, due to the way the Internet protocol reacts to trace-route measurements. In addition, there are other measurement deficiencies arising from resolving address ambiguity. There is also no scalable measurement platform with enough vantage points at the network edge to detect the high-degree nodes there. Therefore, it remains unclear whether the Internet router graph has a power-law degree distribution or not.

However, in this chapter we will assume that these graphs really follow the power-law distribution of their node degrees. Even then, the actual graphs and their properties can be tricky to analyze. Furthermore, there are many layers of protection, and the timescale of recovery from node failure matters. In fact, the time it takes routing protocols to converge and the bugs in the control signaling causes more failure than certain high-degree nodes being under attack. But in this chapter, suppose we can just focus on the topological impact of power-law distribution of node degrees.

So, what is a power-law distribution? Many distributions, like the Gaussian and exponential distributions, have a characteristic scale, defined by the mean and standard deviation as shown in Figure 10.1(a). And the probability that a random variable X following such a distribution has a value above a given number x, i.e., the *tail probability*, becomes small very quickly as x becomes large. It is not likely to be far away from the mean. This leads to what is called a **homogeneous network**, defined here as a network where the degrees of the nodes are more or less similar.

In sharp contrast, as shown in Figure 10.1(b), a **long-tail distribution** does not have a characteristic scale. If the node degrees follow a long-tail distribution, we say the graph represent a **scale-free network**. It is an **inhomogeneous network** in its node degrees. The tail probability, $\text{Prob}[X \geq x]$, of a long-tail distribution exhibits the following power-law characteristics with x raised to the power $-\alpha$:

$$\text{Prob}[X \geq x] \approx kx^{-\alpha},$$

where k is a normalization constant, α is the power exponent, and $\approx$ here means "equal" in the limit as x becomes large, i.e., it *eventually* follows a power-law distribution as we keep going down the tail.

The most famous special case of a long-tail distribution is the **Pareto distribution**, with the following tail distribution for $x \geq k$:

$$\text{Prob}[X \geq x] = \left(\frac{x}{k}\right)^{-\alpha}. \tag{10.1}$$

Differentiating $1 - \text{Prob}[X \geq x]$, we see that the probability density function (pdf) for a Pareto distribution also follows the power law, with the power exponent $-(\alpha + 1)$:

$$\text{Prob}[X = x] = p(x) = \alpha k^\alpha x^{-(\alpha+1)}. \tag{10.2}$$

In sharp contrast, for any $x \in (-\infty, \infty)$, the Gaussian pdf is

$$p(x) = \frac{1}{\sqrt{2\pi}\sigma} e^{-\frac{(x-\mu)^2}{2\sigma^2}},$$

where μ is the mean and σ the standard deviation.

Just to get a feel (ignoring the normalization constant) for the difference between the exponential of square and 1 over square: following the Gaussian distribution's shape, we have $e^{-x^2} = 0.018$ when $x = 2$. But following the Pareto distribution's shape, we have $x^{-2} = 0.25$ when $x = 2$, a much larger number.

We can plot either the tail distribution (10.1), or the pdf (10.2), on a log–log scale, and get a straight line for the Pareto distribution. This can be readily seen:

$$\log \text{Prob}[X \geq x] = -\alpha \log x + \alpha \log k.$$

The slope is $-\alpha$, as shown in Figure 10.2. A straight line on a log–log plot is the visual signature of a power-law distribution. It has been reported that the

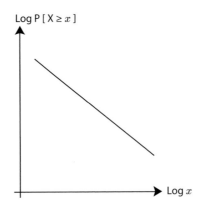

Figure 10.2 Pareto distribution on a log–log plot is a straight line. This is the visual signature of any long-tail, or power-law, distribution.

power exponent is -2.1 for in-degree of webpage graph, -2.4 for out-degree of webpage graph, and -2.38 for router graph.

Just to clarify: a scale-free network is *not* the same as a small world network. As we saw in the last chapter, a network is called a small world if short paths between any pair of nodes exist and can be locally discovered. A network is called scale-free if its node degree distribution follows a power law, such as the Pareto distribution.

- "Scale-free" is a topological property concerning just the node degree distribution. It is not a functionality property. A scale-free network does not have to be a small world.
- "Small world" is a topological *and* functionality property that can arise from different node degree distributions. For example, the Watts–Strogatz graph is a small world but not scale-free, as its node degree distribution does not follow a power law.
- For some social networks, evidence suggests that they are scale-free small-worlds. For example, the last model in Chapter 9, the Generalized Watts–Strogatz model, can generate such networks.

Back to scale-free networks. Researchers have worked out different generative models, but one of them, the preferential attachment model, gained the most attention. As explained in detail in the Advanced Material, preferential attachment generates a type of scale-free network that has highly connected nodes in the center of the network.

In 2000, researchers in statistical physics suggested that, unlike networks with an exponential distribution of node degrees, scale-free networks are robust against random errors, since the chances are that the damaged nodes are not highly connected enough to cause too much damage to the network. But a deliberate attack that specifically removes the most-connected nodes will quickly fragment the network into disconnected pieces. The nodes with large degrees sitting in the center of the network become easy attack targets to break the whole network. Since the Internet graph at router level follows a power-law degree distribution, it, too, must be vulnerable to such attacks on its Achilles' heel.

The argument sounds plausible and the implication alarming, except that it does *not* fit reality.

10.2.2 The Internet's reality

Two networks can have power-law (node) degree distributions and yet have very different features otherwise, with very different implications for functional properties, such as robustness against attacks.

For example, what if the high variability of node degrees happens at the network *edge* rather than at the center? That is unlikely if networks were randomly generated according to the model of preferential attachment, where nodes attach

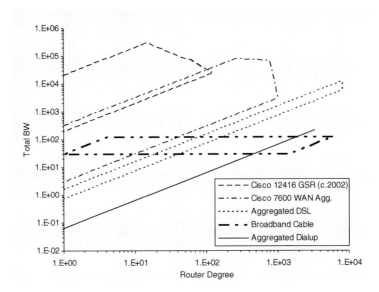

Figure 10.3 Bandwidth vs. connectivity constraint sets for different routers from Alderson et al. (2005). State-of-the-art core routers' capabilities in 2003 are shown. As the degree of the node (i.e., port number per router) goes up, the total bandwidth rises initially, but then drops as more ports force the per-port bandwidth to decrease.

to more popular nodes. In the space of all graphs with power-law degree distributions, degree variability is more likely to arise out of the difference between highly-connected core nodes and sparsely-connected edge nodes.

But what if the unlikely topology is the actual design? There are technical constraints to the router topology. Cisco and Juniper cannot make a router that has *both* a large degree and a large bandwidth per connection. Each router has a limitation on its total bandwidth: the maximum number of packets it can process at each time. So there is an inevitable tradeoff between the number of ports (node degree) and the speed of each port (bandwidth per connection) on the router. This is illustrated in Figure 10.3.

In addition, the Internet takes layers of aggregation to smooth individual users' demand fluctuations through statistical multiplexing. A user's traffic goes through an access network like WiFi, cellular, DSL, or fiber networks, then through a metropolitan network, and, finally, enters the core backbone network. Node bandwidth is related to its placement. The bandwidth per connection in the routers goes up along the way from edge to core. This implies that nodes in the core of the Internet must have large bandwidth per connection, and thus small degree. For example, an AT&T topology in 2003 showed that the maximum degree of a core router was only 68 while the maximum degree of an edge router was 313, almost five times as large.

In summary, access network node degrees have high variability. Core network node degrees do not. Attacks on high-degree nodes can only disrupt access routers

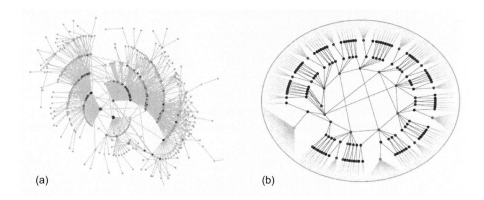

(a)   (b)

Figure 10.4 Achilles' heel or not, from Alderson et al. (2005): (a) is a typical topology from preferential attachment mechanism; (b) is a typical topology of the real Internet. Both satisfy a power-law degree distribution, but the left one has an Achilles' heel whereas the right one does not.

and do not disrupt the entire network. Attacks on medium-to-small-degree nodes have a high chance of hitting the access routers because there are many more of them than there are core routers.

Moreover, robustness of a network is not just about the connectivity pattern in the graph. There are also protocols that take care of detecting and mitigating failures even when routers are down, as we will briefly discuss in a homework problem.

To summarize, the flaws of "The Internet has an Achilles' heel" are three-fold.

- (1) Incomplete measurements skew the data.
- (2) A power-law degree distribution does not imply preferential attachment.
- (3) Functional protection sits on top of topological properties. The question of robustness of the Internet has as much to do with protocols as with graphs.

Certainly, there are all kinds of serious attacks on the Internet, including on core routers. But the vulnerability is not because core routers have unusually high degrees compared with the other nodes of the Internet and therefore become an Achilles' heel.

10.2.3   Functionality model

The Internet might be viewed as "self-organizing," but that self-organization is achieved by designs based on constrained optimization and protocol construction. There is a way to more precisely define the tradeoff between the *performance* and the *likelihood* of vastly different topologies all exhibiting power-law degree distributions.

One of the several ways to capture the aggregate throughput of the Internet, as its performance metric, is through the following optimization problem:

$$\begin{aligned}
\text{maximize} \quad & \sum_i x_i \\
\text{subject to} \quad & \sum_i R_{ki} x_i \leq b_k, \quad \forall k \\
& x_i = \rho y_{S_i} y_{D_i}, \quad \forall i \\
\text{variables} \quad & \rho, \{x_i\}.
\end{aligned} \tag{10.3}$$

Here, an end-to-end session i's rate x_i is proportional to the overall traffic demand y_{S_i} at its source node S_i and y_{D_i} at its destination node D_i (i.e., $\mathbf{x}$ is generated by $\mathbf{y}$ in this traffic model), ρ being the proportionality coefficient and the actual optimization variable. Each entry in the routing matrix R_{ki} is 1 if session i passes through router k, and 0 otherwise. The constraint values $\{b_k\}$ capture the router bandwidth–degree feasibility constraint. The resulting optimal value of the above problem is denoted as $P(G)$: the (throughput) *performance* of the given graph G. Clearly, $P(G)$ is determined by the throughput vector $\mathbf{x}$, which is in part determined by the routing matrix $\mathbf{R}$, which is in turn determined by the graph G.

On the other hand, we can define the *likelihood* $S(G)$ of a graph G with node degrees $\{d_i\}$ as

$$S(G) = \frac{s(G)}{s_{max}},$$

where

$$s(G) = \sum_{(i,j)} d_i d_j$$

captures the pairwise connectivity by summing the degree products $d_i d_j$, over all (i, j) node pairs that have a link between them, similar to the metric of modularity in Chapter 8. And s_{max} is simply the maximum $s(G)$ among all the graphs that have the same set of nodes and the same node degrees $\{d_i\}$. These $\{d_i\}$, for example, can follow a power-law distribution.

Now we have $P(G)$ to represent the performance and $S(G)$ to represent the likelihood of a scale-free network. As shown in Figure 10.5, if we were drawing a network from the set of scale-free networks at random, high-performance topologies like the one exhibited by the real Internet are much less likely to be drawn. But the Internet was *not* developed by such a random drawing. It came through constrained-optimization-based design. The performance of the topology generated by (10.3) is two orders of magnitude higher than that of the random graph generated by preferential attachment, even though it is less than one-third as likely to be picked at random.

The poor performance of graphs with highly-connected core nodes is actually easy to see: a router with a large degree cannot support high-bandwidth links, so if it sits in the core of the network, it becomes a performance bottleneck for the whole network. Since "Achilles' heel nodes" would also have been performance-bottleneck nodes, they were simply avoided in the engineering of the Internet. (We will see this network scalability problem again in Chapter 16, when we discuss cloud services and large-scale data centers.)

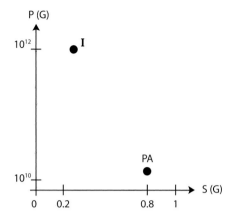

Figure 10.5 Performance vs. likelihood of two topologies, both following a power-law degree distribution. Topology I, which stands for the Internet, is much less likely to be picked up if a graph is drawn at random from the set of graphs a satisfying power-law degree distribution (as measured by $S(G)$). But it also has much higher performance (as measured by total throughput $P(G)$). Topology PA, which stands for Preferential Attachment, is much more likely but has much lower performance.

In summary, the Internet router graph is performance-driven and technologically constrained. Concluding that there is an Achilles' heel is a story that illustrates the risk of overgeneralizing in a "network science" without domain-specific functionality models.

As to the AS graph, the topology can be quite different as it is driven by inter-ISP pricing economics. The webpage graph exhibits yet another topology, increasingly shaped by search-engine optimizers in response to Google's search methods in Chapter 3, and the extremely popular aggregation websites like Wikipedia and YouTube that we saw in Chapters 6 and 7.

10.2.4 Choices of generative models

The story of the Internet's (non-existing) Achilles' heel is part of a bigger picture about generative models of power-law distribution: what might have given rise to these ubiquitous power-law distributions in graphs we find in many areas? It dates back to over a century ago, when the linguist Zipf first documented that the kth most popular word in many languages roughly has a frequency of $1/k$ in its appearance. Since then, Pareto, Yule, and others have documented more of these power laws in the distributions of income and of city size. In the 1950s, there was an interesting debate between Mandelbrot and Simon on the choice of generative models for power law, and on the modeling abilities of fractal/Pareto vs. Poisson/Gaussian. These questions have continued to the present day.

There are several explanatory models of scale-free networks, two of which are well-established and complementary. As illustrated in Figure 10.6, either model

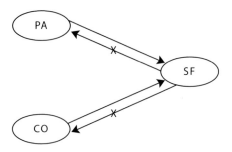

Figure 10.6 A simple logical relationship: Either preferential attachment (PA) or constrained optimization (CO) can lead to a scale-free network (SF), but a network being scale-free does not imply either model is the correct one for that network.

can lead to a scale-free network, but knowing a network is scale-free does not imply that either model is the correct one for that network.

- **Preferential Attachment (PA):** The key idea is that as new nodes are added to a graph, they are more likely to connect to those with a lot of connections already. Conceptually, it says that a self-organizing growth mechanism of graphs leads to a power-law distribution. Mathematically, it turns out that sampling by density leads to a difference equation whose equilibrium satisfies a power law, as we will see in Section 10.4.1.
- **Constrained Optimization (CO):** The key idea is that the graph topology is designed with some objective function and constraints in mind, and the resulting topology shows a power-law distribution. Conceptually, it says that a power law is a natural outcome of constrained optimization. Mathematically, either entropy maximization or isoelastic utility maximization under linear constraints gives rise to a power-law distribution, as we will see in Section 10.4.2. In fact, constrained optimization is not just one single model but a general approach to model functionalities.

The Advanced Material section will continue this discussion with the debate between preferential attachment and constrained optimization as two options of generative models.

10.3   Examples

Suppose we have a network with graph G_1 shown in Figure 10.7(a). We will calculate $S(G_1)$ and $P(G_1)$.

First of all, we have

$$s(G_1) = d_1 d_2 + d_1 d_3 + d_1 d_5 + d_2 d_4 + d_3 d_4$$
$$= (3)(2) + (3)(2) + (3)(1) + (2)(2) + (2)(2) \qquad (10.4)$$
$$= 23.$$

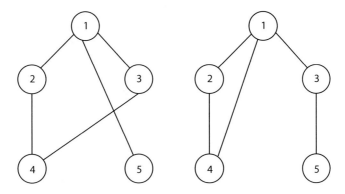

Figure 10.7 Two network topologies G_1 on the left and G_2 on the right, both with node degrees $\{3, 2, 2, 2, 1\}$. G_1 has a smaller S, but a much larger P, than G_2 that maximizes S but suffers from a small P.

Among all the (connected) graphs with node degree distribution the same as G_1, the graph with the greatest $S(G)$ is shown in Figure 10.7(b). Call this graph G_2. A graph that maximizes $S(G)$ generally has more connections between nodes of higher degree. For example, we should swap the 1 in the third term in (10.4) with the 2 in the fifth term. We can readily reason that the largest $s(G)$ must be

$$
\begin{aligned}
s_{max} &= s(G_2) \\
&= d_1 d_2 + d_1 d_3 + d_1 d_4 + d_2 d_4 + d_3 d_5 \\
&= (3)(2) + (3)(2) + (3)(2) + (2)(2) + (2)(1) \\
&= 24.
\end{aligned}
\tag{10.5}
$$

Therefore, $S(G_1) = \frac{s(G_1)}{s_{max}} = \frac{23}{24}$.

Now we turn to calculating $P(G_1)$. Assume we have sessions that wish to connect from each node to each other node, so ten sessions in total. Let session 1 denote the connection between nodes 1 and 2, session 2 denote the connection between nodes 1 and 3, and so forth. Let the node demands be denoted by y_j Gbps, with values $y_1 = 5, y_2 = 2, y_3 = 4, y_4 = 4$, and $y_5 = 2$. Let the bandwidth constraint be $b_k = d_k$ for all k. So $b_1 = 3$ Gbps, $b_2 = b_3 = b_4 = 2$ Gbps, and $b_5 = 1$ Gbps.

To build the routing matrix $\mathbf{R}$, we determine the shortest path between each source–pair destination and write down which routers the path uses. Ties are broken arbitrarily. For example, the shortest path in graph G_1 from node $2 \to 5$ is $2 \to 1 \to 5$. That translates to the 7th column $[1\ 1\ 0\ 0\ 1]^T$ in the routing matrix.

After the above construction, the routing matrix is as follows, with ten sessions, one per column, and five nodes, one per row. For example, the first column corresponds to session 1, from node 1 to node 2; and the first row corresponds to node 1.

$$\mathbf{R} = \begin{bmatrix} 1 & 1 & 1 & 1 & 1 & 0 & 1 & 0 & 1 & 1 \\ 1 & 0 & 1 & 0 & 1 & 1 & 1 & 0 & 0 & 0 \\ 0 & 1 & 0 & 0 & 1 & 0 & 0 & 1 & 1 & 1 \\ 0 & 0 & 1 & 0 & 0 & 1 & 0 & 1 & 0 & 1 \\ 0 & 0 & 0 & 1 & 0 & 0 & 1 & 0 & 1 & 1 \end{bmatrix}. \tag{10.6}$$

The demand vector is found by multiplying appropriate entries of y_k. For example, since session 10 routes from node 4 (S_{10}) to node 5 (D_{10}), we multiply $y_4 \times y_5$ to find the demand constraint $y_{S_{10}} y_{D_{10}}$.

The optimization problem is therefore to

$$\text{maximize} \quad x_1 + x_2 + x_3 + x_4 + x_5 + x_6 + x_7 + x_8 + x_9 + x_{10}$$

$$\text{subject to} \quad \begin{bmatrix} 1 & 1 & 1 & 1 & 1 & 0 & 1 & 0 & 1 & 1 \\ 1 & 0 & 1 & 0 & 1 & 1 & 1 & 0 & 0 & 0 \\ 0 & 1 & 0 & 0 & 1 & 0 & 0 & 1 & 1 & 1 \\ 0 & 0 & 1 & 0 & 0 & 1 & 0 & 1 & 0 & 1 \\ 0 & 0 & 0 & 1 & 0 & 0 & 1 & 0 & 1 & 1 \end{bmatrix} \begin{bmatrix} x_1 \\ x_2 \\ x_3 \\ x_4 \\ x_5 \\ x_6 \\ x_7 \\ x_8 \\ x_9 \\ x_{10} \end{bmatrix} \leq \begin{bmatrix} 3 \\ 2 \\ 2 \\ 2 \\ 1 \end{bmatrix}$$

$$\mathbf{x} = \rho [10 \ 20 \ 20 \ 10 \ 8 \ 8 \ 4 \ 16 \ 8 \ 8]^T$$

variables $\quad \rho, \mathbf{x}.$

Solving this numerically, we find that

$$\mathbf{x}^* = \begin{bmatrix} 0.33 & 0.67 & 0.67 & 0.33 & 0.27 & 0.27 & 0.13 & 0.53 & 0.27 & 0.27 \end{bmatrix}^T \text{ Gbps,}$$

and $\rho^* = 0.033$. The maximized objective function value gives $P(G_1) = 3.73$ Gbps.

Now, we repeat the procedure to find $S(G_2)$ and $P(G_2)$. By definition, $S(G_2) = 1$. For $P(G_2)$, the routing matrix changes, and the optimization problem becomes

$$\text{maximize} \quad x_1 + x_2 + x_3 + x_4 + x_5 + x_6 + x_7 + x_8 + x_9 + x_{10}$$

$$\text{subject to} \quad \begin{bmatrix} 1 & 1 & 1 & 1 & 1 & 0 & 1 & 1 & 0 & 1 \\ 1 & 0 & 0 & 0 & 1 & 1 & 1 & 0 & 0 & 0 \\ 0 & 1 & 0 & 1 & 1 & 0 & 1 & 1 & 1 & 1 \\ 0 & 0 & 1 & 0 & 0 & 1 & 0 & 1 & 0 & 1 \\ 0 & 0 & 0 & 1 & 0 & 0 & 1 & 0 & 1 & 1 \end{bmatrix} \begin{bmatrix} x_1 \\ x_2 \\ x_3 \\ x_4 \\ x_5 \\ x_6 \\ x_7 \\ x_8 \\ x_9 \\ x_{10} \end{bmatrix} \leq \begin{bmatrix} 3 \\ 2 \\ 2 \\ 2 \\ 1 \end{bmatrix}$$

$$\mathbf{x} = \rho [10 \ 20 \ 20 \ 10 \ 8 \ 8 \ 4 \ 16 \ 8 \ 8]^T$$

variables $\quad \rho, \mathbf{x}$

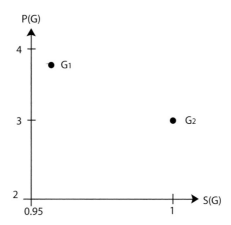

Figure 10.8 $P(G)$ versus $S(G)$ for the simple topology in our small numerical example. The less likely graph is the one with much higher total throughput.

Solving this numerically, we find that

$$\mathbf{x}^* = \begin{bmatrix} 0.27 & 0.54 & 0.54 & 0.27 & 0.22 & 0.22 & 0.11 & 0.43 & 0.22 & 0.22 \end{bmatrix}^T \text{Gbps},$$

and $\rho^* = 0.027$. Now the total throughput is $P(G_2) = 3.03$ Gbps.

We compare graph G_1's and graph G_2's performance metrics with their likelihood metrics in Figure 10.8. We are constrained to a small example that allows us to run through a detailed numerical illustration. So the dynamic range of $S(G)$ is small. But still, we see that, compared with G_2, G_1 has a smaller likelihood of getting picked if a graph is randomly drawn from the set of all the graphs with the given degree distributions, but it has a much higher performance in terms of total throughput.

10.4   Advanced Material

Underlying the Achilles-heel-Internet conclusion is the following chain of logical implications: networks being scale-free implies that they arise from the mechanism of preferential attachment, and preferential attachment always leads to an Achilles heel. Where is the logical fallacy in this chain of reasoning? While preferential attachment generates scale-free networks, scale-free networks do not imply the necessity of preferential attachment. There are other mechanisms that generate scale-free networks just like preferential attachment, but lead to other properties that fit reality better. In the case of the Internet's router graph, preferential attachment necessarily leads to performance bottlenecks, whereas the Internet must avoid such bottlenecks because having core routers with large degree implies that the per-port bandwidth must be small. We now walk through both the preferential-attachment model and the alternative of constrained-optimization model.

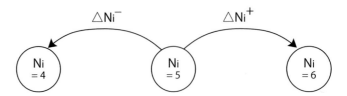

Figure 10.9 An illustration of a key step in the derivation that preferential attachment leads to heavy-tail degree distribution. The number of nodes with in-degree i is denoted by N_i, and it can go up or down as a new node arrives.

10.4.1   Preferential attachment generates a power law

Suppose a graph starts with just one node with a link back to itself. At each of the discrete timeslots, a new node shows up. With probability $\theta \in [0, 1]$, it picks an existing node in proportion to the in-degree of each node. With probability $\bar{\theta} = 1 - \theta$, it picks an existing node randomly (similar to the randomization component of PageRank). Then it connects to the node picked. The bigger θ, the more prominent is the preferential attachment effect: a new node is more likely to connect to a popular node. The smaller θ, the more prominent is the random attachment effect.

Now, look at the evolution of $N_i(t)$, the number of nodes with in-degree i when there are t nodes in the network (t is also the timeslot counter since we add one node each time). We will see that attaching to more popular nodes naturally makes them even more popular, which makes them even more likely to be attached to by future new nodes. This leads to the power-law distribution.

Mathematically, this can be readily reasoned. N_i increases by 1 if the new node picks an existing node with $i - 1$ in-degree, thus pushing its in-degree to i. The expected increase in N_i is

$$\Delta N_i^+ = \theta \frac{(i-1)N_{i-1}}{t} + \bar{\theta}\frac{N_{i-1}}{t} = \frac{\theta(i-1)N_{i-1} + \bar{\theta}N_{i-1}}{t}.$$

The key observation is that $(i - 1)$ multiplies N_{i-1} in the first term. This is the source of the "more popular gets more popular" phenomenon.

On the other hand, N_i decreases by 1 if the new node picks an existing node with i in-degree, thus pushing its in-degree to $i+1$. The expected decrease in N_i is

$$\Delta N_i^- = \frac{\theta i N_i + \bar{\theta}N_i}{t}.$$

The net change in N_i is the difference of the above two expressions, also illustrated in Figure 10.9:

$$\Delta N_i = \Delta N_i^+ - \Delta N_i^- = \frac{\theta((i-1)N_{i-1} - iN_i) + \bar{\theta}(N_{i-1} - N_i)}{t}. \qquad (10.7)$$

At the equilibrium of the growth rates of nodes with different degrees, we can express the number of nodes with i in-degrees as follows:

$$N_i(t) = p_i t,$$

where $p_i \in [0, 1]$ is the proportion of nodes with in-degree i. So the net change of N_i with respect to t is just p_i:

$$\frac{\partial N_i(t)}{\partial t} = p_i.$$

But we just showed that the same quantity can be expressed as in (10.7) too. Setting $p_i =$(10.7) and simplifying the equation, we have

$$p_i(1 + \bar{\theta} + i\theta) = p_{i-1}(\bar{\theta} + (i - 1)\theta). \tag{10.8}$$

This provides the relation between p_i and i that we are looking for as we try to construct the distribution of node degrees. The key observation is that $p_i i$ shows up on the left side, and $p_{i-1}(i - 1)$ on the right.

From this balance equation (10.8), we can examine the ratio p_i/p_{i-1}:

$$\frac{p_i}{p_{i-1}} = 1 - \frac{1 + \theta}{1 + \bar{\theta} + i\theta}.$$

Since we care about the tail of the distribution, we focus on the asymptote where i becomes large and dominates the denominator:

$$\frac{p_i}{p_{i-1}} \approx 1 - \left(\frac{1 + \theta}{i\theta}\right). \tag{10.9}$$

Let us see what kind of distribution of $\{p_i\}$ satisfies the above equation. Maybe a power-law distribution? If the distribution of $\{p_i\}$ follows a power law,

$$k_1 i^{-(1+\theta)/\theta}, \tag{10.10}$$

where k_1 is a normalization constant independent of i, then the asymptote (10.9) is indeed satisfied, since for large enough i, we have

$$\frac{p_i}{p_{i-1}} = \left(\frac{i-1}{i}\right)^{\frac{1+\theta}{\theta}} = \left(1 - \frac{1}{i}\right)^{\frac{1+\theta}{\theta}} \approx 1 - \left(\frac{1+\theta}{\theta}\right)\left(\frac{1}{i}\right),$$

where the approximation $\approx$ is by Taylor's expansion. We can also verify that (10.10) is also the only way to satisfy this asymptote.

In addition to the probability distribution $\{p_i\}$, the tail of the distribution $q_j = \sum_{i \geq j} p_i$ also follows a power law and is proportional to

$$k_2 j^{-1/\theta}, \tag{10.11}$$

where k_2 is another normalization constant independent of j.

For example, for the power exponent to be -2, θ should be 0.5: follow preferential attachment as much as random attachment.

10.4.2   Constrained optimization generates a power law

While preferential attachment is a plausible model that can generate a power-law distribution, it is not the only one. Another major generative model, especially for engineered networks, is constrained optimization. The power-law distribution of node degrees then follows as the consequence of a design that maximizes some objective function subject to technological and economic constraints. The optimization model for the Internet's router topology earlier in this chapter is such an example.

There are more examples, mathematically simple and precise. We will go through one now (maximizing entropy subject to linear cost constraints), and save another for homework (maximizing utility functions subject to linear resource constraints). We can also turn a constrained optimization into a multi-objective optimization, where, instead of forming a constraint, we push it into the objective function. Either way, it is a model of a tradeoff: the tradeoff between *performance* and resource *cost*.

Suppose we want to explain the empirically observed power-law distribution of word lengths in a code or a language. For simplicity of exposition, assume each word has a different length. A word with length i has a probability p_i of being used, and the cost of the word is the number of bits needed to describe it: $\log i$. The average cost is $\sum_i p_i \log i$. The information conveyed by the code is often measured by its entropy: $-\sum_i p_i \log p_i$. We now have the following constrained optimization, where C is a given upper bound on the average cost:

$$
\begin{aligned}
\text{maximize} \quad & -\sum_i p_i \log p_i \\
\text{subject to} \quad & \sum_i p_i \log i \leq C \\
\text{variables} \quad & \{p_i\}.
\end{aligned}
\tag{10.12}
$$

We can normalize **p** later.

We can figure out the structure of the optimizer $\{p_i^*\}$ to the above problem by looking at the Lagrange dual problem, an important approach that we will motivate and explain in detail in Chapter 14. But briefly, we weight the constraints with a Lagrange multiplier λ, and hope that

- when this multiplier is chosen properly, maximizing a weighted sum of the objective function and the constraint will be equivalent to solving the original constrained optimization problem; and
- even when this multiplier is not picked to be exactly the "right" ones, useful structures of the optimal solution can still be revealed.

First we form the weighted sum L, with a Lagrange multiplier $\lambda > 0$:

$$
L(\mathbf{p}, \lambda) = -\sum_i p_i \log p_i - \lambda \left(\sum_i p_i \log i - C \right).
$$

From Lagrange duality theory, it is known that the optimizer $\{p_i^*\}$ to (10.12) must be a maximizer of L. So, taking the partial derivative of L with respect to p_j, we have

$$-\frac{\partial L}{\partial p_j} = \frac{\partial(p_j \log p_j)}{\partial p_j} + \lambda \frac{\partial(p_j \log j)}{\partial p_j}$$
$$= \log p_j + 1 + \lambda \log j.$$

Setting it to zero at $\mathbf{p}^*$, we have

$$p_j^* = \exp(-\lambda \log j - 1) = \frac{1}{e} j^{-\lambda}, \ \forall j,$$

a power-law distribution with a power exponent $-\lambda$, just as we had hoped for.

10.4.3 Broader implications

Which is the "true" explanation of a power law: preferential attachment (PA) or constrained optimization (CO)? There is no universally true explanation for all networks. If the network at hand has some elements of design or optimization, CO is more insightful. If not, PA is the more natural explanation. Since they are "just" explanatory models and both options give rise to the same power law, either could work for that purpose, and it does not matter which we choose. The only way to differentiate between the two is to compare attributes beyond just the power law.

Here lies a relatively under-explored topic, as compared with gathering data and plotting the frequency on a log–log scale to detect a straight line: what kind of *predictions* can generative models make about topological or functionality properties of a given network? In the case of Internet topology, empirical data show that CO correctly predicts the Internet topology, whereas PA does not.

This is part of an even bigger picture of the debate between **Self-Organized Criticality** (SOC) and **Highly Optimized Tolerance** (HOT) models in the study of networks and complexity. In short, in SOC theory, complex behaviors emerge from the dynamics of a system evolving through "meta-stable" states into a "critical" state. In HOT theory, complex behaviors emerge from constrained design aiming at high efficiency and robustness against designed-for uncertainty (but has high sensitivity to design flaws and unanticipated perturbations). If the system exhibits non-generic, structured configurations, HOT is a better model. A fundamental watershed is that the source of randomness is topological in SOC but functional in HOT.

Power laws can be found in very different types of networks:

- biological networks, e.g., brain, species population;
- human or business networks, e.g., webpage, AS, citation, city growth, income distribution;

- technology networks, e.g., the Internet at the router level; and
- a mixture of natural and engineered networks, e.g., forest fire, language.

The third and fourth types of networks tend to have strong constraints and key objectives, and HOT tends to explain them better. Even the first and second types of networks often have constraints, e.g., how many friends one individual can keep is upper bounded by a human being's ability to keep track of friends, thus cutting off the long tail of the degree distribution.

The real question is whether the long tail is still there in the regime that matters to the network functionality under study. However, in some areas in economics and sociology, we cannot easily run controlled, reproducible experiments to falsify a theory, and there are too many self-consistent yet mutually incompatible theories to differentiate using historical data.

There are other ways to generate power laws beyond the two in this chapter. This begs the question: if power laws are so universally observed and easily generated, maybe we should no longer be surprised by discovering them, just like we are not surprised when observing that the sums of many independent random variables (with finite means and variances) roughly follow the Gaussian distribution. This is especially true when networks having power laws can still have diagonally opposite behaviors in properties that matter, such as resilience against attack (in the case of router topology) and peering economics (in the case of AS topology).

Before leaving this chapter, we highlight two interesting messages. First, this and the previous chapter are mostly about *reverse engineering* of network topologies, and we saw the use of optimization in explanatory models. In Chapter 14, we will also see reverse-engineering of network protocols.

Second, we see the importance of domain-specific knowledge that can correct misleading conclusions drawn from a generic network science, the importance of reality checking and falsification against generic topological properties, and the importance of protocol specifics on top of the generic principle of self-organization.

Summary

Box 10 Generative models of scale-free networks

A long tail distribution of node degrees in scale-free networks can be explained by preferential attachment or constrained optimization, but does not imply that either must be true. These two models lead to different conclusions about the robustness of the network. Robustness is as much about the protocols and functionality models as the graph and topological models.

Further Reading

There is a vast literature on power-law and long-tail distributions, and on scale-free graphs found in biological, social, and technological networks.

1. The following paper suggested the existence of an Achilles' heel in the Internet's router level topology:

R. Albert, H. Jeong, and A.-L. Barabasi, "Error and attack tolerance of complex networks," *Nature*, vol. 406, pp. 378–382, 2000.

2. The following paper refuted the above paper in theory and through data. It also developed an optimization model of the Internet's router level topology that we followed in this chapter:

D. Alderson, L. Li, W. Willinger, and J. C. Doyle, "Understanding Internet topology: Principles, models, and validation," *IEEE/ACM Transactions on Networking*, vol. 13, no. 6, pp. 1205–1218, 2005.

3. The following paper summarizes many of the misconceptions in both the collection and the interpretation of Internet measurement data, especially at the AS level. The same special issue of this journal also contains other relevant articles.

M. Roughan, W. Willinger, O. Maennel, D. Perouli, and R. Bush, "10 lessons from 10 years of measuring and modeling the Internet's autonomous systems," *IEEE Journal of Selected Areas in Communications*, vol. 29, no. 9, pp. 1–12, September 2011.

4. The following survey paper traces the history back to Pareto's and Zipf's study of power-law distributions, Yule's treatment of preferential attachment, and the Mandelbrot vs. Simon debate on generative models of power-law distributions in the 1950s:

M. Mitzenmacher, "A brief history of generative models for power-law and lognormal distributions," *Internet Mathematics*, vol. 1, no. 2, pp. 226–249, April 2003.

5. The following best-seller offers many interesting insights to modeling via Gaussian vs. modeling via lon- tail distributions:

N. N. Taleb, *The Black Swan*, 2nd edn., Random House, 2010.

Problems

10.1 *Probability distributions* ⋆

The **log-normal distribution** has a probability density function given by $f(x) = \frac{1}{x\sqrt{2\pi\sigma^2}}e^{-\frac{(\ln x - \mu)^2}{2\sigma^2}}$ and cumulative density function given by $\Phi(\frac{\ln x - \mu}{\sigma})$, where Φ is the cumulative density function of the normal distribution. Parameters μ and σ are the mean and standard deviation of the corresponding normal distribution.

Plot the probability density functions of the following distributions on domain $[1,5]$ with granularity 0.01 (all on the same graph for contrast), first using a linear and then using a (natural) log–log scale:

- Pareto distribution with $x_m = 1, \alpha = 1$;
- normal distribution with $\mu = 1, \sigma = 1$; and
- log-normal distribution with $\mu = 1, \sigma = 1$.

Does the tail of the log-normal distribution look like the normal distribution or the Pareto distribution?

10.2 *Utility maximization under linear constraints* ★★

We will examine another generative model for power-law distributions. Consider the case of nodes communicating to each other over a set of shared links. Each node's utility is a function of the rate it receives. Specifically, we choose the α-fair (also called isoelastic) utility function, where $\alpha \geq 0$ is a fairness parameter. We will talk much more about these functions in Chapter 11. We can formulate this as

$$\underset{\mathbf{x}}{\text{maximize}} \quad \sum_j \frac{x_j^{1-\alpha}}{1-\alpha}$$

$$\text{subject to} \quad \mathbf{Ax} \leq \mathbf{b}$$

$$\mathbf{x} \geq 0,$$

where b_i is the capacity of link i, x_j is the rate of session j, and $\mathbf{A}$ is a binary routing matrix between sessions and links:

$$A_{ij} = \begin{cases} 1 & \text{if session } j \text{ is present on link } i \\ 0 & \text{otherwise.} \end{cases}$$

Show that x_j follows a power law (for fixed α), and give an intuition for your answer.

10.3 *Preferential attachment* ★★

Recall the preferential attachment model, where, with probability θ, a new node attaches to another node proportional to its in-degree. Specifically, the probability of the new node attaching to node k is $\frac{d_k}{\sum_j d_j}$, where d_k is the in-degree of node k. And with probability $1 - \theta$, a new node attaches to one of the existing nodes at random.

Run a simulation of preferential attachment with $\theta = 0.5$ for 500 time steps. Plot p_i versus i. Does it follow the power-law distribution as expected?

10.4   *Backup routing topology design* ★★

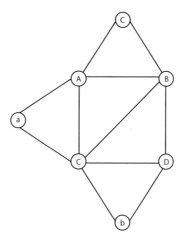

Figure 10.10 A network with 4 routers A, B, C, and D, and 3 end hosts, a, b, and c.

Consider the topology in Figure 10.10. Nodes a, b, and c are end hosts and nodes A, B, C, and D are routers. Each link is 10 Mbps. There are two sessions going on: node a sends to node c, and node b sends to node c. Each session can split traffic across multiple paths. A link's bandwidth is shared equally between the two sessions if they happen to go through the same link.

(a) For a fixed source and destination, *node-disjoint paths* are paths that do not share any nodes. How many node-disjoint paths are there between a and c? Between b and c? If a and b split their traffic evenly across their disjoint paths, how much bandwidth are a and b able to send to c concurrently?

(b) If router A fails, what happens? Repeat (a).

(c) If routers A and B both fail, what happens? Repeat (a).

10.5   *Wavelength assignment in optical networks* ★★★

In an **optical network** for the Internet backbone, each link has a number of wavelengths. An end-to-end path is called a **lightpath**. A lightpath must be assigned the same wavelength on all of the links along its route. And no two lightpaths can be assigned the same wavelength on any link. For a given graph G and routing, we want to find an assignment that uses the smallest number of wavelengths subject to these constraints. This is called the **wavelength assignment** problem.

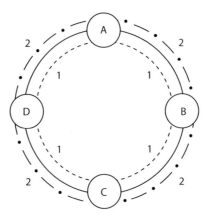

Figure 10.11 An illustration of the wavelength assignment problem in an optical ring network. There are four nodes, A, B, C, and D, in this small example, and two lightpaths, labeled 1 and 2, on each link between two adjacent nodes.

(a) The well-studied graph-coloring problem asks for the most efficient (using the smallest number of colors) assignment of one color to each node of a graph so that no adjacent nodes have the same color. Show that the wavelength assignment problem on G is equivalent to the graph coloring problem on a related graph $\tilde{G}$. What is this $\tilde{G}$?

(b) Many optical networks are rings, like in Figure 10.11. There are always two node-disjoint paths between any pair of source-destination nodes to enhance reliability. Let L be the maximum number of lightpaths on any link. Show that for any set of lightpaths requests, at most $2L - 1$ wavelengths are required, by constructing a greedy algorithm of wavelength assignment on a ring.

11 Why do AT&T and Verizon Wireless charge me $10 a GB?

11.1 A Short Answer

Almost all of our utility bills are based on the amount we consume: water, electricity, gas, etc. But even though wireless cellular capacity is expensive to provide and difficult to crank up, consumers in some countries like the USA have been enjoying flat-rate buffets for mobile Internet access for many years. Can a restaurant keep offering buffets with the same price if its customers keep doubling their appetites every year? Or will it have to stop at some point?

In April 2010, AT&T announced its usage-based pricing for 3G data users. This was followed in March 2011 by Verizon Wireless for its iPhone and iPad users, and in June 2011 for all of its 3G data users. In July 2011, AT&T started charging fixed broadband users on U-Verse services on the basis of usage too. In March 2012, AT&T announced that those existing customers on unlimited cellular data plans will see their connection speeds throttled significantly once the usage exceeds 5 GB, effectively ending the unlimited data plan. The LTE data plans from both AT&T and Verizon Wireless for the "new iPad" launched soon after no longer offered any type of unlimited data options. In June 2012, Verizon Wireless updated their cellular pricing plans. A customer could have unlimited voice and text in exchange for turning an unlimited data plan to usage-based. AT&T followed with a similar move one month later. What a reversal going from limited voice and unlimited data to unlimited voice and limited data. Similar measures have been pursued, or are being considered, in many other countries around the world for 3G, 4G, and even wired broadband networks.

How much is 1 GB of content? If you watch 15 minutes of medium-resolution YouTube videos a day, and do nothing else with your Internet access, that is about 1 GB a month. If you stream one standard-definition movie, it is about 2 GB. With the proliferation of capacity-hungry apps, high-resolution video content, and cloud services (we will discuss cloud and video networking in Chapters 16 and 17, respectively), more users will consume more GBs as months go by. This year's heavy users will become a "normal" user in just a couple of years' time. With the 4G LTE speed much higher than that of 3G (we will look into the details of speed calculation in Chapter 19), many of these GBs will be consumed on mobile devices and fall into the $10/GB bracket. Those who are used to flat-rate, buffet-style pricing will naturally find this quite annoying. And if

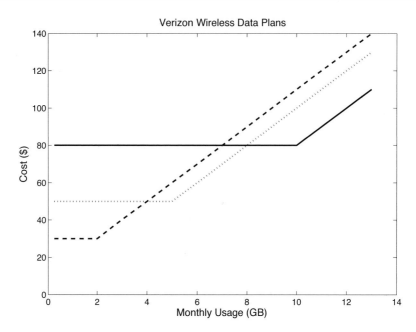

Figure 11.1 Verizon Wireless' data plan options in spring 2012. The plans have a flat-rate component then a usage-based component, e.g., $10 per GB, beyond that.

content consumption is suppressed as a result (which does not have to be the case, as we will see in the next chapter), usage pricing will influence the entire industry ecosystem, including consumers, network providers, content providers, app developers, device manufacturers, and advertisers.

Yet we will see that there are several strong reasons, including those in the interests of consumers, that support usage-based pricing as a better alternative to flat-rate pricing. Whether $10/GB is the right price or not is another matter. We will investigate the pros and cons of usage-based pricing from all these angles.

Despite the different names attached to them, there are two common characteristics of these usage-based pricing plans.

- Charge based on total monthly usage. It does *not* matter when you use it, where you use it, or what you use it for.
- There is a baseline under which the charge is still flat-rate. Then a single straight line with one slope, as the usage grows. The actual numbers in Verizon Wireless cellular data plans in 2012 are shown in Figure 11.1.

11.1.1   Factors behind pricing-plan design

Charging based on consumption probably should have sounded intuitive. That is how most utilities and commodities are charged. But to those who are used to flat-rate Internet connectivity, it represents a radical break. There are two typical precursors to the introduction of usage pricing.

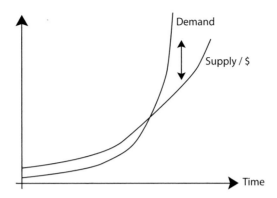

Figure 11.2 The trend of demand and supply/$ of wireless cellular capacity over time. Demand has caught up with supply (per dollar of cost) in recent years, through turning points such as the introduction of iPhone by AT&T in the USA in June 2007, which caused a 50-fold jump in cellular data demand. More importantly, demand is projected to keep growing at a faster pace than supply/$.

- Network usage has surged across many demographics and is projected to climb even higher and faster, e.g., after an ISP introduces iPhones, Android smartphones, and iPads. These devices dramatically enhance the mobile Internet experience and offer many bandwidth-intensive applications and multimedia content. (A more proper word is "capacity-intensive," but we stick to the convention in this field of using the term "bandwidth" in this and the next chapter.) An ISP's profit is the difference between revenue and cost. While demand is rapidly increasing, revenue also needs to catch up with the cost of supporting the rising demand.
- Government regulation allows pricing practices that match cost. There are other regulatory issues that we will discuss soon, but allowing the monthly bill to be proportional to the amount of usage is among the least controversial ones.

So why did the Internet Service Providers (**ISPs**), also called carriers, in countries like the USA avoid usage pricing for many years? There were several reasons, including the following two.

- As the Internet market picked up, each carrier had to fight to capture its market share. A flat-rate scheme is the simplest and easiest one to increase both the overall market acceptance and a particular carrier's market share.
- The growth in the supply of capacity per dollar (of capital and operational expenditure) could still match the growth in demand for capacity.

Then why did the US carriers change to usage pricing during 2010-2012?

- As illustrated in Figure 11.2, the growth rate of demand is outpacing the growth rate of supply/$, and the gap between the two curves is projected

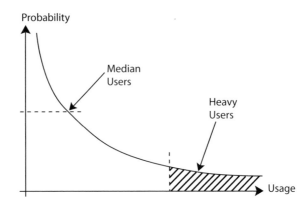

Figure 11.3 Distribution of users' capacity demand, with a long tail. The tail users dictate an ISP's cost structure in both capital expenditure and operational expenditure. If the revenue model is based on the median user, the mismatch between cost and revenue will grow as the tail becomes longer.

to widen even further in the coming years. This we call the "Jobs' inequality of capacity." Once the device suppliers and application communities, such as Apple and iOS app developers, figured out how to make it easy and attractive for users to consume mobile Internet capacity, innovation in those spaces proceeded faster than the supply side can keep up with. Cisco predicts that the mobile Internet demand will keep doubling every year. That is more than 64 times after five years. No technology can double the supply/$ each year forever.

- If we look at the distribution of capacity demand, the tail of that distribution, shown in Figure 11.3, is often the dominant factor in an ISP's cost structure. That tail has always been long, but is getting longer and longer now. If the ISP still collects revenue based on the median user, the difference between cost and revenue will be too big.

One way or another, the cost of building and operating a network must be paid by someone. Usage pricing based on monthly consumption, however, is not the only way to tackle the above issues. ISPs have other choices, such as the following.

- Increase the flat rate for everyone as demand increases. With a sufficiently high flat rate, the revenue collected will be adequate. But clearly this creates affordability and fairness issues.
- Cap heavy users' traffic. Once you exceed a cap, you can no longer use the network. Or the speed will be throttled to the point that the quality of service becomes too low for practical use once the cap has been exceeded. This is actually a special case of usage pricing: the pricing slope becomes infinite beyond the baseline.
- Slow down certain classes of traffic. For example, for a period of time, Comcast throttled BitTorrent users, who often had massive amounts of file sharing

and movie downloads using the popular P2P service. This may raise concerns on network neutrality.

- Offload some of the cellular traffic to open and non-metered WiFi networks operating in unlicensed frequency bands. But as we will see in Chapters 18 and 19, mobility support, intereference management, coverage holes, and backhaul capacity limitation can all become bottlenecks to this solution.
- Implement smarter versions of usage pricing, as discussed in the next chapter.

Most of the ISPs have realized the problem and started pursuing the usage-pricing solution. What are the criteria that we can use to compare alternative solutions to the exploding demand for mobile data? There are too many to list here, but the top ones include the following.

- *Economic viability*: As profit-seeking companies, ISPs need to first recover cost and then maximize profit. Their profit margins are in general declining, as many other transportation businesses have seen in the past. Mobile and wireless networks are bright spots that they need to seize.
- *Fairness*: Consumer A should not have to use her money to subsidize the lifestyle of consumer B.
- *Consumer choice*: Consumers should be able to choose among alternatives, e.g., spend more money to get premium services, or receive standard services with a cost saving.

Along all of the above lines, usage pricing makes more sense than fixed pricing, although it can be further enhanced with more intelligence as we will describe in the next chapter. The key advantages of usage pricing are listed below and will be analyzed in detail in the next section.

- Usage pricing produces less "waste" and matches cost.
- Usage pricing does not force light users to subsidize heavy users.
- Usage pricing helps with better differentiation in the quality of using the Internet.

11.1.2 Network neutrality debates

Before we move to the next section, it is worthwhile to mention "network neutrality," a central policy debate especially in the USA. Counter-productive to useful dialogues, this "hot" phrase has very different meanings to different people. Usually there are three layers of meanings.

- *Access/choice*: Consumers should have access to all the services offered over the Internet, and a choice of how they consume capacity on the Internet.
- *Competition/no monopoly*: ISPs should have no monopoly power and the marketplace needs to have sufficient competition.
- *Equality/no discrimination*: All traffic and all users should be treated the same. This may actually contradict the requirement of access/choice.

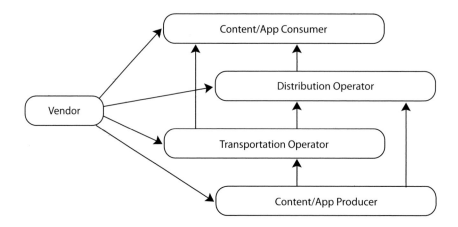

Figure 11.4 Five-party interactions in the industry. Content/app producers include YouTube and Deja, transportation operators include AT&T and Comcast, distribution operators include Akamai and BitTorrent. "Shortcuts" have been created in the traditional food chain, from content/app producers directly to distribution operators, and from transportation operators directly to consumers, further complicating the industry interaction. There is also often a lack of information visibility or incentives for higher efficiency across the boundaries of these parties.

While the last point might sound like an ideal target, it is sometimes neither feasible nor helpful to carry it out. There are four types of "no discrimination," depending on what "discrimination" means.

- *Service limitation*: Because of vertical integration, an ISP also becomes a content owner, possibly blocking access to other content. Or an ISP could block access to voice-call apps on iPhones in order to generate more revenue for its own voice business.
- *Protocol-based discrimination*: Certain protocols generate a significant amount of heavy traffic, e.g., BitTorrent, and get blocked.
- *Differentiation of consumer behaviors*: Usage pricing is one of the simplest ways to correlate pricing with consumer behavior; if consumer A takes up more capacity, she pays more.
- *Traffic management and quality-of-service provisioning*: Examples include maintaining more than one queue in a router, scheduling traffic with weighted fair queuing, or prioritizing emergency traffic like healthcare-monitor signals over non-essential software updates. (We will discuss some of these quality-of-service mechanisms in Chapter 17.)

While neutrality against service limitations is essential, neutrality against protocol-discrimination is debatable, neutrality against consumer behavior differentiation is harmful, and neutrality against traffic management is downright impossible: if having more than one queue is anti-neutral, then the Internet has never been and never will be neutral.

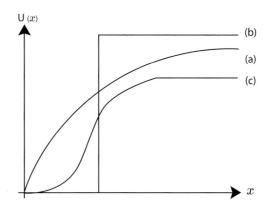

Figure 11.5 Three examples of utility-function shapes: (a) concave, (b) discontinuous, and (c) sigmoidal. Eventually utility functions all become concave as marginal returns diminish. Maximizing concave and smooth utility functions is mathematically easier than maximizing sigmoidal or discontinuous utility functions.

In fact, a naive view of "equality" harms the tenet of providing access and choices to consumers, often viewed as a more important component of neutrality. As summarized by the Canadian Radio, Television and Communications office: "Economic practices are the most transparent Internet traffic management practices," and we should "match consumer usage with willingness to pay, thus putting users in control and allowing market forces to work."

There is much more to the network-neutrality debate than we have space for in this chapter. This debate is further complicated by the fairness and efficiency issues arising out of the five-party interactions shown in Figure 11.4. We will now turn to some basic modeling language about these interactions.

11.2   A Long Answer

11.2.1   Utility maximization model

In order to proceed further to understand Internet access pricing, we need to build some model of consumer demand. The **utility function** is a common modeling tool in economics to capture "how happy" a user would be if a certain amount of resource is allocated. In Chapters 1 and 2, we saw payoff functions in games. Utility functions are a generalization of these. They further lead to models of strategic thinking by users, which are based on assumptions in the expected utility theory and its many extensions.

A typical utility function is shown as curve (a), a logarithmic function, in Figure 11.5. We denote the utility function of session i as $U_i(x_i)$, where x_i is some performance metric like throughput. Maximizing the *sum* of utilities across all users, $\sum_i U_i(x_i)$, is referred to as **social welfare maximization**. It is that theme of scalarization of vectors again, as we saw in voting theory in Chapter 6 and will see again in fairness definition in Chapter 20.

Where do these utility function models come from? One source is human subject experiments. For example, researchers may run tests with a focus group, trying different rates, delays, and jitters of voice calls, and ask them to quantify how happy they are in each case. This leads to utility models for various multimedia applications.

Another is **demand elasticity** modeling, which relies on observed consumer behavior. Given a utility function $U(x)$, we also have an associated **demand function**, capturing the volume of demand as a function of the price offered. Think about the following **net utility maximization**: a user picks the x that maximizes the difference between utility and total price paid:

$$U(x) - px,$$

where p is the unit price. If U is a concave function, it is easy to solve this optimization over one variable: just take the derivative with respect to x and let it be 0: $U'(x) = p$. Since U' is invertible, we can write x, the resulting demand, as a function of p. We call U'^{-1} the demand function D, and it is always a decreasing function, with a higher price inducing a lower demand:

$$x = U'^{-1}(p) = D(p).$$

So, a utility function determines the corresponding demand function. It also determines the demand elasticity, defined as the (normalized) price sensitivity of demand:

$$\eta = -\frac{\partial D(p)/\partial p}{D(p)/p}.$$

For example, if utility is logarithmic, then the demand function is $x = 1/p$ and the elasticity is 1.

The third ground on which we pick utility models is **fairness**. We will later devote a whole chapter to fairness of resource allocation. At this point, we will just bring up one approach. There is a class of utility functions called α-**fair utility functions**, parameterized by a positive number $\alpha \geq 0$, and if you maximize them you will get optimizers that satisfy the definition of α-**fairness**.

Now the details. We call a feasible resource vector $\mathbf{x}$ α-fair, if any other feasible resource vector $\mathbf{y}$ satisfies the following condition:

$$\sum_i \frac{x_i - y_i}{x_i^\alpha} \leq 0. \tag{11.1}$$

That means, roughly speaking, that the (normalized) deviation from $\mathbf{x}$ does not pay off.

It turns out that if you maximize the following function parameterized by $\alpha \in [0, \infty)$,

$$U_\alpha(x) = \begin{cases} x^{1-\alpha}/(1-\alpha) & \alpha \neq 1 \\ \log x & \alpha = 1, \end{cases} \tag{11.2}$$

the optimizer is α-fair. This result says that you can choose utility models by looking at which fairness notion you would like to impose. As you will validate in a homework problem, for a fixed α, the demand elasticity η is a constant, $1/\alpha$, and therefore independent of price. So α-fair utility functions are also called **isoelastic** utility functions. The smaller α, the more elastic the demand. Three values of α are particularly useful.

- $\alpha = 0$ simply maximizes the sum of resources and is often unfair because some user i may receive 0 resource.
- $\alpha = 1$ is called **proportional fairness**: just look at (11.1) with $\alpha = 1$.
- $\alpha \to \infty$ is called **max-min fairness**: you cannot increase some x_i without reducing some other x_j that is already smaller than x_i. This is a "stronger" requirement than Pareto efficiency we saw in Chapter 1.

We have not really justified *why* bigger α is more fair. Also, if **x** evolves over time, this fairness metric is only achieved by the optimizer at the *equilibrium*. We will come back to these limitations in later chapters.

However constructed, this utility function could be a function of all kinds of metrics of interest. Here, we are primarily interested in utility either as a function of data rate (in bits per second) or of data volume (in bytes). But in general it could also be a function that depends on delay, jitter, distortion, energy, etc.

Utility functions are increasing functions ($U' \geq 0$). They are often assumed to be smooth (e.g., continuous and differentiable) and concave ($U'' \leq 0$), even though that does not have to be the case. In some cases, the utility is 0 when x is below a threshold and a constant otherwise, leading to a discontinuous utility function, like curve (b) in Figure 11.5. In other cases, it starts out as a convex function: not only is the user happier as x becomes larger, but also the rate at which her happiness rises increases. But, after an inflection point, it becomes concave: larger x is still better, but the incremental happiness for each unit of additional x drops as x becomes larger. Such functions are called sigmoidal functions, as shown in curve (c) in Figure 11.5 and mentioned in the Bass model in Chapter 7. Due to the principle of **diminishing marginal return**, utility functions eventually become concave, and even possibly flat, for sufficiently large x.

11.2.2 Tragedy of the commons

The positive network effect is often summarized as follows: the benefit of having one more node in a network goes up as the square of the size of the network. The underlying assumptions are that the benefits increase proportionally to the number of links in the network, and that everyone is connected to pretty much everyone else.

There is also a famous negative network effect: "tragedy of the commons." This concept was sketched by Lloyd in 1833 and made popular by Hardin in 1968. Its essence can be captured through a net-utility-maximization argument. Consider a group of N farmers sharing a common parcel of land to feed their cows. If there are too many cows, the land will be overgrazed and eventually all their cows will die. Should a farmer get a new cow? The benefit of having a new cow goes entirely to him, whereas the cost of overgrazing is shared by all N farmers, say, $1/N$ of the cost. So each farmer has the incentive to acquire more cows, even though this collectively leads to overgrazing, the worst case scenario for everyone. This is one more example of negative externality, since it is not represented by pricing signals. It is another example of those mutually destructive phenomena like the Nash equilibrium in the prisoner's dilemma in Chapter 1.

One solution is to charge each farmer a cost proportional to N, to compensate for the inadequate incentive. This amounts to changing the net utility calculation of each farmer from

$$\text{maximize } U(x) - x$$

to

$$\text{maximize } U(x) - Nx.$$

This process of assigning the right price is called *internalizing* the negative externality. If the utility function is logarithmic, the demand drops from $x^* = D(1)$ to $x^* = D(1/N)$ now.

We will see in Chapter 14 how TCP essentially uses congestion pricing to internalize the externality of congestion in the Internet, and in Chapter 15 how P2P protocols use tit-for-tat to internalize the externality of free riders in file-sharing networks.

11.2.3   Comparison between flat-rate and usage-based pricing

There have been many studies comparing usage-based pricing with flat-rate pricing, from those in the late 1980s for the broadband network plan called ISDN to empirical studies like the UC Berkeley INDEX experiment in 1998. The main conclusions can be summarized in several graphs.

These graphs chart the demand function $D(p) = U'^{-1}(p)$, as a function of unit price p. (In the economics literature, we usually plot p vs. $D(p)$, which is the inverse demand function.) For simplicity, let us say the demand functions are linear.

In Figure 11.6, we see that if the charge is usage-based and the price is p_u, the incremental cost for the ISP to provide this much capacity, then the corresponding demand is $x_u = D(p_u)$ (where u stands for "usage.")

- Since $D^{-1} = U'$, the utility to the user is the area (integration) under the inverse demand curve, i.e., area A+B.
- The cost to achieve that utility is the rectangle $p_u x_u$, i.e., area B.
- So the **user surplus**, defined as the difference between the utility achieved and the cost incurred, is area A.

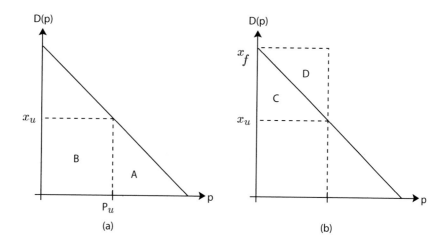

Figure 11.6 Usage-pricing illustrations. (a) The surplus (when utility bigger than cost) is area A, since the utility equals area $A + B$ and the cost equals area B. (b) Flat-rate pricing creates waste (when cost bigger than utility) and reduces the consumer surplus by area D.

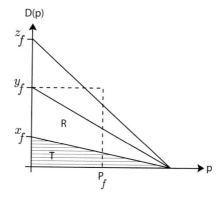

Figure 11.7 Flat-rate pricing penalizes light users in favor of heavy users. The average user's demand curve intercepts the y-axis at y_f, whereas the light user's intercepts at x_f and the heavy user's at z_f. Light users receive negative surplus and subsidize the heavy user's utility.

In contrast, under flat-rate pricing instead of usage-based pricing, the user will consume all the way to x_f (where f stands for "free") as if the price is 0. The extra utility achieved compared with the case of usage pricing is area C, whereas the extra cost required to achieve that extra utility is area C+D. This creates a waste, i.e., negative surplus, of area D. No network service is actually free, so "free" is not a particularly efficient pricing signal.

Now we consider three types of users: an average user's demand curve is shown in the middle of Figure 11.7. A light user's demand curve is the lower one, and

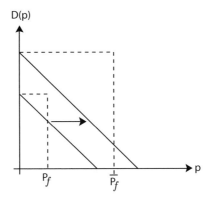

Figure 11.8 Flat-rate pricing discourages the adoption of higher-quality services by light users. A service upgrade is represented by pushing the demand curve to the right as a parallel line. To recover cost, the flat price p_f also increases to $\bar{p}_f$. To some consumers, the surplus can be bigger for a lower quality service than for a higher-quality service.

a heavy user's the higher one. In order for the ISP to recover cost, it has to set the flat-rate charge so that the revenue based on the average user is large enough, i.e., set p_f so that the area R is large enough to recover capacity costs. The utility to a user is the triangle's area, e.g., the shaded triangle T in Figure 11.7 for a light user. Clearly, the surplus for a light user can be negative, even when that for a heavy user is positive. Light users subsidize heavy users under flat-rate pricing.

This becomes even more of a problem if the ISP wants to offer an upgrade, say, in the Internet access speed. This shifts the demand curve to the right, as there will be higher demand if the price remains the same while the service is enhanced. It shifts the cost-recovering flat-rate price too, from p_f to $\bar{p}_f$. It is possible that, for a light user shown in Figure 11.8, the difference between utility (the triangle area) and the cost (the rectangle area) is bigger under the lower-speed service than under the higher-speed service. So the light user sticks to the worse service. This counter-intuitive behavior is due to the fact that recovering cost through a flat rate is inflexible.

11.3   Examples

We consider a simple numerical example with just one user, before moving into the more general case with many users sharing a link with a fixed capacity in the Advanced Material. The user has a weighted (base-10) log utility function, with a positive multiplicative weight σ:

$$U(x) = \sigma \log x.$$

The usage-based price is the combination of a baseline flat rate and a usage fee:

$$p(x) = g + hx.$$

More accurately speaking, it should be $g + h(x - x_0)$, where x_0 is the baseline usage covered by the flat rate, but we will ignore this to simplify the presentation without losing the essence of the result.

The questions we face are as follows. (1) Given (g, h) from the ISP, what would the user's demand $x^*(g, h)$ be? (2) Should the ISP charge a lower flat-rate g dollars, but a steeper slope h dollars per GB? Or should it charge the other way around?

The first question is readily answered from the demand function for a weighted log utility function. Differentiating

$$\sigma \log x - (g + hx)$$

with respect to x, and setting that to zero, we have

$$x^* = \frac{\sigma}{h}.$$

The second question depends on the market power of the ISP. If it is a *monopoly* ISP with price-setting power, it can push the user's net utility to 0:

$$\sigma \log x = g + hx.$$

Using the expression for x^*, we have

$$\sigma \log \left(\frac{\sigma}{h} \right) = g + h\frac{\sigma}{h}.$$

This means we can express the flat rate fee as a function of the usage-based fee:

$$g = \sigma \left(\log \left(\frac{\sigma}{h} \right) - 1 \right).$$

In the Advanced Material, we will see how a common h across all users can be set to prevent the sum of user demand from exceeding a fixed capacity. For now, we can explore some typical h values. Suppose the user's utility level is $\sigma = 100$.

- If $h = \$2/\text{GB}$, g is \$70, and the total revenue to the ISP is $g + hx^* = \$170$, with the flat-rate component $g/(g + hx^*) = 41\%$ of the revenue.
- If $h = \$5/\text{GB}$, g is \$30, and the total revenue to the ISP is $g + hx^* = \$130$, with the flat-rate component $g/(g + hx^*) = 23\%$ of the revenue.

For this user's demand elasticity, it is better for the ISP to charge a higher flat rate fee g and a shallower slope of the usage fee h.

In this example, you probably have spotted the trend: a smaller h means that g can afford to be higher and x will go up, to the point that the total revenue sees a net increase. So we might as well make h arbitrarily close to 0. This impractical conclusion is an artifact of three simplifying assumptions in the example.

- The ISP, as a monopoly, has complete price-setting power.

- There is only one bottleneck link.
- There is no capacity constraint or cost, so it is always advantageous to the ISP to increase demand.

In the next section, we will still keep the first two assumptions but eliminate the third one, which leads to a more insightful analysis. Then, in considering congestion control in Chapter 14, we will further remove the other two assumption to study the problem on a different timescale of machine-protocol reaction to pricing signals.

11.4   Advanced Material

11.4.1   Structure of usage prices

The congestion constraint faced by an ISP stems from the peak aggregate consumer data rate. In contrast, the access price is based on the volume of data consumed over a specified time period t. Pricing data volume is equivalent to pricing the *average* data rate over time t. Therefore an ISP faces a mismatch, because its revenue is accrued on the average data rate but the congestion cost is incurred on the peak data rate. This mismatch will be addressed in the next chapter.

For now, we note that the difference between the peak data rate and the average data rate is reduced when measured over smaller time periods. Consider a unit time interval t that is sufficiently small that the peak data-rate demand of a consumer in that time interval is a close approximation to the average data rate in that interval.

Let f index the consumers' data flows, and let the data rate for flow f in the interval $[(t-1), t]$ be given by x_f^t. The data-volume consumption over time T is then given by $x_f = \sum_{t=1}^{T} x_f^t$, and the capacity constraint C applies at every time instant t on a single bottleneck link (often the access link) across all the flows:

$$\sum_f x_f^t \leq C, \ \forall t.$$

The shape of the utility function depends on the application's performance sensitivity to varying data rates, and the utility level represents the consumer's need for the application or content. This motivates us to assume that the consumer's utility level varies in time, but the shape of the utility function does not. Let

$$\sigma_f^t U_f(x_f^t)$$

be the utility to a consumer associated with flow f at time instant t, with factor σ_f^t denoting the time dependence of the consumer's utility level, leaving the utility shape U_f independent of time t.

Faced with time-varying consumer utilities, the ISP can charge time-dependent, flow-dependent prices $p_f^t(x_f^t)$, as a function of the allocated data rate x_f^t. Consumers maximize the net utility for each flow f:

$$\text{maximize} \quad \sigma_f^t U_f(x_f^t) - p_f^t(x_f^t)$$
$$\text{variable} \quad x_f^t.$$

The most common form of the price p is a flat-rate baseline g, followed by a linear increase as a function of data rate with slope h, like what we saw in Figure 11.1:

$$p_f^t(x_f^t) = g_f^t + h_f^t x_f^t.$$

The flat price g_f^t is fixed for the duration of the time interval, irrespective of the allocated data rate. The usage-based component is based on a price h_f^t per unit data consumption.

The demand function for this form of the price is

$$D_f^t(g_f^t, h_f^t) = \begin{cases} U_f'^{-1}(h_f^t/\sigma_f^t) & \text{if } g_f^t + h_f^t y_f^t \leq \sigma_f^t U_f(y_f^t) \\ 0 & \text{otherwise.} \end{cases}$$

The condition $\sigma_f^t U_f(x_f^t) - g_f^t - h_f^t x_f^t \geq 0$ ensures that consumers have non-negative net utilities, by making g sufficiently small. To simplify the notation, we often use $D_f^t(h_f^t) = U_f'^{-1}(h_f^t/\sigma_f^t)$, with the implicit assumption that the flat price is low enough to ensure non-negative net utilities.

Now we can define the revenue-maximization problem for a monopoly ISP: maximize the total revenue subject to the capacity constraint and the consumer-demand model:

$$\begin{aligned}
\text{maximize} \quad & \sum_t \sum_f (g_f^t + h_f^t x_f^t) \\
\text{subject to} \quad & \sum_f x_f^t \leq C, \ \forall t \\
& x_f^t = U_f'^{-1}(h_f^t/\sigma_f^t), \ \forall t, f \\
& \sigma_f^t U_f(x_f^t) - g_f^t - h_f^t x_f^t \geq 0, \ \forall t, f \\
\text{variables} \quad & \{g_f^t, h_f^t, x_f^t\}.
\end{aligned} \tag{11.3}$$

The variables $\{g_f^t, h_f^t\}$ are controlled by the ISP, and $\{x_f^t\}$ are the reactions from the users to the ISP prices. In this formulation, many ingredients are still missing, including the following two.

- We still have not incorporated any routing matrix that couples the distributed demands in more interesting ways. In Chapter 14 we will bring routing into the formulation.
- Since we assumed a monopoly ISP market, the ISP has complete pricing power, which is not the case in reality. We will see an example of a competitive ISP market, the other end of the abstraction of ISP market power, in the next chapter.

Obviously, ISP revenue increases with a higher flat-fee component g_f^t, which can be set so that the consumer net utility is zero. The revenue from each flow is then $g_f^t + h_f^t x_f^t = \sigma_f^t U_f(x_f^t)$, which can be realized by any combination of flat and usage fee that can support a data rate of x_f^t.

If the usage fee h_f^t is such that the consumer demand $D_f^t(h_f^t)$ is greater than the ISP provisioned data rate x_f^t, then packets have to be dropped. However, the ISP can avoid packet drops by setting a sufficiently high usage price to reduce the consumer demand so that the aggregate demand is within the available capacity. It follows that $x_f^t = D_f^t(h_f^t)$.

Therefore, the ISP's revenue-maximization problem (11.3) simplifies to

$$
\begin{array}{ll}
\text{maximize} & \sum_t \sum_f \sigma_f^t U_f(D_f^t(h_f^t)) \\
\text{subject to} & \sum_f D_f^t(h_f^t/\sigma_f^t) \le C, \ \forall t \\
\text{variables} & \{h_f^t\}.
\end{array}
\tag{11.4}
$$

The capacity inequality should be achieved with equality at optimal prices. It suffices to have the optimal usage fee h^t be the *same* across all flows f, since it will be used to avoid capacity waste in the sum of demands across all the flows (an argument that will be rigorously developed in a homework problem). The optimal flat fee g_f^t, however, is flow-dependent, allowing the ISP to fully extract the consumer net-utility.

Therefore, an optimal pricing scheme that achieves the maximum in (11.4) is given by the following: for each t, the per-unit usage price h^t is set such that the resulting demands fully utilize the link capacity C on the (only) bottleneck link in the network:

$$
\sum_f x_f^t = \sum_f D_f^t(h^t) = C,
$$

and the flat-rate baseline prices $\{g_f^t\}$ are set such that the maximum revenue is generated:

$$
g_f^t = \sigma_f^t U_f(x_f^t) - h^t x_f^t.
$$

Let R_F^* be the revenue from the flat component of the optimal price, and R_S^* the revenue from the usage component. In a homework problem, by using the above solution structure, we can derive the ratio between the flat and usage components. In the special case where utility functions are α-fair with $\alpha_f = \alpha$ for all f, the ratio of flat-rate revenue to usage-dependent revenue becomes

$$
\frac{R_F^*}{R_S^*} = \frac{\alpha}{1-\alpha}.
$$

This reveals that usage-dependent revenue dominates with linear utilities ($\alpha \to 0$), while revenue from flat-rate components dominates with log utilities ($\alpha \to 1$). The flat price is a significant component in the extraction of a consumer's net utility if her price sensitivity is low.

Consider a monopoly ISP providing connectivity service to ten flows over an access link of capacity $C = 10$ Mbps. We generate the utility levels $\{\sigma_f^t\}$ randomly within the range $[\sigma_0, \sigma_1]$.

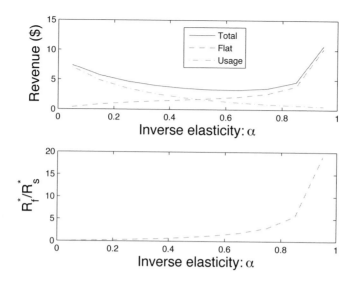

Figure 11.9 Comparison of revenue from flat and usage-based pricing at different consumer demand elasticities. All users have α-fair utility functions with the same α. As α increases and the demand elasticity $\eta = 1/\alpha$ drops, the flat-rate component's fraction of the overall revenue to the (monopoly) ISP increases.

The upper graph in Figure 11.9 illustrates the average revenue received by the monopolist ISP.

- The flat component of the revenue, which enables the monopoly ISP to completely extract the consumer net utility, increases with a decreasing elasticity of demand.
- The usage component of the revenue decreases with a decreasing elasticity of demand.

The lower graph in Figure 11.9 plots the ratio of the flat component to the usage component of the revenue, demonstrating the increased reliance on the revenue from the flat price at low demand elasticity.

The ISP's pricing flexibility, in practice, is restricted along *time* and across *flows*.

- Fixing $\{g, h\}$ to be the same for all times t leads to a tradeoff between oversubscribing link capacity and dropping packets on the one hand, and underutilizing link capacity and losing revenue on the other hand.
- Fixing $\{g, h\}$ to be the same for all the flows leads to a loss of revenue, which can be mitigated by allowing nonlinear pricing: discounted per-GB pricing at high volume to encourage higher utilization. This is called the second-order price discrimination in economics, and can be found in how the baseline charges are structured in AT&T and Verizon Wireless' usage-based mobile data plans.

So far we have focused only on "how much" to charge. In the next chapter, we will continue to discuss three other questions of consumer pricing on the Internet: "how to charge," "what to charge," and "whom to charge."

Summary

┌───┐

Box 11 Economic models of mobile data consumption

Improper pricing creates a tragedy of the commons. As the growth rate of capacity demand surpasses that of capacity supply per dollar, flat rates are no longer sustainable. Compared with flat-rate schemes, charging based on usage reduces waste, avoids light-users subsidizing heavy-users, and provides incentives to shift to higher-quality, more expensive services. Consumers' behavior can be modeled through net utility maximization, where utility functions lead to demand functions and demand elasticities.

└───┘

Further Reading

There is growing research literature on network economics, including Internet access pricing.

1. An excellent edited volume published back in 1998, based on a workshop in 1995, has many insightful chapters still relevant today, such as those written by Clark, by Kelly, and by Varian:
L. W. McKnight and J. P. Bailey, Eds., *Internet Economics*, The MIT Press, 1998.

2. A more recent survey article of the subject can be found at
J. Walrand, "Economic models of communication networks," in *Performance Modeling and Engineering*, Z. Liu and C. Xia, Eds., Springer, 2008.

3. Our basic intuition on flat-rate and usage-based pricing follows the summary of the INDEX experiment in the late 1990s:
R. J. Edell and P. Varaiya, "Providing Internet access: What we learn from INDEX," *IEEE Infocom Keynote*, 1999.

4. The following paper has led to many diverse opinions on how to understand and mitigate the tragedy of the commons:
G. Hardin, "The tragedy of the commons," *Science*, vol. 162, pp. 1243–1248, 1968.

5. We have assumed a lot on how people make decisions based on pricing signals. These assumptions often do not hold. The following book offers an accessible and insightful survey.

D. Kahneman, *Thinking, Fast and Slow*, FSG Publisher, 2011.

Problems

11.1 *Demand elasticity and α-fair utility function* ⋆

(a) Plot α-fair utility functions with $\alpha = 0, \frac{1}{5}, \frac{1}{2}, 1, 2, 5$, and 100 on the same graph, and compare them.

(b) Derive the demand and the demand elasticity as functions of price, if the utility function is $\arctan(x)$.

(c) Repeat (b) for α-fairness utility functions.

11.2 *Optimizing for different utility functions* ⋆⋆

Solve the utility maximization problem in the Examples section, but this time with a weighted α-fair utility functions

$$U(x) = \sigma U_\alpha(x)$$

where $\alpha = 0.5$. Then solve it again for $\alpha = 2$. Compare the results.

11.3 *Demand vs. supply curves* ⋆⋆

In general, the demand $D(p)$ and **supply** $S(p)$ as a function of price p can be defined as follows:

$$D(p) = U'^{-1}(p),$$
$$S(p) = C'^{-1}(p),$$

where $U(x)$ is the utility of the buyer as a function of the amount purchased and $C(x)$ is the cost of the producer as a function of the amount produced. Figure 11.10 gives an illustration, with capital latters indicating the sizes of the corresponding areas.

(a) Assume $U(0) = 0$, show that $U(x_0) = A + B + C$. Therefore, the buyer's net utility is $U(x_0) - p_0 x_0 = A$.

(b) Assume $C(0) = 0$, show that $C(x_0) = C$. Therefore, the seller's profit is $p_0 x_0 - C(x_0) = B$.

(c) Let $x^* = \min\{D(p^*), S(p^*)\}$. At which price p^* does the social welfare, defined here as $U(x^*) - C(x^*)$, take the maximum value?

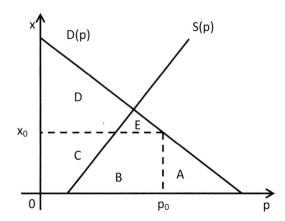

Figure 11.10 A demand curve $D(p)$ and a supply curve $S(p)$ as functions of price.

11.4 *Flat component vs. usage component* ★★★

As in the Advanced Material, the simplified revenue-maximization problem for the monopoly ISP is

$$
\begin{aligned}
&\text{maximize} && \textstyle\sum_{t,f} \sigma_f^t U_f(D_f^t(h_f^t)) \\
&\text{subject to} && \textstyle\sum_f D_f^t(h_f^t) \le C, \ \forall t \\
&\text{variables} && \{h_f^t\}.
\end{aligned}
\tag{11.6}
$$

Let $h_f^{t\,*}$ be the usage price that solves the above maximization problem. The resulting volume of consumption is

$$
x_f^{t\,*} = D_f^t(h_f^{t\,*}),
$$

and the flat-rate price is

$$
g_f^{t\,*} = \sigma_f^t U_f(D_f^t(h_f^{t\,*})) - h_f^{t\,*} x_f^{t\,*}.
$$

(a) Define $R_F^* = \sum_{t,f} g_f^{t\,*}$ as the revenue from the flat-rate component and $R_S^* = \sum_{t,f} h_f^{t\,*} x_f^{t\,*}$ as the revenue from the usage component. Prove that

$$
\frac{R_F^*}{R_S^*} = \frac{\sum_{f,t} \sigma_f^t U_f(x_f^{t\,*})}{\sum_{t,f} \sigma_f^t U_f'(x_f^{t\,*}) x_f^{t\,*}} - 1.
$$

(Hint: Use $\sigma_f^t U_f'(D_f^t(h_f^t)) = h_f^t$.)

(b) Show that if $U_f(x)$ is an α-fair utility function ($\alpha \neq 1$) for all flows f, we have a simpler expression:

$$
\frac{R_F^*}{R_S^*} = \frac{\alpha}{1 - \alpha}.
$$

(c) Argue that $h_f^{t\,*} = h^{t*}$, i.e., the optimal price per unit of usage is independent of the flow f. (For this part you may want to wait till you have seen Lagrange duality in the Advanced Material in the next chapter.)

11.5   *Braess' paradox* $\star\star\star$

Consider a road network as illustrated in Figure 11.11(a), on which 3000 drivers wish to travel from node Start to node End. Denote by x the number of travelers passing through the link Start $\to$ A, and by y the number of travelers passing through the link $B \to$ End. The travel time of each link in minutes is labeled next to the corresponding link. Suppose everyone chooses her route from Start to End to minimize the total travel time.

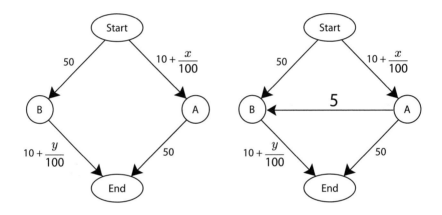

Figure 11.11 Braess paradox: adding a link hurts performance.

(a) What is the resulting traffic and the total travel time for each commuter?

(b) Suppose the government built a shortcut from node A and B with travel time labeled as illustrated in Figure 11.11(b). What is the resulting traffic and the total traveling time for each commuter?

(c) This is the famous **Braess' paradox**. Suggest a way to avoid it.

12 How can I pay less for each GB?

12.1   A Short Answer

ISPs charging consumers on the basis of usage is just one corner of the overall landscape of Internet economics. We will pick consumers' monthly bills to focus on in this chapter, but there are many other key questions.

- The formation of the Internet is driven in part by economic considerations. Different ISPs form peering and transit relationships that are based on business and political decisions as much as on technical ones.
- The invention, adoption, and failure of Internet technologies are driven by the economics of vendor competition and consumer adoption.
- The investment of network infrastructure, from purchasing wireless licensed spectrum to deploying triple-play broadband access, is driven by the economics of capital expenditure, operational expenditure, and returns on investment.

The economics of the Internet are interesting because the technology–economics interactions are *bidirectional*: economic forces shape the evolution of technology, while disruptive technologies can rewrite the balance of economic equations. This field is also challenging to study because of the lack of publicly available data on ISPs' cost structures and the difficulty of collecting well-calibrated consumer data.

12.1.1   Smart data pricing

There is a rapidly growing research field and industry practice on network access pricing. What we described on usage pricing in the last chapter, in the form of tiered and then metered/throttled plans, is just a starter. A few possibilities of **Smart Data Pricing** (SDP) are listed below.

- The hourly-rate model, e.g., Mobinil in Egypt charges data connection by the number of hours of usage.
- Expected-capacity pricing, which relies on resource allocation driven by the needs of different sessions rather than just the byte-counts. Characterizing a session's needs, however, can be tricky, even after a period of performance observation during trials.

- Priority pricing, where you can pay more to get a higher speed, such as the priority pass service by SingTel in Singapore. A turbo mode of anti-throttling is also being considered in the USA for heavy users whose speed is throttled once usage exceeds some threshold. In an elaborate form, priority pricing may even take the form of an auction where the price reflects the negative externality imposed on other users by boosting your speed. Paris metro pricing adds another interesting variation.
- Two-sided pricing, where an ISP charges either the content consumers or the content producers, or both. It is used by Telus in Canada and TDC in Denmark. This can also become an application-dependent pricing method.
- Location-dependent pricing, which is also used in the transportation industry in certain cities, e.g., downtown London and Singapore.
- Time-dependent pricing, which is also used in certain utility industries e.g., energy networks, and will be elaborated in this chapter.

In static pricing, time periods and the associated prices are predetermined and do not vary except over very long timescales like months and years. In dynamic pricing, network access prices are continuously adjusted to reflect the state of the network. We will see that congestion control in Chapter 14 can be interpreted as a type of dynamic pricing.

In this chapter, we bring up several topics that illustrate some central themes in the field. One is charging that is based on *when* the Internet is used, and the other is *differentiating service qualities* by simply charging different prices. We will also explore the question of "whom to charge" through two-sided pricing. These are some of the possibilities to help the entire network ecosystem, from consumers to ISPs, and from content providers to advertisers, move from the shadow of $10/GB to win-win solutions. In a win-win,

- ISPs generate more revenue, lowers cost, and reduces churn;
- consumers pay less per GB of data;
- content providers attract more eyeballs; and
- vendors sell more innovative software and hardware.

12.1.2 Time-dependent pricing

Pricing based just on monthly bandwidth usage still leaves a timescale mismatch: ISP revenue is based on monthly usage, but peak-hour congestion dominates its cost structure. Ideally, ISPs would like bandwidth consumption to be spread evenly over all the hours of a day. **Time-Dependent Pricing** (TDP) charges a user according to not just "how much" bandwidth is consumed but also "when" it is consumed, as opposed to Time-Independent usage Pricing (TIP), which considers only monthly consumption amounts. For example, the day-time (counted as part of minutes used) and evening-weekend-time (free) differentiation, long practiced by wireless operators for cellular voice services, is a simple two-period TDP scheme. Multimedia downloads, file sharing, social media updates, data backup, and software downloads, and even some streaming applications all have various degrees of time elasticity.

As an idea as old as the cyclic patterns of peak and off-peak demand, TDP has been used in transportation and energy industries. Now it has the potential to even out time-of-day fluctuations in (mobile) data consumption: when data plans were unlimited, $1 a GB was infinitely expensive, but now with $10 a GB becoming the norm, $8 a GB suddenly looks like a bargain. As a pricing practice that does not differentiate in terms of traffic type, protocol, or user class, it also sits lower on the radar screen of the network neutrality debate. TDP time-multiplexes traffic demands. It is a counterpart to spatial multiplexing in Chapter 13 and to frequency multiplexing in Chapter 18.

Much of the pricing innovation in recent years has occurred outside the USA. Network operators in highly competitive markets, e.g., in India and Africa, have adopted innovative dynamic pricing for voice calls.

- The African operator, MTN, started "dynamic tariffing," a congestion-based pricing scheme in which the cost of a call is adjusted every hour in each network cell depending on the level of usage. Using this pricing scheme, instead of a large peak demand around 8 pm, MTN Uganda found that many of its customers were waiting to take advantage of cheaper call rates.
- A similar congestion-dependent pricing scheme for voice calls was also launched in India by Uninor. It offers discounts to its customers' calls that depend on the network traffic condition in the location of the call's initiation.
- Orange has been offering "happy hours" data plans during the hours of 8–9am, 12–1pm, 4–5pm, and 10–11pm.

We have to face two questions here. Can we effectively parameterize **delay sensitivity** in setting the right prices? Are users willing to defer their Internet traffic in exchange for a reduced monthly bill? Ultimately, it is the ratio between demand elasticity and delay sensitivity (for each user and each application) that determines how much can time-dependent pricing help.

12.2   A Long Answer

12.2.1   Thinking about TDP

Usage-based pricing schemes use penalties to limit network congestion by reducing demand from heavy users. However, they cannot prevent the peak demand by many users from concentrating during the same time periods. ISPs must provision their network in proportion to these peak demands, leading to a timescale mismatch: ISP revenue is based on monthly usage, but peak-hour congestion dominates its cost structure. Empirical usage data from typical ISPs shows large fluctuations even on the timescale of a few minutes. Thus, usage can be significantly evened out if a TDP induces users to time-shift their demand. However, a simple two-period, time-dependent pricing scheme (e.g., different prices for the day and the night) is inadequate as it can incentivize only the highly price-sensitive users to shift some of their non-critical traffic. Such schemes often end up creating two peaks; one during the day and one at night.

In general, all static pricing schemes suffer from their inability to adapt prices in real time to respond to the usage patterns, and hence fail to exploit the large range of delay tolerance as many types of applications proliferate.

Dynamic pricing, on the other hand, is better equipped to overcome these issues and does not require pre-classification of hours into peak and off-peak periods. However, the current dynamic time-dependent pricing schemes have been explored mainly for voice traffic, which is very different from data in terms of its delay sensitivity, activity patterns, and typical duration. Unlike voice calls, certain classes of mobile data traffic offer greater delay tolerance, e.g., cloud services and multimedia downloads. They can be completed either pre-emptively or in small chunks whenever the congestion conditions are mild. Users of such applications can therefore be incentivized to shift their usage with optimized, time-dependent prices for their mobile data traffic.

Time-dependent pricing can be further extended to **congestion-dependent pricing** by shrinking the timescale of price adaptation. Instead of on a timescale of hours or minutes, the prices may change every several seconds when channel conditions or mobility results in rapidly changing congestion conditions. Similar ideas have been implemented as spot pricing in energy networks and cloud networks. Even during busy hours and over heavily used spectra, there are occasional periods of time with little usage, which we call **flashy whitespaces**. ISPs can offer low spot prices in these less congested timeslots, enabling cost-conscious users to wait for these low prices. In such cases, we need an "auto-pilot" mode, where the device makes decisions and the user need not be bothered in real time (once she has pre-configured her usage requirements and expectations, e.g., the maximum monthly bill, which applications can be delayed up to x minutes, which applications should never be deferred, etc.) In Chapters 16 and 17, we will also see some options of scheduling methods to provide differentiated qualities of service. Now, pushing the auto-pilot TDP approach further, ISPs can even offer intelligent flat-rate data plans. Users may pay a flat rate in exchange for automated scheduling of their traffic.

A schematic of TDP is shown in Figure 12.1, with user-profiling and price-determination as the computational modules, and the feedback loop constructed through a user interface, users' reactions, and network measurement. A TDP system is like a traffic shaper that time-shifts bandwidth demands so that the *statistical multiplexing effect* of a network can be most effectively leveraged. And the degree of freedom in this shaper is the pricing signals that change consumer behavior.

12.2.2   Modeling TDP

Pictorially, an ISP uses TDP to even out the "peaks" and "valleys" in bandwidth consumption over a day. The ISP's problem is then to set its prices to

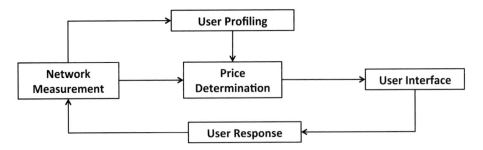

Figure 12.1 A simplified schematic of the TDP architecture. The core computational module is price optimization that takes into account the prediction of user reaction through the user profiling module. The user interface needs to be user-friendly, and must allow both user-generated decisions and auto-pilot decisions on behalf of the user.

balance capacity costs and costs due to price incentives, given its estimates of user behavior and the willingness to defer sessions at different prices. A user is modeled here as a set of application sessions, each with a **waiting function**: the willingness to defer that session for some amount of time under a price incentive for doing so.

Waiting functions are functions of both the reward and the amount of time deferred: $w(p, t)$, where p is the reward offered to the user for waiting t amount of time and the function w maps those to a probability of waiting. Each application of each user in each period has a different waiting function; the users' willingness to defer applications depends on the type of application being deferred, the user's patience, and the period that the application is being deferred from. For instance, I might be generally more willing to defer downloading a cloud synchronization than streaming a YouTube video, and more willing to defer a YouTube video for five minutes at 8pm than at 11am.

For a constant p, $w(p, t)$ should decrease in t, the duration of the wait. And for a constant t, $w(p, t)$ should be increasing in p, the price incentive. Following the principle of diminishing marginal returns, we can say that w is concave in p. For waiting functions to be useful, we need some method for estimating them. To make this estimation easier, we parameterize the waiting functions. Each waiting function has the same form but a different parameter; our job is then to estimate the waiting function parameters. Without going into the many methods of machine learning that can be deployed, here is a simple example. We might take

$$w(p, t) = \frac{p}{(t + 1)^{\beta}},$$

with the parameter $\beta \geq 0$ specifying how users' willingness to shift their traffic falls off with time. With a large β, users do not want to defer for a long time; with a smaller β, users are more willing to defer their traffic. Here are some β values for different applications, estimated from consumer surveys we conducted in the USA and India: 2.027 for YouTube streaming in the USA and 2.796 in India, 0.6355 for video download in the USA and 1.486 in India.

In practice, it is impossible to estimate waiting function parameters for each type of application for each user during each period because there are too many of them. Instead, we use an aggregate approach: several applications for some users are assumed to have the same waiting function. For instance, I might be (almost) equally willing to defer streaming a YouTube video at 9pm as my friend is willing to defer watching Hulu at 10pm. We then have the same waiting function parameters.

Now we know what parameters need to be estimated: the waiting-function parameters in each period and the fraction of traffic that they correspond to. The next step is to do the actual estimation. We can estimate using the difference between traffic before TDP and traffic after TDP. This difference must be equal to the amount of traffic "deferred in" less the amount "deferred out." We are in a position to calculate the amount deferred in: it is the sum of the amount deferred from each period, which can be easily calculated using the waiting functions in those periods. Given the period in which an application session originated, we know how long the traffic was deferred and what reward was offered. We can then write the difference between traffic volumes before and after TDP as a function of the parameters to be estimated and of the rewards offered. In a consumer trial, we can offer a range of rewards and observe the corresponding difference in traffic before and after TDP. We can then choose the waiting-function parameters to fit with the data we observe (e.g., using least squares).

Then, given the estimation of waiting functions, the ISP needs to decide the price per time period. The ISP's decision can also be formulated in terms of "rewards," i.e., price discounts, defined as the difference between TIP and the optimal TDP prices. When determining optimal prices, an ISP tries to balance the cost of meeting a high peak demand, e.g., the capital expenditure of capacity expansion, with the cost of offering price discounts for users to move some of their demand to later times. The optimization variables are these rewards that give users the incentives to defer bandwidth consumption. This module will be our focus in the next two sections.

12.2.3 Implementing TDP

Taking a pricing idea from its mathematical foundation to a deployment in operational networks involves many steps, from theory development, computer simulations, testbed emulations, proof-of-concept implementations, small-scale user trials, to, eventually, large-scale commercial trials. For example, in the case of TDP for mobile (and wireline) networks, an end-to-end system was implemented and consumer trials were carried out at Princeton University in 2011-2012. The trial system involves many components, all parts of the feedback loop in the TDP schematic in Figure 12.1:

- graphical user interfaces connected into the operating systems of iPhones, iPads, Android phones, tablets, and Windows PCs;
- traffic measurement and mobility pattern analysis, partly on end-user devices for privacy reasons and partly in network gateways;
- the computational engines of waiting function estimation and price optimization, again split across end-user devices, network gateways, and servers; and
- a database of usage, traffic, prices, and network conditions on servers;
- software that takes the output of these computational engines to control the scheduling of the application sessions.

There are several challenging aspects embedded in the above list. For example, there are choices to be carefully made in the architectural division of computation and storage among the end user devices, some gateway machines, and the ISPs' in-network management systems. Another example is on the human-computer interface: the graphic user interface must be very intuitive, providing rich features and yet remain as user-friendly as possible. We want to empower users with the freedom to choose and yet not to bother them with the burden of choices. A variety of interface design can provide more pricing certainty and session-scheduling automation to the consumers, so that (most of) the benefits of dynamic pricing can be realized with a minimal amount of uncertainty and complexity.

Initial results from this trial of 50 users indicate that TDP can be effective in reducing the network load, as measured by the *peak-to-average ratio* (PAR). This metric is defined as the peak amount of traffic in one hour, divided by the average traffic over that day. Figure 12.2 shows the distribution of daily PARs both before and after TDP was introduced. The maximum PAR decreases by 30% with TDP, and approximately 20% of the PARs before TDP are larger than the maximum PAR with TDP. TDP significantly reduced the peak-to-average ratio, flattening bandwidth demand over a day. At the same time, averaged over weeks, there was also a 107% increase in the total data consumption. This is an example of the *Macy's sales day effect*, where discounts during limited timeslots attract heavier-than-usual traffic.

12.3   Examples

We present two numerical examples first, before turning to a general formulation of the price optimization module in the Advanced Material.

12.3.1   An illustration

Let us walk through a simulation to visualize TDP, using the β of 0.59 for cloud synchronization, 0.64 for multimedia download, and 2.03 for YouTube streaming. The usage distribution of the different traffic classes was taken from recent

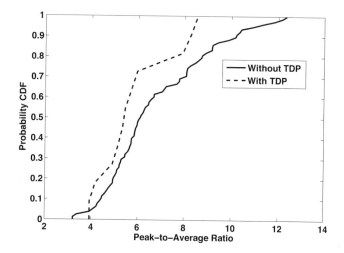

Figure 12.2 A cumulative distribution chart, using empirical data from a Princeton trial of TDP in 2012, shows that TDP reduces the peak-to-average ratios by about 30%.

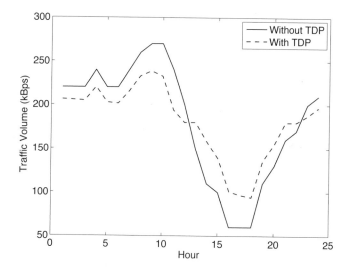

Figure 12.3 TDP and TIP traffic patterns for a sample mix of three classes of traffic with different delay tolerances. The aggregate traffic volume with TDP is flatter than that without.

estimates, and the TIP data estimates were taken from empirical data from an ISP. We consider a system with 100 users and 24 one-hour time periods in each day. The ISP's marginal cost of exceeding capacity is set to $0.30 per MB.

The results of the simulation are shown in Figure 12.3, which gives the demand patterns before and after the use of TDP. It demonstrates that TDP incentivizes

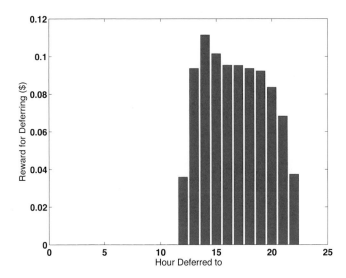

Figure 12.4 Rewards offered at different periods, computed by solving the time-dependent-price optimization that minimizes the overall ISP cost and incorporates user reaction predicted through the estimated waiting functions.

users to shift their traffic, which brings the peaks and valleys closer and improves the smoothness of the demand over time. The daily cost per user decreases from $0.21 with TIP to $0.16 with TDP, a 24% savings.

Figure 12.4 shows the optimal rewards (incentives) awarded for different times of a day. As expected, all hours with positive rewards are at or under capacity with TDP. Rewards are slightly higher in hours 14 and 15 than in subsequent under-capacity hours; hours 14 and 15 represent the under-capacity times closest to the high-usage hours 1–13.

To quantify traffic's unevenness over 24 hours, we define the *residue spread* as the area between a given traffic profile and the ideal one, where the total usage remains the same but with usage constant (i.e., "flattened") across periods in 24 hours. The residue spread decreases 44.8% (from 502.8 MB to 280.3 MB) with TDP. Overused periods closer in time to underused ones have the greatest traffic reduction; users more easily defer for shorter durations. Although TDP does help to even out traffic profiles, some users are impatient and some sessions are simply too time-sensitive to be deferred; thus the usage will never be perfectly flat.

You would expect that when exceeding capacity is expensive, the ISP will offer large rewards to even out demand. Figure 12.5 shows the residue spread with TDP versus the logarithm of a, the weight on the cost of exceeding capacity relative to the cost of handing out rewards. The residue spread decreases sharply for $a \in [0.1, 10]$, then levels out for $a \geq 10$. For $a \geq 10$, demand never exceeds capacity because the cost of exceeding capacity is too big.

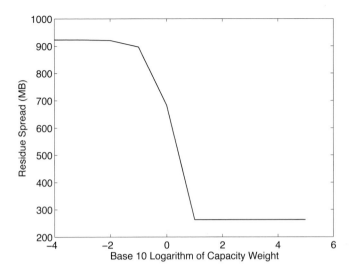

Figure 12.5 The residue spread for different costs of exceeding capacity. As the cost of exceeding capacity rises, the incentives for delaying traffic becomes bigger, and the residue spread lowers. In this example, the ISP cannot entirely evens out traffic, even at the very high cost of exceeding capacity.

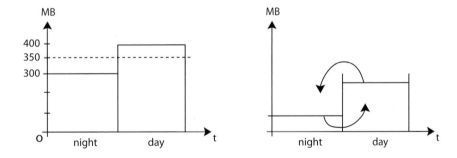

Figure 12.6 A small, two-period, numerical example illustrating TDP over a single bottleneck link, where the link capacity can handle 300 MB of demand. The day-time demand exceeds capacity while night-time demand under-utilizes it. TDP helps provide pricing incentives for users to move some of their day-time demand to night-time. In general, we need to keep track of the demand shifted away from a given period and the demand shifted into that period.

12.3.2   A small example

As illustrated in Figure 12.6, we now walk through a small, much simplified numerical example that bridges the general discussion we just had to the symbolic representation in the Advanced Material. Suppose we have just two periods (e.g., night and day) and we are trying to determine the optimal prices for these periods.

First, we must characterize the types of traffic found during these two periods. Consider just two types of applications, one which users are very impatient to

	Night	Day
Email	200 MB	200 MB
File Downloads	100 MB	200 MB

Table 12.1 The volume of email and file download traffic during the night and day before TDP.

	Shift to night	Shift to day
Email	$\frac{p_n}{4}$ probability, $200\frac{p_n}{4}$ shifted	$\frac{p_d}{4}$ probability, $200\frac{p_d}{4}$ shifted
File downloads	$\frac{p_n}{2}$ probability, $200\frac{p_n}{2}$ shifted	$\frac{p_d}{2}$ probability, $100\frac{p_d}{2}$ shifted

Table 12.2 The probabilities of shifting and the expected amounts of shifting of email and file download traffic to night-time or day-time.

defer and one for which they are more patient. For instance, we could have email or Twitter (impatient) and movie downloads or cloud synchronization (patient).

The volumes of traffic taken up by these two applications before TDP are shown in Table 12.1. Let us say the ISP is at present charging users \$0.01/MB (\$10/GB) both during the night and during the day.

We now need to quantify users' willingness to shift some of their traffic from one period to the other. This willingness is of course influenced by the reward offered in the other period. Let p_n be the reward during the night and p_d the reward during the day. We express users' "willingness to wait" as the probability that the user will wait. To simplify the presentation, let us say this probability is proportional to the reward reaped by waiting. The probabilities of shifting and the corresponding expected amount of traffic shifted are summarized in Table 12.2.

We now formulate the optimization problem. First, consider the cost of offering rewards, which is just the reward per unit of traffic times the amount shifted into a given period. This cost can thus be expressed as

$$p_n \left(200\frac{p_n}{4} + 200\frac{p_n}{2} \right) + p_d \left(200\frac{p_d}{4} + 100\frac{p_d}{2} \right) = 150p_n^2 + 100p_d^2. \qquad (12.1)$$

Next, we quantify the cost of exceeding capacity. We model this cost as linear and assume that the capacity is 350 MB. Thus, the ISP exceeds capacity during the day but not during the night. Compared with the number of users and the volume of traffic in the previous example, the scale is much smaller here. Let us assume that, for each GB of demand over the capacity, the ISP faces a cost of \$1.

We now need to find expressions for the volumes of traffic during the night and day under TDP. Consider the volume of traffic during the night. Before TDP, it is $200 + 100 = 300$ MB. The amount of traffic shifted into the night is

$$200\frac{p_n}{4} + 200\frac{p_n}{2} = 150p_n,$$

and the amount of traffic shifted from the night into the day is

$$200\frac{p_d}{4} + 100\frac{p_d}{2} = 100p_d.$$

Thus, the amount of traffic during the night under TDP, in MB, is

$$300 + 150p_n - 100p_d,$$

and the cost of exceeding capacity during the night is

$$\max\{0, 300 + 150p_n - 100p_d - 350\} = \max\{0, -50 + 150p_n - 100p_d\}.$$

We can find a similar expression for the cost of exceeding capacity during the day using the same line of reasoning.

Finally, our optimization problem is to minimize the following objective function:

$$150p_n^2 + 100p_d^2 + \max\{0, -50 + 150p_n - 100p_d\} + \max\{0, 50 - 150p_n + 100p_d\},$$

over two non-negative variables, p_n and p_d, i.e., the per-unit reward amount, or equivalently, the time-dependent prices.

On solving this optimization problem, we obtain $p_d = 0$, $p_n = 0.33$. Thus, the ISP discounts prices during the night by $3.33/GB. Intuitively, this makes sense: rewards during the day should be lower, so that users are induced to shift their traffic from the day into the night. Indeed, with this pricing scheme, the ISP operates at capacity, with a traffic volume of 350 MB during both the day and the night.

12.4   Advanced Material

12.4.1   TDP price-optimization: A more general formulation

Computing the optimal prices per time period, with user reaction anticipated through waiting-function estimation, is a key part of TDP. We consider a simple version of this problem formulation that generalizes the small numerical example we just saw.

Let X_i denote demand in period i under TIP. The phrase "originally in period i" means that under TIP, this session occurs in period i. Suppose that the ISP divides the day into n periods, and that its network has a single bottleneck link of capacity C, where C is the total amount of traffic that the link can carry in one time period. This link is likely the aggregation link out of the access network, which has less capacity than the aggregate demand and is often oversubscribed by a factor of five or more. The cost of exceeding capacity in each period i, capturing both the impact of customer complaints and the expenses for capacity expansion, is denoted by $f(x_i - C)$, where x_i is the usage in period i. This cost is often modeled as piecewise-linear and convex.

Each period i runs from time $i - 1$ to i. A typical period lasts say 10–30 minutes. Sessions begin at the start of the period. The time between periods i and k is given by $i - k$, which is the number $b \in [1, n]$, $b \equiv i - k \pmod{n}$. If $k > i$, $i - k$ is the time between period k on one day and period i on the next day.

For each session j originally in period i, define the waiting function $w_j(p, t)$ that measures the user's willingness to wait t amount of time, given reward p. Each session j has capacity requirement v_j, so $v_j w_j(p, t)$ is the amount of session j deferred by time t with reward p. To simplify the problem representation, we assume that each session does not defer more than once. To ensure that $w_j \in [0, 1]$ and that the calculated usage deferred out of a period is not greater than demand under TIP, we normalize the w_j by dividing it by the sum over possible times deferred t of $w_j(P, t)$. Here, P is the maximum possible reward offered. The notation $j \in k$ refers to a session j originally in period k (in the absence of TDP).

Now we are ready to state the optimization problem. First, consider the cost of paying rewards in a given period i. The amount of usage deferred into period i is $\sum_{k \neq i} y_{k,i}$, where $y_{k,i}$ is the amount of usage deferred from period k to period i. Consider a session $j \in k$. The amount of usage in session j deferred from period k to period i is $v_j w_j(p_i, i - k)$, since such sessions are deferred by $i - k$ amount of time. Thus,

$$y_{k,i} = \sum_{j \in k} v_j w_j(p_i, i - k),$$

and the ISP's total cost of rewarding all sessions in period i is

$$p_i \sum_{k \neq i} \sum_{j \in k} v_j w_j(p_i, i - k).$$

Now consider the cost of exceeding capacity. It is $af(x_i - C)$, with the weight constant a, but how much is x_i? It is the original amount minus the amount moved out of period i by TDP and plus the amount moved into period i by TDP:

$$x_i = X_i - \sum_{j \in i} v_j \sum_{k=1, k \neq i}^{n} w_j(p_k, k - i) + \sum_{k=1, k \neq i}^{n} \sum_{j \in k} v_j w_j(p_i, i - k).$$

The ISP's total cost function for period i is then

$$p_i \sum_{k \neq i} \sum_{j \in k} v_j w_j(p_i, i - k) + af(x_i - C).$$

Summing over all periods, indexed by i, yields the desired formulation:

$$\text{minimize} \quad \sum_{i=1}^{n} p_i \left(\sum_{k=1, k \neq i}^{n} \sum_{j \in k} v_j w_j(p_i, i - k) \right) + a \sum_{i=1}^{n} f(x_i - C)$$

$$\text{subject to} \quad x_i = X_i - \sum_{j \in i} v_j \sum_{k \neq i}^{n} w_j(p_k, k - i) + \sum_{k \neq i} \sum_{j \in k} v_j w_j(p_i, i - k), \forall i$$

$$\text{variables} \quad p_i \geq 0, \quad \forall i.$$

If the $w(p, t)$ are increasing and concave in p, and f is convex, the ISP's price-, or equivalently, reward-optimization problem is a convex optimization problem that we defined in Chapter 4.

We can carry out several extensions to this problem representation. For example, incorporate stochastic arrivals of new sessions, model the rise of overall traffic (say, over a 24-hour period) due to the attraction of low-priced period, and allow both fixed-size and fixed-duration application sessions.

12.4.2 Paris metro pricing

We have been investigating "how to charge" and "how much to charge." There is also the related question of "what to charge." You would think that a different service, e.g., express delivery, would command a different price. That is true, but it also works the other way: a different price may lead to a different service. This is the idea of "pricing-induced quality differentiation." Whenever a service's quality depends on how crowded its consumers are, we can simply use higher prices to reduce demand in certain portions of the service. This creates a new category of service tier.

Pricing changes consumer behavior and network congestion, and if different congestion levels imply different service grades, i.e., if utility depends on utilization, we can complete the feedback loop where different prices automatically lead to different services. This line of reasoning is exemplified in the **Paris metro pricing**, a thought experiment that Odlyzko presented in 1998 for Internet pricing (and used in the 1970s-1980s in Paris metro services).

Consider a metro (i.e., subway or underground train) service where two passenger cars are identical, but are charged differently: car A's charge is twice as much as car B's. You might protest: how can they charge differently for the exact same service? Well, they are *no longer* the same service as soon as consumers react to the different prices. Compared with car A, car B will be more crowded as the price is lower and the demand is thus higher. Since the congestion level is a key component of the quality of service in riding a metro car, we can call car A first-class and car B coach-class, and their different prices self-justify. This phenomenon applies to all cases where utility depends on utilization.

Even though Paris metro pricing is not yet widely implemented on the Internet, it illustrates an interesting feedback loop between price-setters (a metro company, or an ISP) and intelligent agents (metro riders choosing between cars, or iPhones'

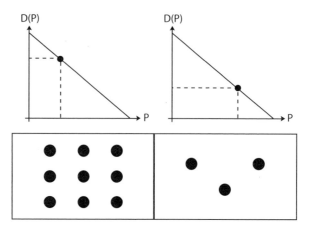

Figure 12.7 Paris metro pricing creates service differentiation through price differentiation. The left graph shows that a lower price increases the demand and the utilization, thus reducing the utility for those in that service tier. The right graph shows that a higher price reduces the demand and utilization, thus increasing the utility for those in that service tier.

network interface cards choosing between WiFi and 3G connections) reacting to the prices. Let us examine Paris metro pricing from an efficiency point of view, and discover a downside of resource pooling.

Suppose an ISP creates two tiers of services on a link: tier A and tier B, and has two possible prices to offer: p_1 and p_2, with $p_2 > p_1$. Demands D_A and D_B can be such that the following statements are true.

- If the ISP offers p_1, the demand will be $D_A(p_1) + D_B(p_1)$, which is too big and causes too much congestion for tier A users to find the service useful. So tier A drops out at this price p_1 completely, and the demand becomes just $D_B(p_1)$, with a revenue of $p_1 D_B(p_1)$ to the ISP.
- If the ISP offers p_2, the demand will be $D_A(p_2) + D_B(p_2)$, which is clearly smaller than $D_A(p_1) + D_B(p_1)$ since price p_2 is higher than p_1. Let us say it is small enough that tier A users stay. The revenue becomes $p_2(D_A(p_2) + D_B(p_2))$ to the ISP.

So the ISP must choose between the two prices, and the revenue is

$$\max\{p_1 D_B(p_1), p_2(D_A(p_2) + D_B(p_2))\}. \tag{12.2}$$

Now, consider a different scheme. The ISP divides the link into two equal parts. In one part, it sets the price to be p_1 and gets revenue $p_1 D_B(p_1)$. In the other part, it sets the price to be p_2 and gets revenue $p_2(D_A(p_2) + D_B(p_2))$. For any demand functions (D_A, D_B), the prices can be set sufficiently high so that each of these demands can fit into half of the link capacity. Now the revenue becomes the *sum*:

$$p_1 D_B(p_1) + p_2(D_A(p_2) + D_B(p_2)), \tag{12.3}$$

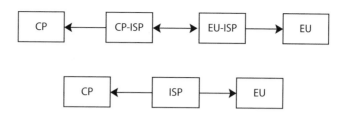

Figure 12.8 An illustration of two-sided pricing, where ISPs charge both the producers and consumers of content and applications. A content-provider-side ISP, CP-ISP, charges content providers, while an end-user-side ISP, EU-ISP, charges EU. The two ISPs may also be the same common ISP.

which is clearly higher than (12.2). A homework problem will further explore Paris metro pricing.

This **anti-resource-pooling** property in revenue maximization through flexible pricing, or the resource-segmentation gain, is the opposite of the principle of statistical multiplexing used in achieving the economy of scale, in smoothing time-varying demand, and in avoiding fragmentation of resources, as we will see in Chapters 13 and 16. In contrast to those chapters, the purpose here is to maximize revenue through multi-class pricing. Anti-resource-pooling turns out to be the right way to go as it provides a more granular control of pricing of different parts of the overall resource.

12.4.3   Two-sided pricing

We conclude this chapter with a brief discussion of "whom to charge?" In **two-sided Internet pricing** models, like in the credit card business, the price of connectivity is shared between content providers (CPs) and end users (EUs). ISPs are just the middle man proving the connectivity between CPs and EUs. A "clearing house" of connectivity exchange market will be a major extension of the 1-800 model of phone-call services in the USA, which charges the callee rather than the caller.

The tradeoff in the resulting economic benefits between CPs and EUs remains to be quantified. Intuitively, end-users' access prices are subsidized and the ISPs have an additional source of revenue. Perhaps more surprisingly, content-providers may also stand to gain from two-sided pricing if subsidizing connectivity to end-users translates into a net revenue gain through a larger amount of consumption. However, the gain to content providers depends on the extent to which content-provider payment translates into end-users' subsidy, and on the demand elasticities of the consumers. The precise gains to the three entities will therefore depend on their respective bargaining powers stemming from their contributions and price sensitivities.

The economic interaction on the flow of data between the EU and the CP includes the peering and transit arrangements between multiple ISPs that operate the links between the CP and the EU, and the access fee charged by the ISPs to the EU and the CP. We can consider an "eyeball ISP", referred to as the EU-ISP, charging an access price of $g + hx$ to the EU; and a "content ISP", referred to as the CP-ISP, charging an access price of $p + qx$ to the CP. The EU-ISP and the CP-ISP can collaborate when charging the access fees to the EU and CP, or act independently. In a homework problem, we will model the collaboration through a representative ISP as shown in Figure 12.8.

Summary

> **Box 12** Smart data pricing
>
> Smart data pricing is a suite of strategies that can help create a win-win across ISPs, consumers, and content and app providers. Charging based on when you use the capacity can time-shift some of the traffic away from the peaks while increasing the total data usage over a 24-hour cycle. Paris metro pricing creates differentiated services through differentiated prices. Two-sided pricing enables the content providers to subsidize consumers at the right pricing.

Further Reading

There is growing literature on Internet access and mobile data pricing, including congestion-, time-, or location-dependent pricing.

1. A short survey of some common topics in transportation networks and the Internet can be found in
F. P. Kelly, "The mathematics of traffic in networks," *Princeton Companion to Mathematics*, Princeton University Press, 2009.

2. Our treatment of time-dependent pricing for Internet access follows
S. Ha, S. Sen, C. Joe-Wong, Y. Im, and M. Chiang, "TUBE: Time-dependent pricing for mobile data," in *Proceedings of ACM Sigcomm*, 2012.

3. The original presentation of the Paris metro pricing method for the Internet can be found at
A. Odlyzko, "Paris metro pricing for the Internet," in *Proceedings of ACM Conference on Electronic Commerce*, 1998.

4. A standard reference on two-sided pricing is
J. C. Rochet and J. Tirole, "Two-sided markets: A progress report," *The RAND Journal of Economics*, vol. 35, no. 3, pp. 645–667, 2006.

5. More persepctives from academia, ISPs, content providers, equipment and software vendors on the challenges and opportunities of data pricing, including practical software implementation and user-friendly interface design, can be found at

Smart Data Pricing (SDP) Forum: `http://scenic.princeton.edu/SDP2012`

Problems

12.1 *Time-dependent pricing* ⋆⋆

An ISP tries to even out the capacity demand over day and night by rewarding its users for delaying their data transmission. Suppose there are just two types of users, type A and type B, which have different levels of willingness to delay their sessions. Originally the demand during day-time is in total $v_{A,day} + v_{B,day} = 14$ GB, which consists of $v_{A,day} = 8$ GB from type A users and $v_{B,day} = 6$ GB from type B users. The demand during night-time is in total $v_{A,night} + v_{B,night} = 5$ GB, which consists of $v_{A,night} = 2$ GB from type A users and $v_{B,night} = 3$ GB from type B users.

Suppose the ISP has capacity $C = 10$ GB and the marginal cost of exceeding capacity is \$1 per GB. It provides a reward of \$$p$ per GB for day-time users to delay their data transfer until night-time. Let $w_A(p)$ and $w_B(p)$ be the proportions of data from type A and type B users, respectively, to be delayed from day-time to night-time:

$$w_A(p) = 1 - \exp\left(-\frac{p}{p_A}\right),$$

$$w_B(p) = 1 - \exp\left(-\frac{p}{p_B}\right),$$

where parameters $p_A = 4$ and $p_B = 2$.

The ISP wishes to find the reward price p^* that minimizes its total cost, i.e., the sum of the cost due to the demand exceeding the capacity and the rewards given out.

(a) What is the formulation of the minimization problem?

(b) Solve p^* numerically by plotting the objective function over the reward price p.

12.2 *User-type estimation* ⋆⋆

Consider the same model as in the above problem, except that now the values $v_{A,day}, v_{B,day}, v_{A,night}$ and $v_{B,night}$ are unknown. Suppose that originally the demand during daytime is in total 17 GB, and after announcing a reward price of \$0.30 per GB the demand during daytime reduces to 15.2 GB in total. What are $v_{A,day}$ and $v_{B,day}$?

12.3   *TDP for smart-grid demand response* ★★

Smart-grid providers often set time-dependent prices for energy usage. This problem considers a simplified example with two periods, the day-time and the night-time. The provider can set different prices for the two periods, and wishes to shift some night-time usage to day-time. The energy provider always offers the full price during the night, and offers a reward of \$p/kWh during the day.

Suppose that with uniform (time-independent) prices, customers vacuum at night, using 0.2 kWh, and also watch TV, using 0.5 kWh, and do laundry, using 2 kWh. During the day, customers use 1 kWh. Suppose the probability of users shifting vacuum usage from the night to the day is

$$1 - \exp\left(-\frac{p}{p_V}\right),$$

(12.4)

where $p_V = 2$. The probability of shifting doing their laundry to the daytime is

$$1 - \exp\left(-\frac{p}{p_L}\right),$$

(12.5)

where $p_L = 3$. Users never shift their TV watching from the night to the day.

Suppose that the electricity provider has a capacity of 2 kWh during the night and 1.5 kWh during the day. The marginal cost of exceeding this capacity is \$1/kWh. In this problem, we ignore the energy cost when the capacity is not exceeded.

(a) Compute the expected amount of vacuum and laundry energy usage (in kWh) that is shifted from the night to the day, as a function of p.

(b) Find the reward p which maximizes the energy provider's profit.

(c) Suppose that if vacuum or laundry usage is shifted from the night to the day, it is shifted by 12 hours. Compute the expected time shift of vacuum and laundry under $p = p^*$, the optimal reward found above.

12.4   *Paris metro pricing* ★★★

Consider a metro system where two kinds of services are provided: service class 1 and service class 2. Let p_1 and p_2 be the one-off fees charged per user when accessing service classes 1 and 2, respectively. Suppose each user is characterized by a valuation parameter $\theta \in [0, 1]$ such that its utility of using service class i is

$$U_\theta(i) = (V - \theta K(Q_i, C_i)) - p_i,$$

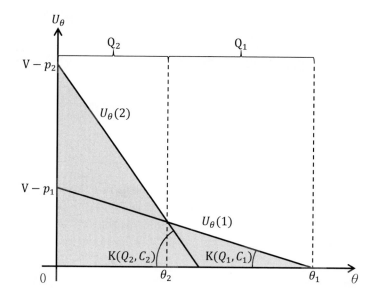

Figure 12.9 An illustration of an equilibrium in Paris metro pricing.

where V is the maximum utility of accessing the service, $K(Q_i, C_i)$ measures the amount of congestion of service class i, given $Q_i \geq 0$ as the proportion of users accessing service class i (with $\sum_i Q_i = 1$), and $C_i \geq 0$ as the proportion of capacity allocated to service class i (with $\sum_i C_i = 1$).

At the equilibrium, i.e., no user changes from her selection, $U_\theta(i)$ is a linear function of θ. Suppose the equilibrium is illustrated as in Figure 12.9.

(a) Let θ_1 be the θ of the user who is indifferent to joining the first service class or opting out of all the services, θ_2 be that of the user who is indifferent to joining the first service class or the second service class, and $F(\theta)$ be the cumulative distribution function of θ. Show that

$$Q_1 = F(\theta_1) - F(\theta_2),$$
$$Q_2 = F(\theta_2),$$
$$V - p_1 = \theta_1 K(Q_1, C_1),$$
$$p_1 - p_2 = \theta_2(K(Q_2, C_2) - K(Q_1, C_1)).$$

(b) Assume that θ is uniformly distributed, i.e., $F(\theta) = \theta$, and that the congestion function is defined as

$$K(Q, C) = \frac{Q}{C}.$$

Solve θ_1 and θ_2 as functions of V, p_1, and p_2.

(Hint: Try $\frac{p_1 - p_2}{V - p_1}$. You may define shorthand notation such as $k = \frac{p_1 - p_2}{V - p_1}$ during the derivation before the formulas become too complicated.)

(For details, see C. K. Chau, Q. Wang, and D. M. Chiu, "On the viability of Paris metro pricing for communication and service networks," in *Proceedings of IEEE Infocom*, 2010.)

12.5   *Two-sided pricing* ⋆⋆

Consider the model where an ISP charges a content provider (CP) a usage price h_{CP} and a flat price g_{CP}, and charges an end user (EU) a usage price h_{EU} and a flat price g_{EU}. Here, for simplicity we assume zero flat price $g_{CP} = g_{EU} = 0$. Let μ be the unit cost of provisioning capacity. The demand functions of the CP and EU, denoted as D_{CP} and D_{EU}, respectively, are given as follows:

$$D_{CP}(h_{CP}) = \begin{cases} x_{CP,max}\left(1 - \frac{h_{CP}}{h_{CP,max}}\right) & \text{if } 0 \le h_{CP} \le h_{CP,max} \\ 0 & \text{if } h_{CP} > h_{CP,max}, \end{cases}$$

$$D_{EU}(h_{EU}) = \begin{cases} x_{EU,max}\left(1 - \frac{h_{EU}}{h_{EU,max}}\right) & \text{if } 0 \le h_{EU} \le h_{EU,max} \\ 0 & \text{if } h_{EU} > h_{EU,max} \end{cases}$$

The parameters are specified as follows:

$$h_{CP,max} = 2\mu,$$
$$h_{EU,max} = 1.5\mu,$$
$$x_{CP,max} = 1,$$
$$x_{EU,max} = 2.$$

The ISP maximizes its profit by solving the following maximization problem

$$\begin{array}{ll} \text{maximize} & (h_{CP} + h_{EU} - \mu)x \\ \text{subject to} & x \le \min\{D_{CP}(h_{CP}), D_{EU}(h_{EU})\} \\ \text{variables} & x \ge 0, h_{CP} \ge 0, h_{EU} \ge 0. \end{array} \qquad (12.8)$$

Find the optimal x^*, h_{CP}^*, and h_{EU}^*.

13 How does traffic get through the Internet?

We have mentioned the Internet many times so far, and all the previous chapters rely on its existence. It is about time to get into the architecture of the Internet, starting with these two chapters on the TCP/IP foundation of the Internet.

13.1   A Short Answer

We will be walking through several core concepts behind the evolution of the Internet, providing the foundation for the next four chapters. So the "short answer" section is going to be longer than the "long answer" section in this chapter.

It is tricky to discuss the historical evolution of technologies like the Internet. Some of what we would like to believe to be the inevitable results from careful design are actually the historical legacy of accidents, or the messy requirements of backward compatibility, incremental deployability, and economic incentives. It is therefore not easy to argue about what could have happened, what could have been alternative paths in the evolution, and what different tradeoffs might have been generated.

13.1.1   Packet switching

The answer to this chapter's question starts with a fundamental idea in designing a network: when your typical users do not really require a *dedicated* resource, you should allow users to *share* resources. The word "user" here is used interchangeably with "session." The logical unit is an application session rather than a physical user or device. For now, assume a session has just one source and one destination, i.e., a **unicast** session.

In the case of routing, the resource lies along an entire path from one end of a communication session, the sender, to the other end, the receiver. We can either dedicate a fixed portion of the resources along the path to each session, or we can mix and match packets from different sessions and also share all the paths. This is the difference between **circuit-switched** and **packet-switched** networks.

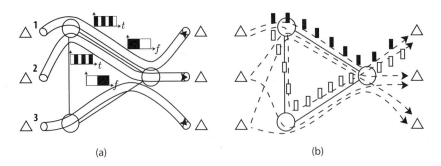

Figure 13.1 A simple network with three interconnected routers and three sessions. (a) Circuit switching: each session gets a dedicated circuit, either a portion of each timeslot t or a fraction of the frequency band f, even when it is not used. (b) Packet switching: each session sends packets along one or more paths (when there are packets to send) and all paths are shared across timeslots and frequency bands.

Before the 1960s, networking was mostly about connecting phone calls in circuit-switched Public Switched Telephone Networks (**PSTNs**). There continued to be active research all the way to the early 2000s, including dynamic routing as you will see in a homework problem.

A revolution, which came to be known as the Internet, started during the 1960s–1970s, witha shift to packet switching as the fundamental paradigm of networking. In the early 1960s, researchers formally developed the idea of chopping up a session's messages into small packets, and sending them along possibly different paths, with each path shared by other sessions. Figure 13.1 contrasts circuit switching with packet switching. Each circuit in circuit switching may occupy either a particular frequency band or a dedicated portion of timeslots. In contrast, in packet switching, there is no dedicated circuit for each session. All sessions have their packets sharing the paths.

In 1969, sponsored by the US Advanced Research Project Agency (ARPA), UCLA and three other institutions put together the first prototype of a packet-switched network, which came to be known as the **ARPANET**. The ARPANET started to grow. In 1974, Cerf and Kahn developed a protocol, i.e., a set of rules for communication among the devices, for packet-switched networks, called the Transmission Control Protocol/Internet Protocol (**TCP/IP**). This protocol enabled scalable connectivity in the ARPANET. From 1985 to 1995, the US National Science Foundation (NSF) took over the next phase of development, sponsoring the creation and operation of an ever-increasing *network of networks* called the **NSFNET**. Starting in the early 1990s, commercial interests and entrepreneurial activities dramatically expanded this inter-connected network of networks. Indeed, by 1994, the World Wide Web and web browser user-interface had matured, and the world quickly moved into commercial applications built on top of this network, known by then as the Internet. Today the Internet has blossomed into an essential part of how people live, work, play, talk, and think.

There are now more Internet-connected devices than people in the world, and it is projected that by 2020 there will be six times as many connected devices as people. It has been a truly amazing five decades of technology development.

The debate between dedicated resource allocation and shared resource allocation runs far and deep. In addition to circuit vs. packet switching here and orthogonal vs. non-orthogonal resource allocation in Chapter 1, we will also see three more special cases of this design choice: client-server vs. peer-to-peer, local storage vs. cloud services, and contention-free scheduling vs. random access in Chapters 15, 16, and 18, respectively.

There is one big advantage of circuit switching, or dedicated resource allocation in general: *guarantee of quality*. As each session gets a circuit devoted to it, throughput and delay performance are accordingly guaranteed, and there is very little jitter (the variance of delay). In contrast, in packet switching, a session's traffic is (possibly) split across different paths, each of which is shared with other sessions. Packets arrive out of order and need to be re-ordered at the receiver. Links may get congested. Throughput and delay performance become uncertain. Internet researchers call this the **best-effort** service that the Internet offers, which is perhaps more accurately described as *no effort* to guarantee performance.

On the other hand, there are two big advantages of packet switching: (1) *ease of connectivity* and (2) *scalability due to efficiency*.

(1) Ease of connectivity is easy to see: there is no need to search for, establish, maintain, and eventually tear down an end-to-end circuit for each session.

(2) Scalability here refers to the ability to take on many diverse types of sessions, some long-duration ones, others short bursts, and to take on many of them. There are two underlying reasons for the efficiency of packet switching, which in turn leads to high scalability. These two reasons correspond to the "many sessions share a path" feature and the "each session can use multiple paths" feature of packet switching, respectively. We call these two features statistical multiplexing and resource pooling.

- **Statistical multiplexing**: packet switching can flexibly map demand of capacity onto supply of capacity. This suits the dynamic, on-demand scenarios with bursty traffic. In particular, when a source is idle and not sending any traffic onto the network, it does not occupy any resources.
- **Resource pooling**: this one takes a little math to demonstrate, as we will in a homework problem. But the basic idea is straightforward: instead of having two sets of resources (e.g., two links' capacities) in isolation, putting them into a single pool lowers the chance that some demand must be turned down because one set of resources is fully utilized.

In the end, the abilities to easily provide connectivity and to scale up with many diverse users won the day, although that was not clear until the early 2000s. In contrast to quality guarantee, which is certainly nice to have, these properties are *essential* to have for a dynamic and large network like the Internet. Once the

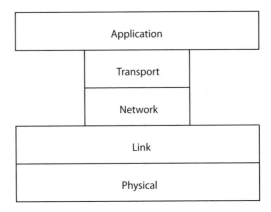

Figure 13.2 Modularization in networking: A typical model of a layered protocol stack. Each layer is in charge of a particular set of tasks, using the service provided by the layer from below and in turn providing a service to the layer above. The horizontal lines that separate the layers represent some kind of limitation of what each layer can see and can do. Over the years, the applications have evolved from file transfer based on command-line inputs to all those applications we experience today. The physical and link layer technologies have evolved from 32 kbps dial-up modem to 10 Gbps optic fibers and 100 Mbps WiFi. The two middle layers, however, dominated by TCP/IP, have remained largely unchanged over the years. They are the "thin waist" of the "hour-glass" model of the protocol stack.

network has grown in an easy and scalable way, we can search for other solutions to take care of quality variation. But you have to grow the network *first*, in terms of the number of users and the types of applications. This is a key reason why IP took over the networking industry and packet switching prevailed, despite alternative designs in protocols (that we will not cover here) like X.25, ATM, frame relay, ISDN, etc.

13.1.2   Layered architecture

Managing a packet-switched network is complicated. There are many tasks involved, and each task's control requires a sequence of communication and computation called a **protocol** to control it. It is a natural practice when engineering such a complex system to break it down into smaller pieces. This process of **modularization** created the **layered protocol stack** for the Internet. The idea of modularizing the design is *not* motivated by efficiency of resource allocation, but by economic viability through the business models of different companies specializing in different layers, and by the robustness regarding unforeseen innovations that may ride on the Internet. This evolvability is further enhanced by the overlay networks that can create new network topologies and functionalities on top of the Internet connectivity, as we will see in Chapter 15.

A typical layered protocol stack is shown in Figure 13.2. TCP/IP sits right in the middle of it in the transport and network layers. Over the short span of Internet evolution, the physical medium's transmission speed has gone up more than 30,000 times, and the applications have gone from command-line-based file transfer to Netflix and Twitter. Yet the Internet itself continued to work, thanks in large part to the "thin waist" of TCP/IP that stayed mostly the same as the applications and the communication media kept changing.

Each layer provides a service to the layer above, and uses a service from the layer below. For example, the transport layer provides an end-to-end connection, running the services of session establishment, packet re-ordering, and congestion control, to the application layer above it that runs applications such as the web, email, and content sharing. In turn, the transport layer takes the service from the network layer below it, including the connectivities established through routing. The link layer is charged with controlling the access to the communication medium, and the physical layer controls the actual transmission of information on the physical medium.

There are functional *overlaps* across layers. For example, the functionality of error control is allocated to many layers: there is error control coding in the physical layer, hop-by-hop retransmission at the link layer, multipath routing for reliability in the network layer, and end-to-end error checking at the transport layer. Functional redundancy is not a bug, it is there by design, paying the price of efficiency reduction for robustness and clear boundaries among the layers.

How should we allocate functionalities among the layers and put them back together at the right interface and timescale? That is the question of **network architecture** that we will continue to explore in later chapters. For example, the horizontal lines in Figure 13.2, denoting the boundaries between protocol layers, are actually very complicated objects. They represent limitations as to what each layer can *do* and can *see*. In the next chapter, we will get a glimpse of some methodologies to understand this architectural decision of "who does what" and "how to glue the modules together."

Just between the transport and network layers, there are already quite a few interesting architectural decisions made in TCP/IP, the dominant special case of the layers 4/3 protocol. First, the transport layer, in charge of end-to-end management, is *connection-oriented* in TCP, whereas the network layer, in charge of connectivity management, runs hop-by-hop *connectionless* routing in IP. As an analogy, calling someone on the phone requires a connection-oriented session to be established first between the caller and the callee. In contrast, sending mail to someone needs only a connectionless session since the recipient does not need to know there is a session coming in. The design choice of connection-oriented TCP and connectionless IP follows the "end-to-end" principle that end-hosts are intelligent and the network is "dumb." Connectivity establishment should be entirely packet switched in the network layer, and end-to-end feedback run by the layer above. But this design choice was *not* the only one that the Internet

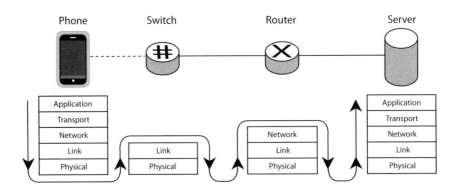

Figure 13.3 Different network elements process up to different layers in the protocol stack. The end-hosts process all the way up to the application layer. Switches that forward frames process up to the link layer. Routers that move datagrams across the network process up to the network layer. In a realistic end-to-end connection, there are usually many more hops.

tried over its decades of evolution, e.g., a connectionless transport layer on top of a connection-oriented network layer is also possible and indeed was once used.

Second, routing in IP is independent of load conditions on the links, whereas congestion control in TCP takes care of managing demand at the end-hosts in response to link loads. In addition to the end-to-end principle, this strategy assumes that rate adaptation at the end hosts is easier to stabilize than route adaptation inside the network.

As we will see in a homework problem, there is also an interesting architectural division-of-labor between the network layer and the link layer below it.

Zooming out of the protocol stack again, the application layer runs applications that generate a sequence of **messages**. Each of these is divided into **segments** at the transport layer, with a layer 4 **header** added in front of the actual content, called **payload**. Then it is passed on to the network layer, which divides and encapsulates the segments as datagrams or **packets**, with a layer 3 header in the front. Each datagram is further passed on to the link layer, which adds another layer 2 header to form a **frame**. This is finally passed on to the physical layer for transmission. These headers are overheads, but they contain useful, sometimes essential, identification and control information. For example, the layer 3 header contains the source node's address and the destination node's address, which are no doubt useful to have in routing. We will examine the impact of these semantic overheads on performance in Chapter 19.

Each network element, e.g., your home gateway, your company's WiFi controller, the central office equipment near the town center, the big router inside the "cloud," runs a subset of the layered protocol stack. Each will decode and read the header information associated with its subset. This is illustrated in Figure 13.3.

13.1.3 Distributed hierarchy

The Internet is not just complex in terms of the number of tasks it has to manage, but also big in terms of the number of users. Hierarchy becomes essential. An end-to-end session, for example, a YouTube streaming session from Google servers to your iPhone may traverse a wireless air-interface, a few links in the cellular core network, and then a sequence of even more links across possibly multiple ISPs in the public Internet.

While *modularization* helps take care of the complexity by "divide and conquer" in terms of functionalities, *hierarchy* helps take care of the large size by "divide and conquer" in terms of the physical span. This is a recurring theme in many chapters. In the current one, we see that the Internet, this network of networks with more than 30,000 **Autonomous Systems** (ASs), has several main hierarchical levels as illustrated in Figure 13.4.

- A few very large ISPs with global footprints are called **tier-1 ISPs**, and they form a *full-mesh* **peering** relationship among themselves: each tier-1 ISP has some connections with each of the other tier-1 ISPs. This full mesh network is sometimes called the Internet backbone. Examples of tier-1 ISPs include AT&T, BT, Level 3, Sprint, etc.
- There are many more tier-2 ISPs with regional footprints. Each tier-1 ISP is connected to some tier-2 ISPs, forming a **customer–provider** relationship. Each of these tier-2 ISPs provides connectivity to many tier-3 ISPs, and this hierarchy continues. The point at which any two ISPs are connected is called the Point of Presence (PoP).
- An ISP of any tier could be providing Internet connectivity directly to consumers. Those ISPs that take traffic only to or from their consumers, but not any transit traffic from other ISPs, are called stub ISPs. Typically, campus, corporate, and rural residential ISPs belong to this group.

Another useful concept in distributed hierarchy is that of a **domain**. Each business entity forms a domain called an AS. There is often a centralized controller within each AS. As we will see later in this chapter, routing *within* an AS and routing *across* ASs follow very different approaches.

Later, in Chapter 15, we will also see how functional and spatial hierarchies combine in building overlay networks.

13.1.4 IP routing

Packet switching, layered architecture, and distributed hierarchy are three fundamental concepts of the Internet. With those topics discussed, we can move on to routing in the Internet.

Transportation networks often offer interesting analogies for communication and social networks. In this case, we can draw a useful analogy from the postal

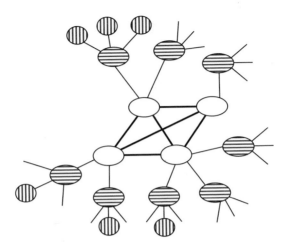

Figure 13.4 Spatial hierarchy in networking: Multiple levels of ISPs and their relationships. Each node in this graph is an ISP, and each link represents a business relationship and physical connections between two ISPs. The four ISPs in the center are tier-1 ISPs, with peering links among themselves. Each of them provides connectivity to many customer ISPs. The stub ISPs at the edge do not provide transit service to any other ISPs. An ISP at any tier may also provide connections to the end-users, which are not shown here.

mail service. In order to route a letter from a sender to the receiver, we need three main functionalities.

- *Addressing.* We first need to attach a unique label to each node in the network, for otherwise we cannot even identify sources and destinations. In the mail system, the label is the postal address, like a street address or mailbox number. Zip codes can quickly zoom you into a subnetwork of the country. In the Internet, we use the **IP address**, a 32-bit number often represented as four decimal numbers separated by dots. Each of these four numbers ranges from 0 to 255 since it is specified by 32/4=8 bits, for example, 127.12.5.88. "Zip codes" here are called subnet masks, for example, 127.12.5.0/24 means that the first 24 bits give the **prefix** of all this subnet's IP addresses: each IP address in this subnet must start with 127.12.5, and can end with any 8 bits. However, in the mail system, an address and a person's ID are separated. In the Internet, an IP address is both an address for establishing connectivity and an identifier of a device. This double loading of functionality onto IP addresses caused various control problems in the Internet.
- *Routing.* Then you have to decide the paths, either one path for each session (single-path routing) or multiple paths for each session (multipath routing). Postal mail uses single-path routing, and routing decides ahead of time which intermediate cities the mail goes through in order to reach, say,

Princeton, NJ, from Stanford, CA. There are two broad classes of routing methods: *metric*-based and *policy*-based routing. Inside an AS, routing is based on some kind of metric, either picking the shortest path between the given source and destination, or distributing the traffic across the paths so that no single path is too loaded. In between the ASs, however, routing is based on policies. For example, AS 1 might suspect there are hackers connected through AS 2, therefore it avoids routing packets along any path traversing AS 2.

- *Forwarding.* Forwarding implements the routing policy. The actual action of forwarding happens each time a packet is received at a router, or each letter is received at an intermediate post office. Some forwarding mechanisms look only at the destination address to decide the next hop, while others, like MultiProtocol Label Switching (**MPLS**), read some labels attached to the packet that explicitly indicate the next hop. In any case, a forwarding decision is made, and one of the egress links connected to the router is picked to send the packet.

Let us look at each of the above in a little more detail now, before focusing the rest of the chapter on just the routing portion.

There are two versions of IP: version 4 and version 6. IPv4 uses 32 bits for addresses, which ran out as of early 2011. IPv6 uses four times as many bits, 128 bits, translating into 2^{128}, about 10^{39}, available addresses. That might sound like a lot, but with the proliferation of Internet-connected devices, we are well on our way to using many of these addresses. One way to upgrade an IPv4 network into IPv6 is to create a "tunnel" between two legacy IPv4 network elements, where IPv6 packets are encapsulated in IPv4 headers.

How are these IP addresses allocated? They used to be given out in blocks, with different block sizes determined by the "class". For example, each class-A address block has a fixed 8-bit prefix, so $2^{32-8} = 2^{24}$ addresses in a class-A block. That is usually given to a national ISP or a large equipment vendor. Lower classes have fewer addresses per block. But this coarse granularity of 8-bit blocks introduced a lot of waste in allocated but unused IP addresses. So the Internet community shifted to Classless InterDomain Routing (**CIDR**), where the granularity does not have to be in multiples of 8 bits.

As a device, you either have a fixed, static IP address assigned to you, or you have to get one dynamically assigned to you by a controller sitting inside the operator of the local network. This controller is called the Dynamic Host Configuration Protocol (**DHCP**) server. A device contacts the DHCP server, receives a currently unused IP address, and returns it to the IP address pool when no longer needed. You may wonder how a device can communicate with a DHCP server in the first place. We will address the protocols involved in Chapter 19. Sometimes the address given to a device within a local network, e.g., a corporate intranet, is different from the one seen by the outside world, and a Network Address Translation (**NAT**) router translates back and forth.

As mentioned, inter-AS routing is very different from intra-AS routing. Border Gateway Protocol (**BGP**) is the dominant protocol for address discovery and reachability for inter-AS routing. It "glues" the Internet together. However, as a policy-based routing protocol, it is a complicated, messy protocol, with many gray areas. We will only briefly describe it in the Advanced Material.

Within an AS, there are two main flavors of metric-based routing protocols: Routing Information Protocol (**RIP**) uses the **distance vector** method, where each node collects information about the distances between itself and other nodes, and Open Shortest Path First (**OSPF**) uses the **linked state** method, where each node tries to construct a global view of the entire network topology. We will focus on the simpler RIP in the next section, saving OSPF for the Advanced Material.

A packet arrives at a router interface, that interface acts as the input port for the packet while another interface acts as the output port. In-between is the switching fabric that physically moves the packet from an input port to the output port. If packets arrive too fast, congestion occurs inside the router. Sometimes it occurs because the intended output ports are occupied, sometimes because the switching fabric is busy, and sometimes because a packet is waiting for its turn at the input port's queue, thus blocking all the packets behind it in the same input queue.

Which output port is the "right" one? That is decided by looking up the forwarding table, which is either stored centrally in the router, or duplicated with one copy at each input port. The forwarding table connects the routing decisions to actual forwarding actions. A common type of forwarding table lists all the destination IP addresses in the Internet, and indicates which output port, thus the next hop router, a packet should go to on the basis of its destination address written in the header. There are too many IP addresses out there, so the forwarding table often groups many addresses into one equivalent class of input.

We are now going to study one member of the intra-AS routing family, and then how forwarding tables are constructed from distributed messages passing among the routers.

13.2   A Long Answer

Consider a directed graph $G = (V, E)$ representing the topology inside an AS, where each node in the node set V is a router, and each link in the link set E is a physical connection from one router i to another router j.

Each link has a cost c_{ij}. It is often a number approximately proportional to the length of the link. If it is 1 for all the links, then minimizing the cost along a path is the same as minimizing the hop count. If it were dynamically reflecting the congestion condition on that link, it would lead to dynamic, load-sensitive routing. But IP does not practice dynamic routing, leaving load sensitivity to TCP congestion control.

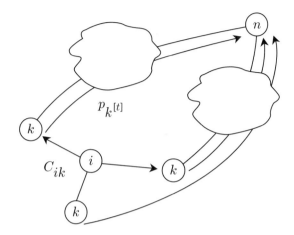

Figure 13.5 Bellman's equation for minimum-cost routing. The minimum cost from a node i to destination n is the smallest, among all its neighbors, of the sums of the cost from i to a neighbor and the cost from that neighbor to n. Node i does not need to know how its neighbors get to n, just the cost of reaching n via each neighbor.

The **shortest path problem** is an important special case of the network flow problem, which is in turn an important special case of linear programming. A more accurate name is the *minimum cost path problem*, since we are actually finding the minimum cost path for any pair of nodes in a given graph. "Minimum cost" would be equivalent to "shortest" only when the link costs are the physical distances. But still, "shortest path problem" sticks, and we will use that term.

The shortest path problem has been studied extensively since the early 1950s, and there are several famous approaches: the Bellman–Ford algorithm, the Dijkstra algorithm, Lagrange duality, etc. We will focus on the **Bellman–Ford algorithm**, because it is simple, distributed, and illustrates the fundamental principle of **dynamic programming**. It also leads to a fully distributed and asynchronous implementation used in RIP, in part of BGP, and in the routing method in the original ARPANET.

First, a little bit of notation. For now, fix one destination node n; we can generalize to multiple destinations readily. Let $p_i[t]$ be the length of the shortest path from node i to destination n using *at most* t links. It is not a coincidence that we are using the time symbol t to capture this spatial definition. We will soon see that t indeed indexes the iterations too.

If node i cannot reach destination n in t hops, we say $p_i[t] = \infty$. Obviously, at initialization $p_i[0] = \infty$ for all nodes i, unless it is the destination n itself.

Here comes the core idea behind the Bellman–Ford algorithm. Obviously, i needs to get to n via some neighbor. And we realize that $p_i[t+1]$ can be *decomposed* into two parts, as illustrated in Figure 13.5.

- The cost c_{ik} of getting from node i to one of its outgoing neighbors k. An outgoing neighbor is a node where there is a link pointing from i to it, and we denote the set of these neighbors for node i as $O(i)$.

- The minimum cost of getting from that neighbor k to the destination n, using at most t hops, since we have already used 1 hop (out of $t + 1$ hops) just to get to node k.

The minimum-cost path picks the neighbor that minimizes the *sum* of the above two costs:

$$p_i[t + 1] = \min_{k \in O(i)} \{c_{ik} + p_k[t]\}. \tag{13.1}$$

Let us assume each node knows the cost to reach its outgoing neighbors. Then, by iteratively updating $p_i[t]$ and passing a vector describing these updated numbers to its neighbors, we can execute (13.1) in a distributed way. No wonder it is called a distance vector routing algorithm. Passing the distance vectors around and initializing the c_{ik} values require a little protocol, as we will see soon with RIP.

A quick detour: there is actually a very broad idea behind the recursive **Bellman equation** (13.1). Optimizing over a sequence of timeslots or spatial points is studied in the research area of dynamic programming. For many system models where the cost we want to minimize is *additive* over time or space, and the dependence between stages of the problem is *memoryless* (the next stage depends only on the current one), we know that (this is a mouthful) the "tail" of the optimal solution to the cost minimization is the optimal solution to the "tail" of the cost minimization. The Bellman Ford algorithm is a special case of this general principle.

13.3   Examples

13.3.1   Centralized Bellman–Ford computation

First, an example on the Bellman–Ford algorithm. Suppose we have a network topology as in Figure 13.6. The negative link weights are just there to show that the Bellman–Ford algorithm can accommodate them, as long as there are no negatively weighted *cycles*, since those can reduce some path costs to negative infinity.

In this small example with four nodes (not counting the destination), we know we can stop after four iterations, since any path traversing five nodes or more will have to go around a cycle, and that can only add to the length, thus never being optimal. But in a real, distributed implementation of a distance vector protocol, we do not know how many nodes there are, so we have to rely on the lack of new messages to determine when it is safe to terminate the algorithm.

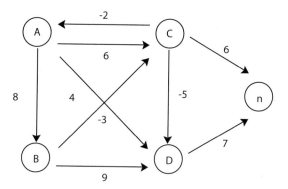

Figure 13.6 An example to illustrate the Bellman–Ford algorithm. Here we want to find out the shortest paths from nodes A, B, C, and D to a common destination node n. There are negatively weighted links just to illustrate that the algorithm can handle them. But there are no negatively weighted cycles, for they would have made the shortest path problem ill-defined.

We try to find the minimum paths from nodes A, B, C, and D to destination n. We initialize distances $p_A[0] = p_B[0] = p_C[0] = p_D[0] = \infty$. And of course it takes zero cost to reach oneself: $p_n[t] = 0$ at all times t.

For $t = 1$, by Bellman's equation, we have

$$p_A[1] = \min\{c_{AB} + p_B[0], c_{AC} + p_C[0], c_{AD} + p_D[0]\}$$
$$= \min\{8 + \infty, 6 + \infty, 4 + \infty\}$$
$$= \infty,$$

$$p_B[1] = \min\{c_{BC} + p_C[0], c_{BD} + p_D[0]\}$$
$$= \min\{-3 + \infty, 9 + \infty\}$$
$$= \infty,$$

$$p_C[1] = \min\{c_{Cn} + p_n[0], c_{CD} + p_D[0], c_{CA} + p_A[0]\}$$
$$= \min\{6 + 0, -5 + \infty, -2 + \infty\}$$
$$= 6,$$

$$p_D[1] = \min\{c_{Dn} + p_n[0]\}$$
$$= 7 + 0$$
$$= 7.$$

Notice that node D has only one outgoing link, so $p_D = 7$ and we do not need to keep calculating it.

Similarly, for $t = 2$, we have

$$p_A[2] = \min\{c_{AB} + p_B[1], c_{AC} + p_C[1], c_{AD} + p_D[1]\}$$
$$= \min\{8 + \infty, 6 + 6, 4 + 7\}$$
$$= 11,$$

$$p_B[2] = \min\{c_{BC} + p_C[1], c_{BD} + p_D[1]\}$$
$$= \min\{-3 + 6, 9 + 7\}$$
$$= 3,$$

$$p_C[2] = \min\{c_{Cn} + p_n[1], c_{CD} + p_D[1], c_{CA} + p_A[1]\}$$
$$= \min\{6 + 0, -5 + 7, -2 + \infty\}$$
$$= 2.$$

For $t = 3$,

$$p_A[3] = \min\{c_{AB} + p_B[2], c_{AC} + p_C[2], c_{AD} + p_D[2]\}$$
$$= \min\{8 + 3, 6 + 2, 4 + 7\}$$
$$= 8,$$

$$p_B[3] = \min\{c_{BC} + p_C[2], c_{BD} + p_D[2]\}$$
$$= \min\{-3 + 2, 9 + 7\}$$
$$= -1,$$

$$p_C[3] = \min\{c_{Cn} + p_n[2], c_{CD} + p_D[2], c_{CA} + p_A[2]\}$$
$$= \min\{6 + 0, -5 + 7, -2 + 11\}$$
$$= 2.$$

For $t = 4$,

$$p_A[4] = \min\{c_{AB} + p_B[3], c_{AC} + p_C[3], c_{AD} + p_D[3]\}$$
$$= \min\{8 - 1, 6 + 2, 4 + 7\}$$
$$= 7,$$

$$p_B[4] = \min\{c_{BC} + p_C[3], c_{BD} + p_D[3]\}$$
$$= \min\{-3 + 2, 9 + 7\}$$
$$= -1,$$

$$p_C[4] = \min\{c_{Cn} + p_n[3], c_{CD} + p_D[3], c_{CA} + p_A[3]\}$$
$$= \min\{6 + 0, -5 + 7, -2 + 8\}$$
$$= 2.$$

We can also readily keep track of the paths taken by each node to reach n: D directly goes to n, C goes through D to reach n, B goes through C, and A goes through B.

13.3.2   Distributed RIP

So far we have assumed centralized computation. But imagine you are one of the nodes trying to figure out how to reach the other nodes in the network. How do you know the cost to reach different nodes and how do you even start?

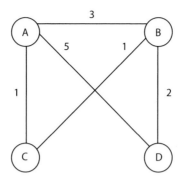

Figure 13.7 An example to illustrate the distributed message passing in RIP, where each node wants to find one of the shortest paths to every other node through message passing. Again, we choose an extremely small network so that we can go through the numerical steps in detail. A major challenge in routing in the Internet is actually the scale of the network, hence the desire for distributed solutions in the first place.

We now describe the message-passing protocol in RIP, which allows the discovery and update of c_{ik} and $p_i[t]$ across the nodes. For simplicity of presentation, assume all the links are bidirectional: if node i can send messages to node j, so can j to i.

The message passed around in distance vector routing protocols has the following format: [NodeID, DestinationID, Cost of MinCost Path].

At the very beginning, iteration 0, each node knows only its own existence, so each node i can only pass around this vector [node i, node i, 0]. But once each node has received the messages from its neighbors, it can update its list of vectors.

There are several key features of the message passing.

- *Short messages*: All the detailed topology information (about who connects to whom and what the link costs are) can be *summarized* into these lists of distance vectors.
- *Local interaction*: Neighbor-to-neighbor message passing is enough to propagate this summary so that end-to-end shortest paths can be discovered. There is no need for broadcasting all the summaries to all the nodes.
- *Local view*: Even when these optimal paths are obtained, still each node has only a local view: it knows only which neighbor to send a packet with a given destination address, but has no idea what the actual end-to-end path looks like. As far as forwarding packets using the knowledge of their destination IP addresses is concerned, it does not need to know, because the routers will forward packets hop by hop.

Here is an example to illustrate this iterative and distributed routing method, for the small network shown in Figure 13.7. We try to find the minimum paths between all nodes. We can collect all the distance vectors [NodeID, DestinationID,

Cost of MinCost Path], together with the next hop decision, into a routing table for each node. Each node stores only the information for its own NodeID. At $t = 0$, we have the following four near-empty tables, one per node:

NodeID	DestinationID	Cost of MinCost Path	Next node
A	A	0	A

NodeID	DestinationID	Cost of MinCost Path	Next node
B	B	0	B

NodeID	DestinationID	Cost of MinCost Path	Next node
C	C	0	C

NodeID	DestinationID	Cost of MinCost Path	Next node
D	D	0	D

At each iteration, each node sends the distance vectors (the routing table above except the next hop information) to its neighbors. For example, at $t = 1$, node B receives messages from nodes A, C, and D. Node B receives $[A, A, 0]$ from node A; $[C, C, 0]$ from node C; and $[D, D, 0]$ from node D.

In terms of Bellman's equation (13.1), this tells node B that $p_A[0] = 0$ for destination A, $p_C[0] = 0$ for destination C, and $p_D[0] = 0$ for destination D. All other distances are infinite. Node B then uses (13.1) to calculate the new distances in the routing table.

Let us work out the calculations for destination A at $t = 1$:

$$p_A[1] = 0,$$

$$p_B[1] = \min\{c_{BA} + p_A[0], c_{BC} + p_C[0], c_{BD} + p_D[0]\}$$
$$= \min\{3 + 0, 1 + \infty, 2 + \infty\}$$
$$= 3,$$

$$p_C[1] = \min\{c_{CA} + p_A[0], c_{CB} + p_B[0]\}$$
$$= \min\{1 + 0, 1 + \infty\}$$
$$= 1,$$

$$p_D[1] = \min\{c_{DA} + p_A[0], c_{DB} + p_B[0]\}$$
$$= \min\{5 + 0, 2 + \infty\}$$
$$= 5.$$

For destination B at $t = 1$, we have

$$p_A[1] = \min\{c_{AB} + p_B[0], c_{AC} + p_C[0], c_{AD} + p_D[0]\}$$
$$= \min\{3 + 0, 1 + \infty, 5 + \infty\}$$
$$= 3,$$

$$p_B[1] = 0,$$

$$p_C[1] = \min\{c_{CA} + p_A[0], c_{CB} + p_B[0]\}$$
$$= \min\{1 + \infty, 1 + 0\}$$
$$= 1,$$

$$p_D[1] = \min\{c_{DA} + p_A[0], c_{DB} + p_B[0]\}$$
$$= \min\{5 + \infty, 2 + 0\}$$
$$= 2.$$

For destination C at $t = 1$, there is no $p_D[1]$ since node C does not realize the existence of node D yet:

$$p_A[1] = \min\{c_{AB} + p_B[0], c_{AC} + p_C[0], c_{AD} + p_D[0]\}$$
$$= \min\{3 + \infty, 1 + 0, 5 + \infty\}$$
$$= 1,$$

$$p_B[1] = \min\{c_{BA} + p_A[0], c_{BC} + p_C[0], c_{BD} + p_D[0]\}$$
$$= \min\{3 + \infty, 1 + 0, 2 + \infty\}$$
$$= 1,$$

$$p_C[1] = 0.$$

For destination D at $t = 1$, there is no $p_C[1]$ because node D does not know the existence of node C yet:

$$p_A[1] = \min\{c_{AB} + p_B[0], c_{AC} + p_C[0], c_{AD} + p_D[0]\}$$
$$= \min\{3 + \infty, 1 + \infty, 5 + 0\}$$
$$= 5,$$

$$p_B[1] = \min\{c_{BA} + p_A[0], c_{BC} + p_C[0], c_{BD} + p_D[0]\}$$
$$= \min\{3 + \infty, 1 + \infty, 2 + 0\}$$
$$= 2,$$

$$p_D[1] = 0.$$

Although each set of calculations organized above is for sending various nodes to a single destination, during the execution of RIP, it is the other way around: each node performs calculations for various destinations. At $t = 1$, each node stores its own table shown below. C and D cannot reach each other because so far the message passing has revealed only zero- or one-hop paths and these two nodes require at least two hops to connect.

NodeID	DestinationID	Cost of MinCost Path	Next node
A	A	0	A
A	B	3	B
A	C	1	C
A	D	5	D

NodeID	DestinationID	Cost of MinCost Path	Next node
B	A	3	A
B	B	0	B
B	C	1	C
B	D	2	D

NodeID	DestinationID	Cost of MinCost Path	Next node
C	A	1	A
C	B	1	B
C	C	0	C

NodeID	DestinationID	Cost of MinCost Path	Next node
D	A	5	A
D	B	2	B
D	D	0	D

Now, at $t = 2$, all nodes send updated distance vectors to their neighbors.

- Node B receives $[A, B, 3], [A, C, 1], [A, D, 5]$ from node A. This tells node B that $p_A[1] = 3$ for destination B, $p_A[1] = 1$ for destination C, and $p_A[1] = 5$ for destination D.
- Node B receives $[C, A, 1], [C, B, 1]$ from node C. This tells node B that $p_C[1] = 1$ for destination A, $p_C[1] = 1$ for destination B, and $p_C[1] = \infty$ for destination D.
- Node B receives $[D, A, 5], [D, B, 2]$ from node D. This tells node B that $p_D[1] = 5$ for destination A, $p_D[1] = 2$ for destination B, and $p_D[1] = \infty$ for destination C.

We update all routing tables at $t = 2$. We focus on node A's table to illustrate the derivation. For destination B at $t = 2$, we have

$$p_B[2] = \min\{c_{AB} + p_B[1], c_{AC} + p_C[1], c_{AD} + p_D[1]\}$$
$$= \min\{3 + 0, 1 + 1, 5 + 2\}$$
$$= 2.$$

For destination C at $t = 2$, we have

$$p_C[2] = \min\{c_{AB} + p_B[1], c_{AC} + p_C[1], c_{AD} + p_D[1]\}$$
$$= \min\{3 + 1, 1 + 0, 5 + \infty\}$$
$$= 1.$$

For destination D at $t = 2$, we have the same cost either through node B or directly to D:

$$p_D[2] = \min\{c_{AB} + p_B[1], c_{AC} + p_C[1], c_{AD} + p_D[1]\}$$
$$= \min\{3 + 2, 1 + \infty, 5 + 0\}$$
$$= 5.$$

Each node stores its own routing information in a table at $t = 2$. The above explains the entries in the first table below, and we also see that nodes C and D know about the existence of each other now:

NodeID	DestinationID	Cost of MinCost Path	Next node
A	A	0	A
A	B	2	C
A	C	1	C
A	D	5	B

NodeID	DestinationID	Cost of MinCost Path	Next node
B	A	2	C
B	B	0	B
B	C	1	C
B	D	2	D

NodeID	DestinationID	Cost of MinCost Path	Next node
C	A	1	A
C	B	1	B
C	C	0	C
C	D	3	B

NodeID	DestinationID	Cost of MinCost Path	Next node
D	A	5	A
D	B	2	B
D	C	3	B
D	D	0	D

Similarly, each node stores its own routing information in a table at $t = 3$:

NodeID	DestinationID	Cost of MinCost Path	Next node
A	A	0	A
A	B	2	C
A	C	1	C
A	D	4	C

NodeID	DestinationID	Cost of MinCost Path	Next node
B	A	2	C
B	B	0	B
B	C	1	C
B	D	2	D

NodeID	DestinationID	Cost of MinCost Path	Next node
C	A	1	A
C	B	1	B
C	C	0	C
C	D	3	B

NodeID	DestinationID	Cost of MinCost Path	Next node
D	A	4	B
D	B	2	B
D	C	3	B
D	D	0	D

Further iterations produce no more changes. The routing tables have converged to the right solution through distributed message passing in RIP.

What about link failures? Is distance vector routing robust against events like a link breaking? We will find out in a homework problem.

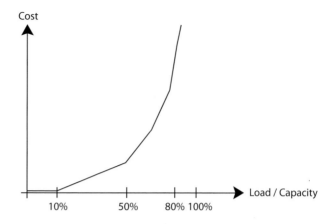

Figure 13.8 A typical cost function $C(f)$: increasing, convex, and piecewise-linear. It rises sharply as the amount of load reaches about 50% of the link capacity, and reaches very high values by the time this loading is 80%.

13.4   Advanced Material

In this section, we will go into further detail of routing within an AS and across ASs. Then, in a homework problem, we will also go through the essential ideas of switching within a small, local area network and a distributed protocol that determines a spanning tree to connect a given set of nodes.

13.4.1   Link state routing: OSPF

Picking the shortest paths is just one metric out of several reasonable ones. Another popular metric is the minimization of link loads, where the load of a link is defined as the percentage of its capacity used. This is carried out through a procedure called **traffic engineering** by the ISPs. The goal is to load-balance the traffic across the paths so that all the link loads are as low as possible under the traffic demand, or at least no one link load becomes too high and forms a bottleneck.

The benchmark of traffic engineering performance is defined by the **multi-commodity flow problem**. It is a basic optimization problem encountered in many networks to design a mapping of flows onto different paths with a given topology $G = (V, E)$.

We assume that each destination n has one flow coming to it from various other nodes in the network. Let f_{ij}^n denote the amount of flow on link (i, j) destined to reach node n, and f_{ij} the sum of load on link (i, j) across all destinations. The objective function is to minimize some cost function: the higher the load, the higher the cost. A typical link cost function is a piecewise-linear, increasing, and convex function, as shown in Figure 13.8. As the load approaches the link capacity, the cost rises sharply. The constraint is simply a flow-conservation

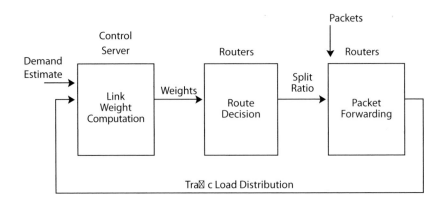

Figure 13.9 Three main modules in link state routing protocols like OSPF. Link weight computation is done in a centralized management server based on an estimation of traffic demand. These link weights are then given to the routers, each autonomously deciding the splitting ratio for load balancing on the basis of these weights. When a packet actually arrives at a router, forwarding to the next hop is done by reading the destination IP address.

equality: the incoming traffic to a node v, plus $D(v, n)$, the traffic that originates at v and destined to reach n, must be equal to the outgoing traffic from node v:

$$
\begin{aligned}
\text{minimize} \quad & \sum_{(i,j)} C(f_{ij}) \\
\text{subject to} \quad & \sum_{j:(v,j)\in E} f_{v,j}^n = \sum_{i:(i,v)\in E} f_{i,v}^n + D(v,n), \quad \text{for all } v \neq n \quad (13.2) \\
& f_{ij} = \sum_t f_{i,j}^t, \quad \text{for all } (i,j) \in E.
\end{aligned}
$$

Using the terminology from Chapter 4, we know this problem (13.2) is a linearly constrained, convex optimization problem. So it is easy to solve, at least through a centralized algorithm. And there are many specialized algorithms to further speed up the computation.

But IP does *not* allow end-to-end tunneling, so it cannot keep track of the f_{ij}^n. This is a design choice driven by the simplicity of network management. Neither does IP allow dynamic routing adaptive to link loads; that is the job of TCP. What actually happens is that the most popular intra-AS routing protocol, OSPF, solves the above problem *indirectly* through link weight optimization. This is illustrated in Figure 13.9 and summarized below.

- *Link weight computation.* A centralized management server collects or estimates the source–destination traffic, i.e., all the $D(v, n)$ in (13.2), every 12 hours or so. This timescale is not dynamic at the same timescale as traffic fluctuation. Then the server computes a set of link weights, one for each link in the network, and uploads this information to all the routers. Each router has a global view of the topology, not just a neighborhood local view as in distance vector routing.

- *Use link weights to split traffic.* Given the weights computed by the central-ized management server, each router constructs many paths to each of the possible destinations. In OSPF, each router constructs just the shortest paths under the given link weights, and splits the incoming traffic equally among all the shortest paths.
- *Forward the packets.* As each packet arrives at the router, the next hop is decided just from the destination IP address, and the traffic splitting de-cided in the step above. It does not matter which the source is and what routers the packet has traversed so far. This is called *destination-based and hop-by-hop forwarding.*

We can compare link state routing, like OSPF, with distance vector routing, like RIP. Link state routing passes detailed messages about each local topology, while distance vector routing passes coarser messages about the global topol-ogy. There are also different tradeoffs among communication overhead and local computation between link state and distance vector protocols.

Is it easy to turn the knob of link weight, and hope the right weights will indirectly induce a traffic distribution $\{f_{ij}^t\}$ solving (13.2)? The answer is no. Picking the link weights for OSPF in order to induce a solution to (13.2) is a computationally intractable problem.

But you do *not* have to use OSPF. There are other members of this family of link state routing protocols, e.g., PEFT, where link weights are used to define the weight for all the paths to be used, not just the shortest paths, but also the longer paths. However, the fraction of traffic put on each path exponentially drops as a function of the path length. If link weights are used this way, it turns out that computing the right link weights becomes a computationally tractable problem. As in mechanism design of games, this is a case of the principle of "design for optimizability:" designing the network protocol so that its operation can be readily optimized.

13.4.2   Inter-AS routing: BGP

Since the Internet is a network of networks, we need all ASs to cooperate with each other so that one AS's customers can reach customers of another AS. Inter-AS routing glues the entire Internet together, but it is messy, because the best path (defined by some distance or weight metric) across the ASs is often *not* chosen due to policy and economic concerns.

Consider the small-scale inter-AS connectivity graph in Figure 13.10. Each AS has a number called the ASN, just like each end host and each interface of a router has an IP address.

BGP governs the routing across the ASs (called an eBGP session) and moves inter-AS packets within an AS (called an iBGP session). Picking the next-hop AS in BGP is *almost* like picking the next-hop router in RIP, as each AS passes a list of *BGP attributes* to neighbor ASs. Part of the attributes is called AS-PATH,

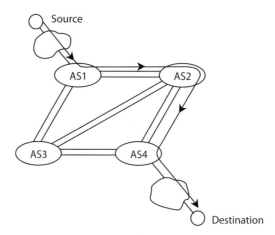

Figure 13.10 An example of BGP routing a session across multiple ASs in the same tier. Each node is an AS, and each link is a peering relationship manifested through some physical connectivity between border routers. Which AS to pick as the next hop depends not just on performance metrics but also on policies relating to economic and security concerns. Which router to pick to forward the packet out of an AS is based on hot-potato routing.

listing the ASs that this AS needs to pass in order to reach a certain range of IP addresses.

Here comes the messy but critical part in eBGP: each AS can have a sequence of *filters* that reprioritize the choices of neighboring ASs to go to in order to reach a certain range of IP addresses. On top of that list of filters is a *local preference policy*: AS 1 might decide that it simply does not want to go through AS 3, so any AS-PATH containing AS 3 is put to the bottom of the priority list. This decision might be due to some security or economic concern: AS 3 may be perceived by AS 1 as non-secure, or AS 1 might not want to send too much traffic along to AS 3 in case it tips the traffic-volume balance between them and results in a new payment contract between these two peering ASs.

Usually there are multiple border routers connecting two ASs. Which one should we use? iBGP uses a simple rule: pick the border router that has the minimum cost to reach (from where the packet is positioned when it enters the AS). This means each AS wants to get rid of the packet as soon as possible. This is called *hot-potato* routing.

There is a lot of detail of BGP beyond the above gist, including an interesting model called the *stable path problem* that crystallizes the BGP stability issues.

But we have to conclude this chapter on the fundamentals of the Internet now. There are actually many other types of routing in communication networks. There are specialized routing protocols in wireless mesh networks and optical networks. There is also much work on routing with a guarantee on the quality of service such as delivery time. Some of these routing protocols are centralized whereas others are distributed. Some have centralized control planes for parameter optimization, yet distributed data planes that forward the actual packets. Some of these protocols are static, whereas others are dynamic as a function of the link loads on a fast timescale.

And there have been years of work both in the research community and in industry on implementing **multicast routing**. We have only been looking at unicast routing: from one sender to one receiver. As the Internet is used increasingly for content distribution and video entertainment delivery, often there are many receivers at the same time. Creating a unicast routing session for each of the receivers is inefficient. But multicasting at the IP layer turns out to be hard to manage, and has often been replaced by an *architectural alternative*: application-layer multicasting in an overlay network. We will pick this up in Chapter 15 on P2P.

Summary

Box 13 Architectural principles of the Internet

Packet switching for scalable connectivity, layered protocols of modularization, and distributed hierarchy with overlay are three fundamental concepts of the Internet. Routing by IP in the network layer is done differently based on whether it is within a domain or across domains. Intra-domain routing comes in two major types: distance vector or link state. Distance vector routing implements the Bellman–Ford algorithm through distributed message passing.

Further Reading

There is a whole library of computer networking and Internet design textbooks, and thousands of papers on all kinds of aspects of routing.

1. The following book provides an accessible and concise overview on Internet routing protocols:

   C. Huitema, *Routing in the Internet*, 2nd edn., Prentice Hall, 1999.

2. Routing is a popular topic in algorithms courses in computer science departments. The following is one of the standard textbooks on the subject:

   T. H. Cormen, C. E. Leiserson, R. L. Rivest, and C. Stein, *Introduction to Algorithms*, 3rd edn., MIT Press, 2009.

3. One of the several standard textbooks on computer networking that covers routing in much more detail is

L. L. Peterson and B. Davie, *Computer Networks: A Systems Approach*, 5th edn., Morgan Kaufmann, 2012.

4. Below is a recent research article that proved the feasibility of efficient computation of the optimal link weights in link state routing with hop-by-hop forwarding:

D. Xu, M. Chiang, and J. Rexford, "Link state routing protocol can achieve optimal traffic engineering," *IEEE/ACM Transactions on Networking*, vol. 19, no. 6, pp. 1717–1730, November 2011.

5. We did not have the time to cover BGP in detail. The following is a relatively recent survey article on BGP:

M. Caesar and J. Rexford, "BGP routing policies in ISP networks," *IEEE Network*, vol. 19, no. 6, pp. 5–11, November 2005.

Problems

13.1 *Packet switching* ⋆

We will quantitatively examine the two benefits of packet switching in this problem.

(a) *Statistical multiplexing*

Suppose you have a 10 Mbps link shared by many users. Each user of the link generates 1 Mbps of data 10% of the time, and is idle 90% of the time.

If we use a circuit-switched network, and the bandwidth allocation is equal among the users, how many users can the link support? Call this number N. Now consider a packet-switched network. Say we have M users in total, and we want the probability of a user being denied service to be less than 1%. Write down the expression that must be solved in the form of $f(M, N) < 0.01$. Solve this numerically for M.

(Hint: Use the binomial distribution's cumulative distribution function.)

(b) *Resource pooling*

We will consider modeling a shared resource and see what happens when both the requests for the resource and the ability to fulfill requests increase. Suppose we have m servers. When a request comes in, an idle server answers the request. If all servers are busy, the request is dropped. The following **Erlang formula** gives the probability of a request being denied, given m servers and E units of traffic:

$$P(E, m) = \frac{\frac{E^m}{m!}}{\sum_{i=0}^{m} \frac{E^i}{i!}}.$$

Calculate $P(3, 2)$. Now calculate $P(6, 4)$. What do you observe? In general, $P(wx, wy) < P(x, y)$, $\forall w > 1$. This is one of the several standard ways to quantify the notion of resource pooling's benefits.

13.2 RIP ⋆⋆

Consider the network with the topology shown in Figure 13.2.

(a) Run an example of RIP on this network to find the minimum paths between all nodes. Show the routing tables at each time step.

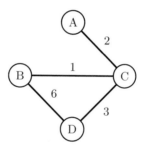

Figure 13.11 A small graph for an example of RIP.

(b) Now the link between A and C fails, resulting in a cost of ∞ for both directions of transmission. B and C immediately detect the link failure and update their own routing tables using the information they already have from their one-hop neighbors. Write down the routing tables for four iterations after the link failure. You need to only show the routing tables that change. What is happening to the paths to A?

(c) Propose a solution to the problem found in (b).

13.3 *Ford–Fulkerson algorithm and the max flow problem* ⋆⋆

You are in the engineering library of Princeton University studying for your final exam, which will take place in 1 hour. You suddenly realize you did not attend a key lecture on power control. Your kind TA offers to send you a video of the lecture. Unfortunately, she lives off in the Graduate College, which is somewhere way off-campus (you do not even know where).

Since you want to get the video as quickly as possible, you decide to split it into many pieces before sending it over the Princeton network. Suppose the Princeton network has the pipe capacities given in Figure 13.12. How much capacity should you send over each pipe so that you maximize your total rate?

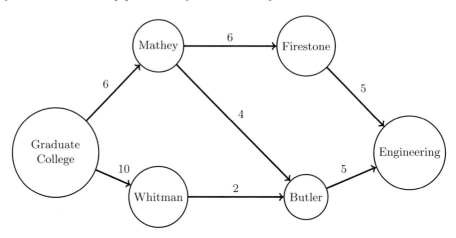

Figure 13.12 A simplified Princeton campus network, with the six nodes abbreviated as G, M, F, E, B, and W.

We will walk through a step-by-step approach for solving this *maximum flow problem*. A useful operation will be generating a *residual graph*. Given a flow, for each link along the flow's path, we draw a *backward link* with the amount of flow, leaving a forward link with the remaining capacity of the link. An example is shown in Figure 13.3.

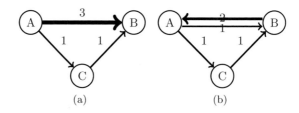

Figure 13.13 An example of drawing the residual graph. We decide to push two units of flow from A to B in graph (a), which gives us the residual graph in (b).

(a) Allocate 4 Mbps to the path G–M–B–E. Draw the residual graph.

(b) Allocate 2 Mbps to the path G–W–B–M–F–E on the residual graph from (a). Draw the new residual graph.

(c) Allocate 2 Mbps to the path G–M–F–E on the residual graph from (b). Draw the new residual graph.

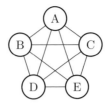

Figure 13.14 A small, full mesh graph to illustrate DAR.

(d) There are no paths remaining on the residual graph from (c), so the algorithm terminates. The capacity allocation is given by the net capacityfrom steps (a), (b), and (c). Draw the graph with the final capacity allocation on each of the links.

This classic algorithm is called the **Ford–Fulkerson algorithm**. More on this in Chapter 16.

13.4 *Dynamic alternative routing* ★★

We did not get a chance to talk about routing in circuit-switched networks. A major brand there is Dynamic Alternative Routing **DAR**, which was adopted by British Telecom in 1996 for their networks.

Suppose that, instead of having fixed paths to route packets from a source to a destination, we want to have the routes change dynamically in response to network conditions. We will consider such a protocol, DAR, in the case of a fully-connected graph.

We want the routing to adapt dynamically to the link utilization and select a new path if the current one is too busy. Each possible session (source–destination pair) has an associated backup node. When a session is initiated, it first tries to send its traffic along the direct link. If the direct link fails because it is full, the session tries to use the two-hop path with the backup node. If the backup path fails too because it is busy, the session fails and selects a new backup node from the network. Clearly, the backup node should not be the same node as the destination of the session.

One possible problem with this scheme is that many sessions may end up using two-hop paths. Then we are not being very efficient, since we are using double the capacity compared with a one-hop path. Therefore, DAR reserves a fraction t_l of each link l for direct, one-hop sessions (sessions that use the direct link between the source and destination, as opposed to the backup path). We call this parameter, $t_l \in (0, 1)$, the trunk reservation coefficient. Each link l in the network has an overall capacity of c_l Mbps and is bidirectional. The non-reserved capacity, $c_l - t_l c_l$, may be used for either direct or indirect sessions.

(a) Suppose we have a full mesh shown in Figure 13.14. Let $c_l = 10$ Mbps, $t_l = 0.1$, $\forall l$. The backup nodes are initialized as in Table 13.1.

Session	Backup node
(B, C)	A
(C, B)	A
(A, C)	E

Table 13.1 Backup nodes for a DAR example.

Link	Unreserved capacity used [session]	Reserved capacity used [session]
(A, B)		
(A, C)		
(A, E)		
(B, C)	9 Mpbs [9 × (B, C)]	1 [1 × (B, C)]
(C, E)		

Table 13.2 The table of link utilizations to be filled out in the DAR example.

The following events occur in sequence.
1. Ten parallel sessions of (B, C) begin.
2. Ten parallel sessions of (C, B) begin.
3. Ten parallel sessions of (A, C) begin.

Assume the sessions last a long time and each session consumes 1 Mbps. Fill in Table 13.2 after the above sequence of events has occurred. Remember that sessions and links are two different concepts. One row has been filled out for you as an example.

(b) Repeat (a) without the trunk reservation scheme.

(c) What is the efficiency of link utilization, i.e., the number of sessions divided by the total network capacity used, under (a) and (b), respectively?

(For more details, see R. J. Gibbens, F. P. Kelly, and P. B. Key, "Dynamic alternative routing," *Routing in Communication Networks*, M. Steenstrup Eds., Prentice Hall, 1995.)

13.5 *Spanning tree* ★ ★ ★

Routing in an inter-connected set of local area networks (**LANs**), like the one in a corporation or on a campus, is much easier than over the entire Internet

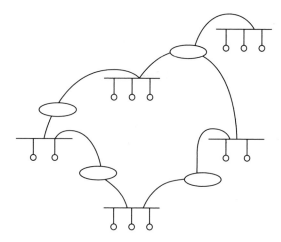

Figure 13.15 An example of local area networks connected by bridges. There are two types of nodes in this graph: each local area network (a line with its hosts represented as small circles) is a node, and each bridge (an oval) is also a node. The links in this graph connect bridges with local area networks. If all the bridges are used, the graph becomes cyclic. Distributed spanning tree protocols discover an acyclic subgraph that includes all the local area networks but not all the links.

globally connecting devices with many different layers 1 and 2 technologies. The connections among LANs are called **bridges**, or **switches**. Each has multiple ports, with one port connecting to one LAN. We could configure the LAN IDs on each port of a bridge, but a more scalable and automated way is for each bridge to listen to each packet that arrives on a port, and copy the source address in that packet's header to the database of hosts reachable from that port.

This "learning bridge" protocol works well, except when there are loops in the graph of LANs and bridges. So we need to build a tree that connects all the LANs without forming any cycles, a so-called **spanning tree**. Then there is only one way to forward a packet from one device to another. (We will later see building multiple trees for multicast overlay in P2P networks in Chapter 15.)

If each link has a weight, say, the distance of the link, then finding the smallest weight spanning tree is the well-studied graph-theoretic problem of the **minimum spanning tree**. If you can add extra intermediate nodes to shorten the total distance in the tree, it is called the **Steiner tree** problem. It turns out that some neurological networks in animals follow the principle of Steiner tree construction.

In this homework problem, however, we tackle a simpler and still important problem of *distributedly* discovering a spanning tree (not necessarily the minimum spanning tree). This protocol was invented by Perlman in 1985. Like link state routing protocols, the spanning tree protocol we see below is an example of achieving a globally consistent view of the topology through only local interactions that eventually propagate throughout the network.

Consider the set of LAN segments (the horizontal lines, each with some devices attached to it) and bridges (the ovals) depicted in Figure 13.15. Clearly there are cycles in the given graph: we can go from one segment through other segments and back to itself without going through a link twice. We want to determine a cycle-free way to provide connectivity among all the segments. One way to arrive at a consistent spanning tree is to have the bridge with the smallest ID number as the root of the tree, and each of the other bridges reach this root bridge through the smallest-hop-count path. That is easy, at least for such small networks, if we have a global view. But how to do that distributedly, with message passing only between neighbors? How do the nodes even agree on which bridge is the root of the tree?

One possibility is to ask each bridge to announce the following message, consisting of three fields, during each timeslot:

- the ID of the bridge believed to be the root,
- the number of hops to reach that root bridge from this bridge, and
- the ID of this bridge.

Now, here come the questions. (a) Initially, each bridge only has local information about itself. What are the messages from the four bridges in Figure 13.15?

(b) Upon receiving a message, each bridge selects the root bridge and discover the way to reach it by appying the following ordered list of criteria.

1. The bridge with a lower ID number wins and becomes the new root bridge (as believed by this bridge).
2. If there are multiple paths to reach the same root bridge, the path with the smallest hop count wins.
3. If there are multiple equal-hop-count paths to reach the same root bridge, the path announced from a bridge with a smaller ID number wins.

Each bridge then updates the root bridge field of the message, and increases the hop count by 1. Then it sends the new message to its neighbors, except of course those neighbors that have a shorter path toward the same root bridge.

Write down the evolution of the messages for the bridges in Figure 13.15. Does it converge to a spanning tree?

(c) Even after convergence, the root bridge keeps sending the message once every regular period. Why is that? Consider what happens when a bridge fails.

(The following book provides a comprehensive survey of layer 2 switching and layer 3 routing protocols. It is also a rare example of a book that is devoted to network protocols and yet maintain a sense of humor; see Chapter 18 for example: R. Perlman, *Interconnections: Bridges, Routers, Switches, and Internetworking Protocols*, 2nd edn., Addison-Wesley, 1999.)

14 Why doesn't the Internet collapse under congestion?

14.1   A Short Answer

14.1.1   Principles of distributed congestion control

When demand exceeds supply, we have congestion. If the supply is fixed, we must reduce the demand to alleviate congestion. Suppose the demand comes from different nodes in a network, we need to coordinate it in a distributed way.

As the demand for capacity in the Internet exceeds the supply every now and then, **congestion control** becomes essential. The timescale of congestion control is on the order of ms, in contrast to shaping consumer behavior through pricing in Chapters 11 and 12. The need for congestion control was realized in October 1986, when the Internet had its first congestion collapse. It took place over a short, three-hop connection between Lawrence Berkeley Lab and UC Berkeley. The normal throughput was 32 kbps (that is right, kbps, not the Mbps numbers we hear these days). That kind of dial-up modem speed was already low enough, but during the congestion event, it dropped all the way down to 40 bps, by almost a factor of 1000.

The main reason was clear as we saw from the last chapter on routing: when users send so many bits per second that their collective load on a link exceeds the capacity of that link, these packets are stored in a buffer and they wait in the queue to be transmitted. But when that wait becomes too long, more incoming packets accumulate in the buffer until the buffer overflows and packets get dropped. This is illustrated in Figure 14.1.

These dropped packets never reach the destination, so the intended receiver never sends an acknowledgement (an ACK packet) back to the sender, as it should do in the **connection-oriented**, end-to-end control in TCP. As mentioned in the last chapter, Internet design evolution considered different divisions of labor between layers 3 and 4, eventually settling on a connection-oriented layer 4 and connectionless layer 3 as the standard configuration. According to TCP, the sender needs to resend the unacknowledged packets. This leads to a vicious cycle, a *positive-feedback loop* that feeds on itself: congestion persists as the same set of senders that caused congestion in the first place keeps resending the dropped packets. Packets keep getting dropped at the congested link, resent from the source, dropped at the congestion link ... Senders need to rethink how they can avoid congestion in the first place, and they need to back off when congestion

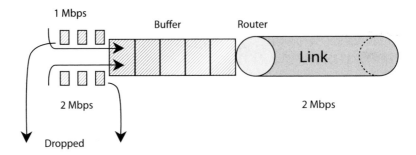

Figure 14.1 An illustration of congestion at one end of a link. Two sessions arrive at the buffer with an aggregate demand of 3 Mbps, but there is only a supply of 2 Mbps in the outgoing link. The buffer is filled up and packets start to get dropped. Which packets get dropped depends on the details of the queue-management protocols.

happens. We need to turn the positive-feedback loop into a *negative*-feedback loop.

That was what Van Jacobson proposed in the first congestion control mechanism added to TCP in 1988, called **TCP Tahoe**. It has been studied extensively since then, and improved significantly several times. But most of the essential ideas in congestion control for the Internet were in TCP Tahoe already.

- *End-to-end control via negative feedback.* We can imagine congestion control within the network where, hop by hop, routers decide for the end hosts at what rates they should send packets. That is actually what another protocol, called Asynchronous Transmission Mode (**ATM**), does to one type of its traffic, the Arbitrary Bit Rate traffic. But TCP congestion control adopts the alternative approach of having an *intelligent edge network* and a *dumb core network*. The rate at which a sender sends packets is decided by the sender itself. But the network provides hints through some feedback information to the senders. Such feedback information can be inferred from the *presence* and *timing* of acknowledgement packets, transmitted from the receiver back to the sender acknowledging the in-order receipt of each packet.

- *Sliding-window-based control.* If a sender must wait for the acknowledgement of a sent packet before it is allowed to send another packet, it can be quite slow. So we pipeline by providing a bigger allowance. Each sender maintains a sliding window called the **congestion window**, with its value denoted by cwnd. If the window size is 5, that means up to five packets can be sent before the sender has to pause and wait for acknowledgement packets to come back from the receiver. For each new acknowledgement packet received by the sender, the window is slid one packet forward and this enables the sending of a new packet, hence the name "sliding window." This way of implementing a restriction on transmission rate introduces the

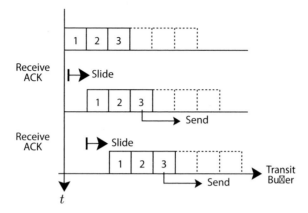

Figure 14.2 An illustration of a sliding window with a fixed size of three. When three packets are outstanding, i.e., have not been acknowledged, transmission has to pause. As each acknowledgement is received, the window is slided by one packet, allowing a new packet to be transmitted.

so-called *self-clocking property* driven by the acknowledgement packets. A picture illustrating the sliding-window operation is shown in Figure 14.2.

- *Additive increase and multiplicative decrease.* We will not have the time to discuss the details of how the cwnd value is initialized as a new TCP connection is established, during the so-called **slow start** phase. We focus on the **congestion avoidance** phase instead. If there is no congestion, cwnd should be allowed to grow, in order to efficiently utilize link capacities. *Increasing* the cwnd value is different from *sliding* the window under the same given cwnd value: cwnd becomes larger in addition to getting slided forward. And in TCP, when cwnd grows, it grows *linearly*: cwnd is increased by 1/cwnd upon receiving each acknowledgement. That means that over one round trip time, cwnd grows by 1 if all ACKs are properly received. This operation is shown in the space–time diagram in Figure 14.3. But if there is congestion, cwnd should be reduced so as to alleviate congestion. And TCP says when cwnd is cut, it is cut *multiplicatively*: cwnd next time is, say, half of its current value. Increasing cwnd additively and decreasing it multiplicatively means that the control of packet injection into the network is conservative. It would have been much more aggressive if it were the other way around: multiplicative increase and additive decrease.

- *Infer congestion by packet loss or delay.* But how do you know whether there is congestion? If you are an iPhone running a TCP connection, you really have no idea what the network topology looks like, what path your packets are taking, which other end hosts share links with you, and which links along the path are congested. You have only a local and noisy view, and yet you have to make an educated guess: is your connection experiencing

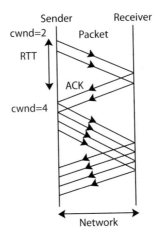

Figure 14.3 A space–time diagram of TCP packets being sent and acknowledged. The horizontal distance between the two vertical lines represents the spatial distance between the sender and the receiver. The vertical axis represents time. As two acknowledgements are received by the sender, the congestion window not only slides, but also increases by 1.

congestion somewhere in the network or not? The early versions of TCP congestion control made an important assumption: if there is a packet loss, there is congestion. This sounds reasonable enough, but sometimes packet loss is caused by a bad channel, like in wireless links, rather than congestion. In addition, often it is a little too late to react to congestion by the time packets are already getting dropped. The first problem has been tackled by many proposals of TCP for wireless. The second problem is largely solved by using packet delay as the congestion feedback signal. Instead of a binary definition of congestion or no congestion, a delay value implies the *degree* of congestion.

- *Estimate packet loss and delay by timers.* Say you agree that packet loss or delay implies congestion, how can you tell whether a packet has been lost and how do you calculate delay? TCP uses two common sense approximations. (1) If the sender waits for a long time and the acknowledgement does not come back, probably the packet has been lost. How long is a "long time"? Say this timeout timer is three times the normal **round trip time** (RTT) between the sender and the receiver. And what is the "normal" RTT? The sender timestamps each packet, and can tell the RTT of that packet once the acknowledgement is received at a later time. This is how the sender calculates a delay for each packet. Then it can calculate a moving-averaged RTT. The smallest RTT over a period of time is approximately the no-congestion, "normal" RTT. (2) Each packet sent has a sequence number, and if the sender hears from the receiver that several, say three, later packets (numbered 10, 11, and 12) have been received but this particular packet 9 is still not yet received, that probably means packet 9 has been lost. Packet 9 may have traversed a different path with a longer RTT (as discussed in IP routing in the last chapter), but if as many as three later packets have already arrived, chances are that packet 9 is not just late but lost.

As mentioned in the last chapter, TCP/IP is the "thin waist" of the Internet layered protocol stack. It glues the functional modules below it, like the physical and link layers, to those above it, like the application layer. (There are alternatives to TCP in this thin waist, such as the connectionless UDP that does not maintain an end-to-end feedback control that we will see in Chapter 17.) As part of that thin waist, the above five elements of congestion control design in TCP led to a great success. The wonderful fact that the Internet has not collapsed, despite the incredible and unstoppable surge of demand, is partially attributable to its congestion control capability.

Starting with TCP Tahoe in 1988 and its slightly modified variant **TCP Reno** in 1990, TCP congestion control had gone through over twenty years of improvement. For example, **TCP Vegas** in 1995 shifted from a loss-based congestion signal to a delay-based congestion signal. **FAST TCP** in 2002 stabilized congestion control to achieve high utilization of link capacity. **CUBIC** in 2005 combined loss- and delay-based congestion signals, and is now the default TCP in the Linux kernel. There have also been many other variants of TCP congestion control proposed over the past two decades.

14.1.2 Loss-based congestion inference

If you think about it, for end-to-end congestion control without any message passing from the network, an end host (like your iPad) really has very little to work with. Estimates of packet loss and calculations of packet delay are pretty much the only two pieces of information it can obtain through time stamping and numbering the packets.

For loss-based congestion control like TCP Reno, a major TCP variant especially for the Windows operating system, the main operations are as follows.

- If all the `cwnd` outstanding packets are received at the receiver properly (i.e., in time and not out of order more than twice), increase the `cwnd` by 1 each RTT.
- Otherwise, decrease it by cutting it in half, e.g., from `cwnd` to 0.5×`cwnd`.

There are also other subtle features like Fast Retransmit and Fast Recovery that we will not have time to get into.

Let us look at an example. For simplicity, let RTT = 1 unit, and assume it is a constant. Actually, RTT is about 50 ms across the USA and varies as the congestion condition changes. Initialize `cwnd` to be 5. Suppose all packets

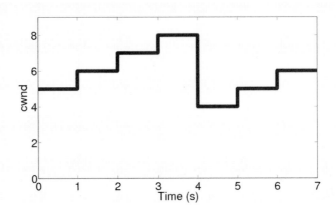

Figure 14.4 A zoomed-in view of cwnd evolution for TCP Reno, with RTT=1 unit of time.

are successfully received and acknowledged (ACK) during each RTT, except at $t = 4$, when a packet loss occurs.

At $t = 0$, cwnd=5, so the sender sends five packets and pauses.

At $t = 1$, the sender has received five ACKs, so it slides the congestion window by five packets and increases cwnd by 1. It sends six packets.

At $t = 2$, the sender has received six ACKs, so it sends seven packets.

At $t = 3$, the sender has received seven ACKs, so it sends eight packets.

At $t = 4$, the sender detects a lost packet. It halves cwnd to four, and sends four packets.

At $t = 5$, the sender has received four ACKs, so it sends five packets.

At $t = 6$, the sender has received five ACKs, so it sends six packets.

Figure 14.4 shows these values of cwnd over time. When there was no packet loss ($t = 0, 1, 2, 3$), cwnd grew linearly. When the packet loss occurred ($t = 4$), cwnd decreased sharply, then began growing linearly again ($t = 5, 6$).

Zooming out, Figure 14.5(a) shows a typical evolution of TCP Reno's cwnd over time. The y-axis is the congestion window size. If you divide that by the RTT and multiply it by the average packet size, you get the actual transmission rate in bps.

14.1.3   Delay-based congestion inference

Now we turn to delay-based congestion control like TCP Vegas. We first have to appreciate that the total RTT is mostly composed of **propagation delay**, the time it takes to just go through the links, and **queueing delay**, the time a packet spends waiting in the queue due to congestion. The heavier the congestion, the longer the wait. So the sender needs to estimate RTT_{min}, the minimum RTT that tells the sender what the delay value should be if there is (almost) no congestion.

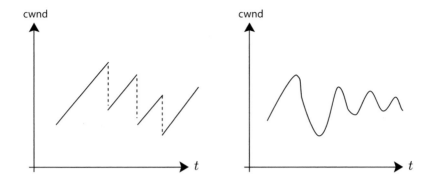

Figure 14.5 Typical evolutions of `cwnd` values in TCP Reno on the left and in TCP Vegas on the right. TCP Reno uses loss as the congestion signal whereas TCP Vegas uses delay as the congestion signal. The zig-zags between overshooting and under-utilizing capacity tend to be smaller in Vegas if the parameters are properly tuned.

Then, upon receiving each acknowledgement, the sender looks at the difference between $cwnd/RTT_{min}$ and $cwnd/RTT_{now}$. It is the difference between the transmission rate (in packets per second) without much congestion delay and that with the current congestion delay.

- If this difference is smaller than a prescribed threshold, say 3, that means there is little congestion, and `cwnd` is increased by 1.
- If the difference is larger than the threshold, that means there is some congestion, and `cwnd` is decreased by 1.
- If the difference is exactly equal to the threshold, `cwnd` stays the same.
- If all sources stop adjusting their `cwnd` values, an equilibrium is reached.

We can compare this congestion control with the power control in Chapter 1: at the equilibrium everyone stops changing their variable simultaneously. We would like to know what exactly the resource allocation is at such an equilibrium, and whether it can be reached through some simple, distributed, iterative algorithm.

Figure 14.5(b) shows a typical evolution of TCP Vegas' `cwnd` over time. You can see that the zig-zag between a rate that is too aggressive (leading to congestion) and one that is overly conservative (leading to under-utilization of link capacities) can be reduced, as compared with TCP Reno. Using delay as a *continuous* signal of congestion is better than using only loss as a *binary* signal, and we will see several arguments for this observation in the next section.

14.2   A Long Answer

Whether distributedly like TCP or through a centralized command system, any protocol trying to control congestion in a network must consider this fundamental

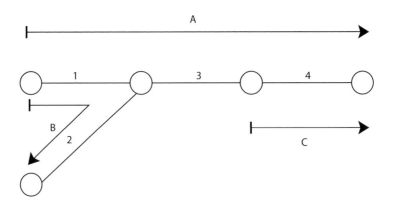

Figure 14.6 A simple network with four links and three sessions. Sessions A and B share link 1, and sessions A and C share link 4. Constrained by the fixed capacities on the four links, it is not trivial to design a distributed algorithm that allocates the capacities in an efficient and fair way among the three competing sessions.

issue: each link l's fixed capacity c_l is shared by multiple sessions, and each of these end-to-end sessions traverses multiple links. We assume each source i has one session and uses a single-path routing. So, "flows," "sessions," and "sources" are interchangeable terms in this chapter. Each link l is shared by a set of sessions $S(l)$, and each session i uses a set of links $L(i)$ along its path decided by IP routing.

Consider the simple example in Figure 14.6. As we will see in Chapter 19, it is often accompanied by control signaling that traverses other paths too. Sometimes, the contents of one session also reside at different locations, e.g., advertisements on a webpage need to be downloaded from a different server than the actual content of the webpage. We ignore these factors here.

In this graph, session A originating from node 1 traverses links 1, 3 and 4. And link 1 is shared by sessions A and B. Even when link 3 is not fully utilized, we cannot just increase session A's rate since the bottleneck link for that session may be link 1.

How can we allocate each link's capacity so that the sessions collectively use as much capacity as they can without causing congestion, and their competition is fairly coordinated? A capacity allocation must first be *feasible* under the link capacity constraints, and then, also be *efficient* and *fair*.

In Figure 14.6, consider each link's capacity as 1 Mbps. One solution that satisfies all four links' capacity constraints, thus constitutes feasible solution, is (0.5, 0.5, 0.5) for the three sessions A, B, and C, in Mbps. In this equal distribution of end-to-end rates, 3 Mbps of the link capacities across all the links is used. For the same capacity utilization, another feasible solution is (1, 0, 0), which starves sessions B and C, and probably is not viewed as a fair allocation. It turns out that a standard notion of fairness, called proportional fairness, would give the allocation $(1/3, 2/3, 2/3)$ to the three competing sessions. This may make intuitive sense to some people, since session A traverses two links that are potential bottlenecks whereas sessions B and C traverse only one such link.

Now we need to write down the problem statement more precisely. We will call this optimization the basic **Network Utility Maximization** (NUM) problem. It is a networked version of the social welfare problem in Chapter 11. We will soon present a distributed solution to this problem in this section, leaving the derivation steps to the Advanced Material. Then we will show that TCP Reno and TCP Vegas actually can be reverse-engineered as solutions to specific NUM problems.

14.2.1   Formulating the NUM problem

We need to address two issues in modeling congestion control: (1) how to measure efficiency and fairness, and (2) how to capture capacity constraint.

(1) How do we measure efficiency and fairness? We use the utility functions introduced in Chapter 11, and we sum up each individual TCP session's utility to the end-user. We model utility as a function of the end-to-end transmission rate of a TCP session here, since we are only adjusting these rates and we assume the application's performance depends only on this rate. Fairness may also be captured by some of these utility functions, like the α-fair utility functions.

(2) How do we represent the link capacity constraint? On each link l, there is a limited capacity c_l in bps. The load must be smaller than c_l. There are several ways to express the load on a link in terms of the variable transmission rates at the sources and the given routing decisions.

We can write the load on link l as the sum of source rates x_i across those sources using this link: $\sum_{i \in S(l)} x_i$. Or, we can use R_{li} as a binary-valued indicator, so that $R_{li} = 1$ if source i's session traverses link l, and $R_{li} = 0$ otherwise. Then the load on link l is simply $\sum_i R_{li} x_i$. In this notation, you can readily see that the constraints

$$\sum_i R_{li} x_i \le c_l, \quad \forall l,$$

are equivalent to the following linear inequalities in matrix notation:

$$\mathbf{Rx} \le \mathbf{c},$$

where $\le$ between two vectors means a component-wise inequality between the corresponding entries of the vectors.

For example, in the network topology in Figure 14.6, the link capacity constraint in matrix form becomes

$$\begin{bmatrix} 1 & 1 & 0 \\ 0 & 1 & 0 \\ 1 & 0 & 0 \\ 1 & 0 & 1 \end{bmatrix} \begin{bmatrix} x_A \\ x_B \\ x_C \end{bmatrix} \le \begin{bmatrix} c_1 \\ c_2 \\ c_3 \\ c_4 \end{bmatrix}.$$

14.2.2   Distributed algorithm solving NUM

Now we have completely specified the link capacity allocation problem that pre-
scribes what congestion control should be solving:

$$
\begin{array}{ll}
\text{maximize} & \sum_i U_i(x_i) \\
\text{subject to} & \mathbf{Rx} \le \mathbf{c} \\
\text{variables} & x_i \ge 0, \ \forall i.
\end{array}
\qquad (14.1)
$$

We refer to this problem as the basic NUM problem. Problem (14.1) is easy to
solve for several reasons.

- It is a convex optimization problem, as defined in Chapter 4. More precisely,
  this time it is maximizing a concave utility function (the sum of all sessions'
  utilities) rather than minimizing a convex cost function. Therefore, it enjoys
  all the benefits of being convex optimization: a locally optimal solution is
  also globally optimal, the duality gap (a concept to be introduced soon) is
  zero, and it can be solved very efficiently, at least in a centralized computer.
- It is also decomposable. **Decomposition** here refers to breaking up one optimization problem into many smaller ones, somehow coordinated so that
  solving them will be equivalent to solving the original one. Why would we
  be interested in having many problems instead of just one? Because such
  a decomposition leads to a distributed algorithm. Each of these smaller
  problems is much easier to solve, often locally at each node in a network.
  And if their coordination can be done without explicit message passing, we
  have a truly distributed way to solve the problem.

Postponing the derivation of decomposition (through the Lagrange dual problem) to the Advanced Material, we have the following iterative solution to (14.1),
consisting of source actions and router actions.

At each of the discrete timeslots $[t]$, the source of each session simply decides
its transmission rate from its demand function, with respect to the current price
along its path. This path price q_i is the sum of link prices p_l along all the links
this session traverses: $q_i = \sum_{l \in L(i)} p_l$. The transmission rate of source i is

$$
x_i[t] = D_i(q_i[t]) = U_i'^{-1}(q_i[t]),
\qquad (14.2)
$$

where D is the demand function as defined in Chapter 11.

Of course, in a sliding-window-based implementation, the source adjusts its
window size cwnd rather than x_i directly. The path price serves as the *congestion
feedback signal* from the network. We hope it can be obtained without explicit
message passing in actual implementations.

At the same time, the router on each link l updates the "price" on that link:

$$
p_l[t] = \{p_l[t-1] + \beta\,(y_l[t] - c_l)\}^+,
\qquad (14.3)
$$

where y_l is the total load on link l: $y_l[t] = \sum_{i \in S(l)} x_i[t]$. And $\{\dots\}^+$ simply says
that if the expression inside the brackets takes on a negative value, then just

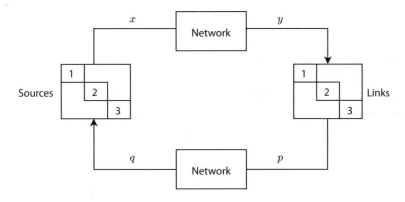

Figure 14.7 The feedback control loop in the distributed solution of NUM. Each source i autonomously adapts its window size (or transmission rate x_i) on the basis of the path-congestion price feedback q_i, while each link l autonomously adapts its congestion price p_l on the basis of its load y_l.

return 0. In this case, it means that the link price is never allowed to be negative. This is an *interpretation* in the language of pricing, not actual money changing hands between network entities like in Chapters 11 and 12. The parameter $\beta \geq 0$ is the **stepsize** that controls the tradeoff between a convergence guarantee and the convergence speed. Think about playing golf: if you hit the ball too hard, even in the right direction, it will fly by the hole, and you will have to hit backward again. Stepsize is like the magnitude of your force. If it is sufficiently small, the above algorithm is guaranteed to converge, and converge to the right solution to (14.1). If it is too small, the guaranteed convergence becomes too slow. In real systems, tuning this parameter is not easy.

The feedback loop in the pair of equations (14.2) and (14.3) is illustrated in Figure 14.7. It makes sense from an economic standpoint. If, at time t, there is more load than there is capacity on link l, then price p_l will go up according to (14.3), and the prices for all the paths containing link l will rise in the next timeslot $t+1$. A higher price will reduce demand according to (14.2), and x_i will drop at all the sources that use link l, helping to restore the balance between demand and supply on link l. This pricing signal balances the elastic demand and the fixed supply of capacities. What is interesting is that it carries out this task distributedly through a network consisting of many links and sessions. Mathematically, these link prices turn out to be the variables in a "mirror image" of the basic NUM problem: the variables in the Lagrange dual problem of NUM.

The above algorithm not only solves the basic NUM (for proper β), but also solves it in a very nice way: fully distributed and intuitively motivated.

- As clearly shown in (14.3), each link needs to measure its own total load only. It does not need to know any other link's condition, or even the load coming from each of the sources using it.

- As clearly shown in (14.2), each source needs to know only the total price along the path it is using. It does not need to know any other path's or source's condition, or even the price per link along the path that it is using. If the path price q_i can be measured locally at each source i without explicit message passing, this would be a completely distributed solution. That is the case for using packet losses as the pricing feedback in TCP Reno, and packet delays as the pricing feedback in TCP Vegas.

14.2.3   Reverse engineering

At this point, it might feel like the first section of this chapter and the current section are somehow disconnected. Are the TCP congestion control protocols implemented on the Internet related to the distributed solution of the basic NUM problem? Roughly a decade after the first TCP congestion control protocol was invented, researchers reverse-engineered these protocols and showed that they actually can be interpreted, approximately, as solutions to NUM. If you describe a protocol to me, I can tell you what the utility function being implicitly maximized is and what the price variables are.

For example, it turns out that TCP Reno implicitly maximizes arctan utilities, with packet losses as the price. And TCP Vegas implicitly maximizes logarithmic utilities, with packet delays as the price. In the Advanced Material, we will present the derivation in the case of TCP Reno, and a similar derivation is in a homework problem for TCP Vegas.

Reverse engineering presents a peculiar viewpoint: give me the solution and I will tell you what the problem being solved by this solution is. You might wonder why I would care about the problem if I already have the solution? Well, discovering the underlying problem being solved provides a rigorous understanding of why the solution works, when it might not work, and how to make it work better. It also leads to new designs: forward engineering using the insights from reverse engineering.

Here is an example of the implications on the properties of TCP derived from reverse engineering. Since TCP Reno implicitly solves NUM with arctan utility, we know that the equilibrium packet-loss rate, i.e., the optimal dual variables, cannot depend on parameters that do not even show up in NUM. In particular, if we double the buffer size, it will not help reduce the equilibrium packet-loss rate, since buffer size does not appear in the NUM problem (14.1). Intuitively, what happens is that increasing the buffer size simply postpones the onset of congestion and packet loss, but does not help with the equilibrium.

Here is another example: since TCP Vegas is reverse-engineered as a solution to log utility maximization, it leads to a proportionally fair allocation of link capacities. Of course, that does not guarantee TCP Vegas will converge at all. But, if it does converge, we obtain proportional fairness.

In addition to the mentality of reverse engineering, we have introduced two important themes in this chapter, themes that run beyond just TCP congestion control in the Internet.

- *A feedback signal can be generated and used in a network for distributed coordination.* Selfish interests of the users may be aligned by pricing signals to achieve a global welfare maximization. In some cases, the pricing signals do not even require explicit message passing, an additional benefit not commonly found in general.
- *A network protocol can be analyzed and designed as a control law.* Properties of Internet protocols can be analyzed through the trajectories of the corresponding control law. This might sound straightforward once it has been stated, but it was an innovative angle when first developed in the late 1990s. It opened the door to thinking about network protocols not just as bottom-up, trial-and-error solutions, but also as the output of a first-principled, top-down design methodology.

14.3   Examples

Consider the network shown in Figure 14.6 again. The routing matrix, where the rows are the links and the columns are the sources, is

$$
\mathbf{R} = \begin{bmatrix} 1 & 1 & 0 \\ 0 & 1 & 0 \\ 1 & 0 & 0 \\ 1 & 0 & 1 \end{bmatrix}.
$$

Routing can also be represented by $\{S(l)\}$ or $\{L(i)\}$. In this example, we have $S(1) = \{A, B\}, S(2) = \{B\}, S(3) = \{A\}, S(4) = \{A, C\}$, and $L(A) = \{1, 3, 4\}, L(B) = \{1, 2\}, L(C) = \{4\}$.

Assume the capacity on all links is 1 Mbps and the utility function is a log function for all sources, i.e., $U_i(x_i) = \log x_i$. Our job is to find the sending rate of each source: x_A, x_B and x_C. The NUM problem is formulated as

$$
\text{maximize} \quad \log(x_A) + \log(x_B) + \log(x_C)
$$

$$
\text{subject to} \quad \begin{bmatrix} 1 & 1 & 0 \\ 0 & 1 & 0 \\ 1 & 0 & 0 \\ 1 & 0 & 1 \end{bmatrix} \begin{bmatrix} x_A \\ x_B \\ x_C \end{bmatrix} \leq \begin{bmatrix} 1 \\ 1 \\ 1 \\ 1 \end{bmatrix}, \tag{14.4}
$$

$$
x_i \geq 0, \ \forall i.
$$

Recall that the source rates and the link prices converge to the distributed solution to the NUM problem by the following iterative updates:

$$
x_i[t] = U'^{-1}(q_i[t]) = \frac{1}{q_i[t]},
$$

$$
p_l[t] = \{p_l[t-1] + \beta(y_l[t] - c_l)\}^+,
$$

where q_i is the path price seen by source i (we step over the time index from $t - 1$ to t here), and y_l is the total load on link l:

$$q_i[t] = \sum_{l \in L(i)} p_l[t - 1],$$

$$y_l[t] = \sum_{i \in S(l)} x_i[t].$$

Let us initialize the source rates to 0 and the link costs to 1, i.e., $x_A[0] = x_B[0] = x_C[0] = 0$ and $p_1[0] = p_2[0] = p_3[0] = p_4[0] = 1$. Let stepsize $\beta = 1$. At $t = 1$, we first update the source rates. Since

$$q_A[1] = p_1[0] + p_3[0] + p_4[0] = 1 + 1 + 1 = 3,$$
$$q_B[1] = p_1[0] + p_2[0] = 1 + 1 = 2,$$
$$q_C[1] = p_4[0] = 1,$$

we have, in Mbps,

$$x_A[1] = \frac{1}{q_A[1]} = 0.333,$$

$$x_B[1] = \frac{1}{q_B[1]} = 0.5,$$

$$x_C[1] = \frac{1}{q_C[1]} = 1.$$

We then update the link prices. Since the link loads are, in Mbps,

$$y_1[1] = x_A[1] + x_B[1] = 0.333 + 0.5 = 0.833,$$
$$y_2[1] = x_B[1] = 0.5,$$
$$y_3[1] = x_A[1] = 0.333,$$
$$y_4[1] = x_A[1] + x_C[1] = 0.333 + 1 = 1.33,$$

we have

$$p_1[1] = [p_1[0] + y_1[1] - c]^+ = [1 + 0.833 - 1]^+ = 0.833,$$
$$p_2[1] = [p_2[0] + y_2[1] - c]^+ = [1 + 0.5 - 1]^+ = 0.5,$$
$$p_3[1] = [p_3[0] + y_3[1] - c]^+ = [1 + 0.333 - 1]^+ = 0.333,$$
$$p_4[1] = [p_4[0] + y_4[1] - c]^+ = [1 + 1.33 - 1]^+ = 1.33.$$

At $t = 2$, we update the source rates. Since

$$q_A[2] = p_1[1] + p_3[1] + p_4[1] = 0.833 + 0.333 + 1.33 = 2.5,$$
$$q_B[2] = p_1[1] + p_2[1] = 0.833 + 0.5 = 1.33,$$
$$q_C[2] = p_4[1] = 1.33,$$

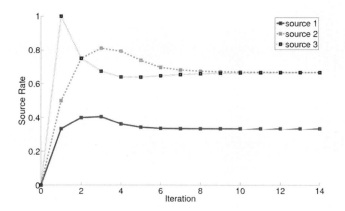

Figure 14.8 Source rates converge over time in a congestion control example.

we have, in Mbps,

$$x_A[2] = \frac{1}{q_A[2]} = 0.4,$$

$$x_B[2] = \frac{1}{q_B[2]} = 0.75,$$

$$x_C[2] = \frac{1}{q_C[2]} = 0.75.$$

We then update the link prices. Since the link loads are, in Mbps,

$$y_1[2] = x_A[2] + x_B[2] = 0.4 + 0.75 = 1.15,$$
$$y_2[2] = x_B[2] = 0.75,$$
$$y_3[2] = x_A[2] = 0.4,$$
$$y_4[2] = x_A[2] + x_C[2] = 0.4 + 0.75 = 1.15,$$

we have

$$p_1[2] = [p_1[1] + y_1[2] - c]^+ = [0.833 + 1.15 - 1]^+ = 0.983,$$
$$p_2[2] = [p_2[1] + y_2[2] - c]^+ = [0.5 + 0.75 - 1]^+ = 0.25,$$
$$p_3[2] = [p_3[1] + y_3[2] - c]^+ = [0.333 + 0.4 - 1]^+ = 0,$$
$$p_4[2] = [p_4[1] + y_4[2] - c]^+ = [1.33 + 1.15 - 1]^+ = 1.48.$$

These iterations continue. We plot their evolution over time in Figures 14.8 and 14.9.

We see that an equilibrium is reached after about ten iterations. At this point, the source rates are

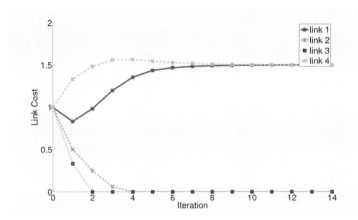

Figure 14.9 Link prices converge over time in a congestion control example.

$$x_A^* = 0.33 \text{ Mbps},$$
$$x_B^* = 0.67 \text{ Mpbs},$$
$$x_C^* = 0.67 \text{ Mbps}.$$

It makes sense that, for proportional fairness, the session that takes up more network resources is given a lower rate: session 1 traverses twice as many bottleneck links and receives half as much the allocated rate.

As a sanity check, let us make sure that the equilibrium values satisfy the constraints in (14.4). Obviously, all the x_i^* are non-negative. To check the link capacity constraint, we see that

$$\begin{bmatrix} 1 & 1 & 0 \\ 0 & 1 & 0 \\ 1 & 0 & 0 \\ 1 & 0 & 1 \end{bmatrix} \begin{bmatrix} x_A \\ x_B \\ x_C \end{bmatrix} = \begin{bmatrix} 1 & 1 & 0 \\ 0 & 1 & 0 \\ 1 & 0 & 0 \\ 1 & 0 & 1 \end{bmatrix} \begin{bmatrix} 0.33 \\ 0.67 \\ 0.67 \end{bmatrix} = \begin{bmatrix} 1 \\ 0.67 \\ 0.33 \\ 1 \end{bmatrix} \leq \begin{bmatrix} 1 \\ 1 \\ 1 \\ 1 \end{bmatrix}.$$

At equilibrium, the link prices are

$$p_1^* = 1.5,$$
$$p_2^* = 0,$$
$$p_3^* = 0,$$
$$p_4^* = 1.5.$$

Again, it makes sense that links 2 and 3 have 0 price at equilibrium, since their capacity is not fully utilized at equilibrium, as constrained by the way the sessions are routed. Conversely, links 1 and 4 have strictly positive prices at equilibrium, and by virtue of the complementary slackness property in the Advanced Material, we know that their link capacities must be fully utilized at equilibrium, i.e., links 1 and 4 are the bottlenecks.

14.4   Advanced Material

14.4.1   Decomposition of NUM

In this subsection, we derive the solution (14.2) and (14.3) to the basic NUM problem (14.1). The objective function is already decoupled across the sessions indexed by i: each term U_i in the sum utility depends only on the rate x_i for that session i. So we just need to decouple the constraints. It is precisely this set of linear capacity constraints that couples the sessions together through the given routing matrix $\mathbf{R}$.

The decomposition method we will use is called **dual decomposition**, since it actually solves the **Lagrange dual problem** of NUM. Given any optimization problem, we can derive a "mirror" problem called the dual problem. Sometimes the dual problem's optimized objective value equals that of the original primal problem. And at all times, it provides a performance bound to that of the original problem. The dual problem can sometimes be solved much faster, and in our case, solved in a distributed way.

The first step in deriving the Lagrange dual problem is to write down the **Lagrangian**: the sum of the original objective function and a weighted sum of the constraints $(c_l - \sum_{i \in S(l)} x_i \geq 0)$. The positive weights are called **Lagrange multipliers** $\mathbf{p}$, which can be interpreted as the link prices:

$$L(\mathbf{x}, \mathbf{p}) = \sum_i U_i(x_i) + \sum_l p_l \left(c_l - \sum_{i \in S(l)} x_i \right). \tag{14.5}$$

The intuition is that we change a constrained optimization to a much easier, *unconstrained* one, by moving the constraints up to augment the objective function. The hope is that if we set the right weights $\mathbf{p}$, we can still get the original problem's solution.

Next, we group everything related to the variables $\mathbf{x}$ together, in order to try to extract some structure from the Lagrangian:

$$L(\mathbf{x}, \mathbf{p}) = \sum_i U_i(x_i) - \sum_l \sum_{i \in S(l)} p_l x_i + \sum_l c_l p_l.$$

Now something almost magical happens: we can rewrite the double summation above by reversing the order of summation: first sum over l for a given i and then sum over all i:

$$\sum_i \sum_{l \in L(i)} p_l x_i,$$

which allows us to rewrite the part of L involving the rate variables $\mathbf{x}$ as follows:

$$L(\mathbf{x}, \mathbf{p}) = \sum_i \left[U_i(x_i) - \left(\sum_{l \in L(i)} p_l \right) x_i \right] + \sum_l c_l p_l.$$

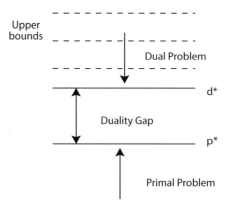

Figure 14.10 Suppose we have a maximization problem, which we will call the primal problem, with an optimized objective function's value p^*. There is a corresponding Lagrange dual problem, which is a minimization problem, with an optimized objective function's value d^*. Any feasible solution in the dual problem generates an upper bound on the primal problem's p^*. This is called weak duality. The tightest bound is d^*, which may still have a gap from p^*. If there is no gap, as is the case when the primal problem is convex optimization (and satisfies some technical conditions), we say the strong duality property holds.

For example, for the network in Figure 14.6, we have $L(x_A, x_B, x_C, p_1, p_2, p_3, p_4) =$

$$U_A(x_A) - (p_1 + p_3 + p_4)x_A + U_B(x_B) - (p_1 + p_2)x_B + U_C(x_C) - p_4 x_C +$$
$$c_1 p_1 + c_2 p_2 + c_3 p_3 + c_4 p_4.$$

We denote the path price (the sum of prices along the links used by session i) as q_i, so $q_i = \sum_{l \in L(i)} p_l$. Then we have a decomposed Lagrangian: the maximization over $\mathbf{x}$ can be independently carried out by each source i (see the square bracket within $\sum_i$ below):

$$L(\mathbf{x}, \mathbf{p}) = \sum_i [U_i(x_i) - q_i x_i] + \sum_l c_l p_l.$$

Suppose now we maximize the Lagrangian over the original variables $\mathbf{x}$. This was our plan in the first place: to turn a constrained optimization into an unconstrained one. Of course the maximizer and the maximized L value depend on what Lagrange multipliers $\mathbf{p}$ we used. So, we have to denote the resulting value as a function of $\mathbf{p}$:

$$g(\mathbf{p}) = \max_{\mathbf{x}} L(\mathbf{x}, \mathbf{p}).$$

This function $g(\mathbf{p})$ is called the **Lagrange dual function**.

It turns out that no matter what $\mathbf{p}$ we use (as long as they are non-negative), $g(\mathbf{p})$ is *always* a performance bound. It is an upper bound on the maximum $U^* = \sum_i U_i(x_i^*)$ of the original NUM problem. This is easy to see. Consider the

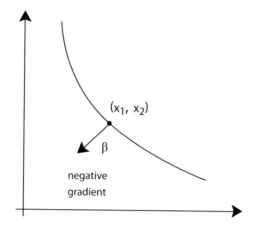

Figure 14.11 Suppose we want to minimize a function of two variables as shown here. The gradient algorithm moves from the current point (x_1, x_2) along the direction of the negative gradient of the function, with a certain stepsize β. Sometimes, the gradient can be computed distributedly.

maximizer of the NUM problem $\mathbf{x}^*$. It must be a feasible vector and satisfies the inequality constraint. Also $\mathbf{p}$ is non-negative. So the Lagrangian L (14.5) must be at least as large as U^* when $\mathbf{x} = \mathbf{x}^*$. Since the Lagrange dual function g is the largest Lagrangian over all $\mathbf{x}$, it must also be no smaller than U^*. This is called the **weak duality** property, which actually holds for all optimization problems.

How about tightening this bound $g(\mathbf{p})$, by picking the best $\mathbf{p}$? We call the resulting problem the Lagrange dual problem, and give the name Lagrange dual variables to $\mathbf{p}$ now:

$$\text{minimize}_{\mathbf{p}} \ g(\mathbf{p}).$$

As illustrated in Figure 14.10, if this tightening generates the exact answer to the original optimization, we say the optimal **duality gap** is zero, and the property of **strong duality** holds. Strong duality is not always true. Together with some technical conditions, the original problem being a convex optimization problem is a sufficient condition for strong duality to hold. This is another reason why convex optimization is easy.

Applying the above dual decomposition to break up one problem into many smaller problems to NUM, we see that the first step is maximizing over $\mathbf{x}$ for a given $\mathbf{p}$. This is nothing other than the net utility maximization we saw in Chapter 11, selfishly and distributedly carried out at each source now:

$$x_i^*(\mathbf{p}) = \text{argmax} \ [U_i(x_i) - q_i x_i].$$

We obtain exactly (14.2).

The second step, minimizing the Lagrange dual problem's objective function $g(\mathbf{p}) = L(\mathbf{x}^*(\mathbf{p}), \mathbf{p})$, over $\mathbf{p}$, can be carried out by the gradient method. Going down along the direction of the negative gradient with a stepsize β, as illustrated in Figure 14.11, we have

$$\mathbf{p}[t+1] = \mathbf{p}[t] - \beta\,(\text{gradient of } g(\mathbf{p}) \text{ at } \mathbf{p}[t])\,.$$

It turns out that for a linearly constrained, concave maximization problem like NUM, the constraint function itself $c_l - \sum_{i \in S(l)} x_i = c_l - y_l$, is the gradient for each p_l. So all we need to do is multiply the gradient by a stepsize β (and then make sure it is never negative):

$$p_l[t] = \left\{ p_l[t-1] - \beta\left(c_l - \sum_{i \in S(l)} x_i^*(\mathbf{p}) \right) \right\}^+,$$

which is exactly (14.3).

Since strong duality holds for NUM, solving the Lagrange dual problem is equivalent to solving the original problem. This concludes the derivation of (14.2) and (14.3) as a distributed solution algorithm to (14.1).

These optimized primal variables (the rate vector $\mathbf{x}^*$) and dual variables (the link price vector $\mathbf{p}^*$) also satisfy other useful properties, including the following **complementary slackness** property. If primal constraint l is slack, that is, $\sum_{i \in S(l)} x_i^* < c_l$, the corresponding optimal dual variable p_l^* (the optimal link congestion price) must be 0, i.e., the non-negativity constraint in the dual problem is not slack. Conversely, if the optimal link congestion price $p_l^* > 0$ for some link l, we must have $\sum_{i \in S(l)} x_i^* = c_l$, i.e., link l is a bottleneck link at equilibrium.

14.4.2   Reverse engineering

We mentioned earlier that if you give me a TCP congestion control protocol, I can return to you a corresponding NUM problem implicitly solved by it, with the utility function completely specified, where the source rates (or window sizes) are the variables, and the pricing signals are the Lagrange dual variables. We now illustrate the main steps in this reverse engineering approach for TCP Reno.

The first step in the derivation is to write down the evolution of the congestion window size, denoted now by w, as specified by the given protocol. Each time an in-order acknowledgement is received at the source, the window size grows by $1/w$ above its current size. Therefore, if every packet is properly received, the window size increases by 1 after one RTT. But each time a packet loss is detected, the window size is halved. Therefore, the net change to the window size $w[t]$ is

$$x[t](1 - q[t])\frac{1}{w[t]} - x[t]q[t]\frac{w[t]}{2}, \tag{14.6}$$

where we omitted the subscript i for notational simplicity.

Now let RTT be d and assume it is constant (even though it obviously varies in time depending on the congestion condition). Since $x = w/d$, (14.6) leads to the following difference equation:

$$x[t+1] = x[t] + \frac{1-q[t]}{d^2} - \frac{1}{2}q[t]x^2[t].$$

By definition, x no longer changes at an equilibrium, which means $\frac{1-q}{d^2} = \frac{1}{2}qx^2$. This equilibrium condition gives us an equation connecting q with x:

$$q = \frac{2}{x^2 d^2 + 2}.$$

From the demand function's definition, we know $U_i'(x_i) = q_i$. So, if we integrate the above expression in x, we recover the utility function:

$$U(x) = \frac{\sqrt{2}}{d}\arctan(\sqrt{1/2}x_i d).$$

In summary, TCP Reno's equilibrium solves NUM with arctan utility, with the help of packet loss as the Lagrange variables (i.e., the pricing feedback). We can also verify that complementary slackness is satisfied: if a primal constraint is slack, i.e., the demand is strictly less than the capacity on a link at equilibrium, there will be neither loss nor queueing delay. Conversely, if a dual constraint is slack, i.e., there is loss or queueing delay on a link, its capacity must be fully utilized.

Now, the packet-loss rate q along a path is not actually equal to the *sum* of loss rates on the links along the path. It is $1 - \prod_l(1 - p_l)$, i.e., 1 minus the probability that no packet is lost on any link. But when the loss rate is small, $q_i = 1 - \prod_l(1 - p_l) \approx \sum_{l \in L(i)} p_l$ holds pretty well as an approximation.

We have made quite a few other assumptions implicitly along the way.

- We focused only on the equilibrium behavior. But a protocol may not converge. Its equilibrium behavior may be desirable, but an equilibrium may never be reached.
- We ignored the actual queueing dynamics inside the queues.
- We ignored the propagation delay it takes for packets and acknowledgements to travel through the network.
- We assumed that there is a fixed set of sessions sharing the network capacities, each going on forever.

Many of these assumptions have been taken away and stronger results obtained over the years. What is somewhat surprising is that, even with some of these assumptions, the theoretical prediction from the NUM analysis works quite well when compared with actual TCP operations.

Further, the optimization model of congestion control has led to forward engineering of new TCP variants that are provably stable. **Stability** here means that the trajectory of a protocol's variables converges to the desired equilibrium, such as the solution to an NUM with a properly chosen utility function. Some of these variants have been demonstrated in lab experiments and then commercialized, including FAST TCP for long-distance and large-volume transmissions, and CUBIC, the default TCP in the Linux kernel.

Summary

Box 14 Distributed congestion control

TCP runs connection-oriented, end-to-end control in the transport layer, including sliding-window-based congestion control. Congestion feedback signals are sent implicitly from the links to the sources as packet loss or delay. This negative feedback can be viewed as pricing signals for distributed coordination of capacity demand. Each congestion control protocol can be reverse-engineerd as implicitly solving a Network Utility Maximization problem and its Lagrange dual problem through a distributed gradient algorithm.

Further Reading

Congestion control modeling and design have been an active research area in networking for more than 20 years.

1. A standard reference book on TCP/IP is
W. R. Stevens, *TCP/IP Illustrated, Vol. 1: The Protocols*, Addison Wesley, 1994.

2. The control dynamic system viewpoint, the optimization model, and the pricing interpretation of TCP were pioneered by Kelly and others in the late 1990s:
F. P. Kelly, A. Maulloo, and D. Tan, "Rate control in communication networks: Shadow prices, proportional fairness, and stability," *Journal of the Operational Research Society*, vol. 49, no. 3, pp. 237–252, March 1998.

3. Reverse engineering TCP protocols into NUM models has been summarized in the following article:
S. H. Low, "A duality model of TCP and queue management system," *IEEE/ACM Transactions on Networking*, vol. 11, no. 4, pp. 525–536, July 2003.

4. The following monograph summarized the major results in congestion control modeling up to 2004, including stochastic session arrivals and departures:
R. Srikant, *The Mathematics of Internet Congestion Control*, Birkhauser, 2004.

5. The following article surveyed the generalization of the modeling approach in this chapter to "layering as optimization decomposition:"

M. Chiang, S. H. Low, A. R. Calderbank, and J. C. Doyle, "Layering as optimization decomposition: A mathematical theory of network architecture," *Proceedings of the IEEE*, vol. 95, no. 1, pp. 255–312, January 2007.

Problems

14.1　*A numerical example of NUM* ⋆⋆

Suppose we have a network whose topology is shown in Figure 14.12.

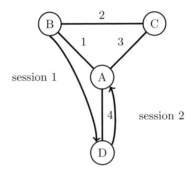

Figure 14.12 A small network for a numerical example of NUM.

(a) Write down the routing matrix $\mathbf{A}$.

(b) Run a numerical example for ten time steps to solve the NUM problem using the link-price and source-rate updates in (14.4) and (14.5). The utility function is a logarithmic function of the source rate. Initialize the link prices to 1, and run the source-rate update step first. Set the step size $\beta = 1$. Plot the source rates and link prices over time. What are the equilibrium values?

(c) Change the step size β and observe the impact on the convergence of the algorithm.

14.2　*TCP slow start* ⋆

We learned the primary mode of operation of TCP Reno, where the congestion window size, denoted as w in the homework problems, increases by 1 for each RTT. Equivalently, w increases by $\frac{1}{w}$ for each ACK received. This operational mode is called *congestion avoidance*.

However, this results in a linear increase of w over time. At the beginning of a TCP connection, we would like to quickly ramp up w before transitioning to congestion avoidance mode. Most TCP protocols have a (somewhat confusingly named) *slow start* phase that accomplishes this. In this mode, w increases by 1 for each ACK received.

(a) If we plot w versus *time*, instead of having a linear increase as in congestion avoidance, what kind of increase do we see in slow start?

(b) Assume w starts at 1. Draw a space–time diagram for the slow start phase over four RTTs.

14.3   *TCP Reno congestion window* ⋆⋆

Recall that the congestion window length changes with time as follows during TCP Reno's congestion avoidance phase.
• If an ACK is received, then increase w by $\frac{1}{w}$.
• If congestion is detected, then decrease w by $\frac{w}{2}$.
  Suppose the probability of failed transmission is p; the probability of a successful transmission is then $1-p$. The transmission rate $x = \frac{w}{RTT}$ packets per second.

(a) Write down the equation for the expected change of w per time step.

(b) At equilibrium, the expected change is 0. Using (a), show that $x_r = \frac{1}{RTT}\sqrt{\frac{2(1-p)}{p}}$.

14.4   *TCP Vegas* ⋆⋆

TCP Vegas tries to anticipate congestion by estimating delay. In this scheme, the congestion window size is changed on the basis of the timings of the ACKs it has received. Specifically, we have

$$w[t+1] = \begin{cases} w[t] + 1 & \text{if } \frac{w[t]}{d} - \frac{w[t]}{D[t]} < \beta \\ w[t] - 1 & \text{if } \frac{w[t]}{d} - \frac{w[t]}{D[t]} > \beta \\ w[t] & \text{otherwise,} \end{cases}$$

where d is the minimum RTT observed historically, $D[t]$ is the RTT observed at time t, and β is a threshold. So, $\frac{w}{d}$ is the expected rate and $\frac{w}{D}$ is the observed rate, and w decreases (increases) if the expected rate is greater (smaller) than the actual rate by β.

Suppose

$$D[t] = \begin{cases} t & \text{if } t \le 4 \\ 4 & \text{otherwise} \end{cases},$$

and $w[1] = 4, \beta = 3$ and $d = 1$. Plot the evolution of w versus time for ten time steps.

14.5   *Reverse engineering TCP Vegas* ★★★

It can be shown that TCP Vegas approximately solves the following weighted log utility maximization problem, where β_i is the protocol parameter in the last problem, and d_i is the no-congestion RTT for session i:

$$\text{maximize} \quad \sum_i \beta_i d_i \log x_i$$

$$\text{subject to} \quad \sum_{i \in S(l)} x_i \leq c_l, \ \forall l \tag{14.7}$$

$$\text{variables} \quad x_i \geq 0, \ \forall i.$$

The links update their prices as follows:

$$p_l[t+1] = \{p_l[t] + \gamma_l \left(y_l[t] - c_l\right)\}^+ . \tag{14.8}$$

As before,

$$L(i) = \text{the set of links used by session } i,$$

$$S(l) = \text{the set of sessions using link } l,$$

$$y_l[t] = \sum_{i \in S(l)} x_i[t], \tag{14.9}$$

$$q_i[t] = \sum_{l \in L(i)} p_l[t].$$

We will show this through several steps.

(a) Define the total backlog on link l from all sessions as $b_l[t]$. Then each link updates its backlog at each time step by $b_l[t+1] = \{b_l[t] + \beta(y_l[t] - c_l)\}^+$. Show that if we define $p_l[t] = \frac{b_l[t]}{c_l}$, the links update their prices as in (14.8).

(b) If a network is trying to solve (14.7), what should the source update rule be? Recall that $x_i[t] = U_i'^{-1}(q_i[t])$.

(c) Recall that each window size in TCP Vegas is updated by:

$$w_i[t+1] = \begin{cases} w_i[t] + 1 & \text{if } \frac{w_i[t]}{d_i} - \frac{w_i[t]}{D_i[t]} < \beta_i \\ w_i[t] - 1 & \text{if } \frac{w_i[t]}{d_i} - \frac{w_i[t]}{D_i[t]} > \beta_i \\ w_i[t] & \text{otherwise} \end{cases}$$

We also know that the backlog on link l from session i is $\frac{x_i[t]}{c_l} b_l[t]$. The congestion window size, w_i, is the sum of the total backlog in the path of i and the bandwidth–delay product, i.e., $w_i[t] = \sum_{l \in L(i)} \frac{x_i[t]}{c_l} b_l[t] + d_i x_i[t]$. Show that the source-rate update rule matches the answer to part (b).

15 How can Skype and BitTorrent be free?

We just went through some of the key concepts behind the TCP/IP thin waist of the Internet protocol stack. We will now go through five more chapters on technology networks, focusing on two major trends: massive amounts of content distribution and the prevalent adoption of mobile wireless technologies.

Scaling up the distribution of content, including video content, can be carried out either through the help of peers or by using large data centers. These two approaches, P2P and cloud, are described in this chapter and the next, respectively. In particular, P2P illustrates a key principle behind the success of the Internet: under-specify protocols governing the operation of a network so that an overlay network can be readily built on top of it for future applications unforeseen by today's experts. It also illustrates the importance of backward compatibility, incremental deployability, and incentive alignment in the evolution of the Internet.

15.1   A Short Answer

Skype allows phone calls between IP-based devices (like laptops, tablets, and smartphones) or between IP devices and normal phones. It is free for IP-to-IP calls. How could that be? Part of the answer is that it uses a peer-to-peer (**P2P**) protocol riding on top of IP networks.

P2P started becoming popular around 1999. For example, Kazaa and Gnutella were widely used P2P file- and music-sharing systems back then. However, incentives were not properly designed in those first-generation P2P systems; there were a lot of *free riders* who did not contribute nearly as much as they consumed.

Skype started in 2001 from Kazaa, and was acquired by eBay for $2.6 billion in 2006 and then by Microsoft for $8 billion in 2011. As of 2010, there were 663 million Skype users worldwide. On any given day there are, on average, 700 million minutes of Skype calls.

BitTorrent started in 2001 as well, and is heavily used for file sharing, including movie sharing. Like Skype, it is free and uses P2P technologies. At one point, P2P was more than half of Internet traffic, and BitTorrent alone in the mid 2000s was about 30% of the Internet traffic. P2P sharing of multimedia content is still very popular today, with over 250 million users just in BitTorrent.

P2P showcases a major success of the evolution of the Internet: make the basic design simple and allow overlay constructions. The architecture of the Internet focuses on providing simple, ubiquitous, stable, and economical connectivities, leaving the rest of the innovations to overlays to be constructed in the future for unforeseeable applications. Different types of applications, unicast as well as multicast, have been built using P2P overlays, including file sharing, video streaming, and on-demand multimedia distribution.

Both Skype and BitTorrent are free (of course the Internet connection from your device might not be free).

- Skype is free in part because it leverages peer capability to locate each other and establish connections. P2P is used for signaling in Skype.
- BitTorrent is free in part because it leverages peer uplink capacities to send chunks of files to each other, without deploying many media servers. (And it is free in part because the content shared sometimes does not incur royalty fees). P2P is used for sharing content in BitTorrent.

Both Skype and BitTorrent are *scalable*. They illustrate a positive network effect whereby each additional node in the network contributes to many other nodes. We can therefore add many more nodes as the network scales up without creating a bottleneck. Of course this assumes the nodes can effectively contribute, and that requires some smart engineering design. As this chapter shows, this "**P2P law**" is a refinement of our intuition about the network effect codified in **Metcalfe's law** (named after the inventor of Ethernet that connects computers in a local area network): the benefit of joining a network grows as the *square* of the number of nodes. One of the underlying assumption is that all connections are equally important. But as we saw in Chapter 9 on triad closures vs. long-range links and in Chapter 10 on long-tail distribution, there is often a "diminishing marginal return" on similar types of links. Metcalfe's law also makes an assumption that each node is basically connected to all the other nodes, or at least the number of neighbors per node grows as a linear function of the network size. In contrast, the P2P law does *not* require that, and shows that the benefit of scalability can be achieved even when each node has only a small number of neighbors at any given time, as long as these are carefully chosen.

Skype's operational details are a commercial secret. BitTorrent is much more transparent, with papers written by the founder explaining its operation. So our treatment of Skype P2P connection management will be thinner than that of BitTorrent's P2P content sharing.

15.1.1 Skype basics

To understand how the technology behind Skype works, we need to understand two major topics: voice over IP (**VoIP**) and P2P. We postpone the discussion of VoIP to Chapter 17 together with multimedia networking in general. This chapter's focus is on P2P.

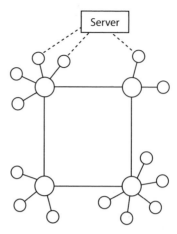

Figure 15.1 A typical topology of Skype. There is a mesh of super nodes (the bigger circles) and a shallow tree of ordinary nodes (smaller circles) rooted at each super node. Super nodes can be users' computers or skype machines. There is also an authentication server (the rectangle) that each node exchanges control messages with first.

Phone calls are intrinsically P2P: a peer calls another peer (as opposed to a server). What is interesting is that Skype uses P2P to discover peers and to traverse firewalls (software and hardware that blocks incoming data connections). As shown in Figure 15.1, Skype's central directory allows a caller to discover the IP address of the callee and then establish an Internet connection. These directories are replicated and distributed in **super nodes** (SNs).

The problem is that sometimes both the caller and the callee are behind firewalls, with a NAT box (see Chapter 13) in between. So the actual IP address is not known to the caller. Those outside of a firewall cannot initiate a call into the firewall.

What happens then is that super nodes have *public* IP addresses, serving as anchors to be reached by anyone and collectively acting as a network of publicly visible relays. The caller first initiates a connection with an SN, and the callee initiates a connection with another SN. Once a connection has been established, two-way communication can happen. The caller then calls her SN, which calls the callee's SN, which then calls the callee. Once a connection between the caller and the callee has been established through these two SNs, they mutually agree to use just a single SN that they both can connect to, thus shortening the communication path.

15.1.2   BitTorrent basics

BitTorrent uses P2P for resource sharing: sharing upload capacities of each peer and sharing the content stored in each peer, so that content sharing can scale

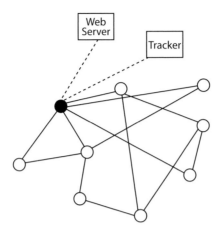

Figure 15.2 A typical topology of BitTorrent. There are actually three topologies: (1) a graph of physical connections among peers and routers, (2) a graph of overlay neighbor relationships among peers, and (3) a graph of peering relationships among peers. Graph (3) is an overlay on graph (2), which is in turn an overlay on (1). This figure shows graph (3). It changes regularly depending on the list of peers provided by the tracker to, say, peer A (the dark circle), as well as the subset of those peers chosen by peer A.

itself. It is designed primarily for **multicasting**: many users all demand the same file at about the same time.

In BitTorrent, each file is divided into small pieces called *chunks*, typically 256 kB, so that pieces of a file can be shared simultaneously. Each peer polls a centralized directory called the *tracker*, which tells a peer a set of 50 (or so) peers with chunks of the file it needs. Then the peer picks five peers with which to exchange file chunks. This set of five peering neighbors is refreshed at the beginning of every timeslot, depending in part on how much a neighbor is helping this peer and in part on randomization.

As shown in Figure 15.3, each individual chunk traverses a *tree* of peers, although the overall peering relationship is a general graph that evolves in time. A **tree** is an undirected graph with only one path from one node to any other node. There are no cycles in a tree. We usually draw a tree with the root node on top and the leaf nodes on the bottom.

We see that the control plane for signaling is somewhat centralized in both Skype and BitTorrent, but the data plane for the actual data transmission is distributed, indeed peer to peer. This is in sharp contrast to the traditional **client–server** architecture, where each of the receivers requests data from a centralized server and do not help each other.

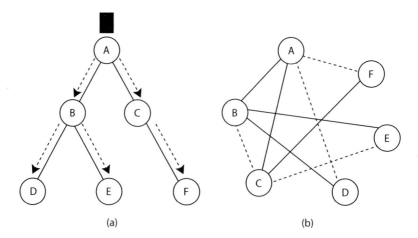

(a)   (b)

Figure 15.3 (a) Each chunk traverses a tree (with the chunk represented by the rectangle and the data transmission in dotted lines), even though (b) the peering relationships form a general graph (where the solid lines represent the current peering relationships and dotted lines represent possible peering relationships in the next timeslot).

15.2   A Long Answer

Before we go into some details of the smart ideas behind Skype and BitTorrent, we highlight two interesting observations.

- P2P is an **overlay network**, as illustrated in Figure 15.4. Given a graph with a node set V and a link set E, $G = (V, E)$, which we call the underlay, if we select a subset of the nodes in V and call that the new node set $\tilde{V}$, and we take some of the *paths* connecting nodes in $\tilde{V}$ as *links* and call that the new link set $\tilde{E}$, we have an overlay graph $\tilde{G} = (\tilde{V}, \tilde{E})$. The Internet itself can be considered as an overlay on top of the PSTN, wireless, and other networks; and online social networks are an overlay on top of the Internet too. The idea of overlay is as powerful as that of layering in giving rise to the success of the Internet. It is evolvable: as long as the Internet provides the basic service of addressing, connectivity, and application interfaces, people can build overlay networks on top of existing ones. For example, multicasting could have been carried out in the network layer through **IP multicast**. And there are indeed protocols for that. But other than within a Local Area Network (see the homework problem in Chapter 13) and IPTV for channelized content (see Chapter 17), IP multicast is rarely used. The management of IP multicast tends to be too complicated. Instead, P2P offers an alternative, overlay-based approach with less overhead.
- P2P is about *scalability*, and, in BitTorrent's case, scalability in *multicasting*. If you consume, you also need to contribute. This upside of the network effect is the opposite of the wireless network interference problem, where

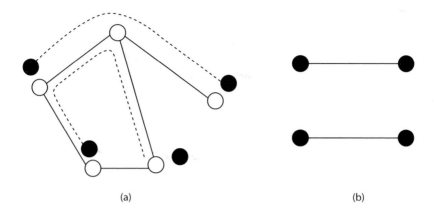

Figure 15.4 Building an overlay network of four nodes (the dark circles) on top of an underlay network of five nodes (the light circles). In the overlay graph (b), the nodes are a subset of the nodes in the underlay graph (a), and each link represents a path (the dotted lines) in the underlay graph. The overlay graph in this example is simply two parallel lines.

one user's signal is other users' interference. Of course, even in BitTorrent, there is the problem of free riders: what if a user only consumes but does not contribute? We will look at BitTorrent's solution next. We will also see in the next chapter another way to provide scalability to the server–client architecture using small switches to build large ones as the data center scales up.

15.2.1   More on Skype

As mentioned before, there are two types of nodes in Skype: super nodes (SNs) and ordinary hosts. An SN must have a public IP address, so that it can help traverse NATs and firewalls. Preferably, it should also have abundant resources, including CPU, memory, and capacities on ingress and egress links. An ordinary host must connect to an SN. Some of your desktops may actually be SNs on Skype networks.

Skype uses an overlay P2P network with two tiers: a mesh of SNs and a shallow tree rooted at each super node. This two-tier structure is mainly for the purpose of firewall traversal in Skype, although we will encounter it again in the Advanced Material for performance optimization in P2P file sharing.

When a Skype client, whether on an ordinary host or on an SN, wants to initiate a call, it must first authenticate itself with the Skype login server, which stores all the usernames and passwords. If the call is between an IP device and a PSTN phone, additional servers and procedures are required (and it is no longer free).

Each ordinary host maintains and updates a host cache, which contains the IP addresses and port numbers of SNs. During login, a host advertises its presence to other hosts, determines whether it is behind a firewall, and discovers which SN to connect to and which public IP address to use. As we will see, compared with BitTorrent, the P2P topology is much more stable in Skype after login is finished.

Once logged in, a user search can be carried out through Skype's global index database. A TCP connection needs to be established for signaling between the caller and callee, and a UDP (or a TCP, if the firewall needs to be traversed) connection established for the actual voice traffic. For conferencing, more steps are needed for connection establishment.

15.2.2   More on BitTorrent

The first smart idea in BitTorrent file sharing is to use *smaller granularity* than the whole file. In this way, each chunk can traverse different trees, and the transmission can be *spatially pipelined*. The advantage of *multi-tree transmission* is similar to the advantage of multipath routing in packet switching, which divides a given message into smaller granularity called packets, and lets them go through possibly different paths. In fact, the richness of the tree topology, compared with the path topology, and the heavy usage of multiple trees make P2P tree selection more robust than IP routing. We can pick peers without too much optimization and still achieve very high overall efficiency for the network. We will see more of this in the Advanced Material.

When we discuss content distribution networks in Chapter 17, we will see that they are similar to deploying peers with large upload and storage capacities. Indeed, the term "peer" in BitTorrent refers to the fact that the node is both a sender and a receiver of content, and, when it acts as a sender, it is a (small) server. In content distribution networks, deciding which content to place on which servers is a key design step. In P2P, this content placement is optimized through the strategy of *rarest chunk first*. When a peer looks at the bitmap and chooses which chunks to download, it should start with the chunks that very few peers have. By equalizing the availability of chunks, this strategy mitigates the problem where most of the peers have most of the chunks, but all must wait for the few rare chunks.

Yet another smart idea in BitTorrent is its peering construction method. The first step is for the tracker to suggest a set of 50 or so potential peers to a new peer. These potential "friends" are recommended based on the content they have and other performance-driven factors like the distance to the new peer. They are also driven by peer churns: which peers are still sending "I am here" messages to the tracker. A list of 50 provides a larger degree of freedom than is actually used by each peer.

The second step is to let the new peer pick, at each time, her actual "friends." These are the peers to exchange chunks with. Usually five peers are picked, and the upload bandwidth is evenly distributed among these five in the next timeslot:

- Four of them are the top four peers in terms of the amount of content received from them by this node in the last timeslot. This is called the **tit-for-tat** strategy.
- The remaining peer is selected at random from the set of 50.

The first feature mitigates the free-rider problem, where a node could contribute but decides not to. The second feature avoids unfairness to those nodes with little upload capacity to contribute. Randomization is also generally a good idea to avoid getting trapped in a locally optimal solution. This is similar to Google's PageRank calculation in Chapter 3, which was 85% topology-driven and 15% randomization.

Now we can summarize the BitTorrent operation, knowing why each step is designed as it is.

---

1. A new peer A receives a .torrent file from one of the BitTorrent web servers, including the name, size, and number of chunks of a particular file, together with the IP address and port number of the corresponding tracker.
2. It then registers with the right tracker. It will also periodically send keep-alive messages to the tracker.
3. The tracker sends to peer A a list of potential peers.
4. Peer A selects a subset (following the tit-for-tat and randomization rules) and establishes connections with these five peers.
5. They exchange bitmaps to indicate which chunks of the content they each have.
6. With chunks selected, they start exchanging chunks among themselves, starting with the rarest chunks.
7. Every now and then, each peer updates its peer list.

15.3 Examples

We will discuss the throughput limit of P2P in the Advanced Material, and games in P2P systems in homework problems. But first, several simple examples to quantify the benefits of P2P.

15.3.1 Back-of-the-envelope bounds

To illustrate the P2P network effect, and how P2P changes the scalability property of file distribution, we run a simple back-of-the-envelope calculation.

First, consider N clients requesting a file of size F bits from a server with an upload capacity of u_s bps. Each of these clients has a download capacity of d_i bps. This is illustrated in Figure 15.5.

- The server needs to send out NF bits, so it takes at least NF/u_s seconds.

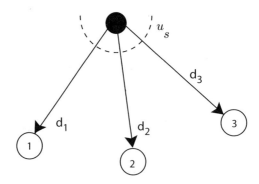

Figure 15.5 A typical server–client star topology. The upload speed of the server is u_s bps, and the download speeds of the clients are $\{d_i\}$ bps.

- All the clients need to receive the file, including the slowest one with a download capacity of d_{min}, and that takes at least F/d_{min} seconds.

So the total download time is the larger of the two numbers above:

$$T = \max\left\{\frac{F}{d_{min}}, \frac{NF}{u_s}\right\}. \qquad (15.1)$$

This would be fine, if we could increase u_s as N becomes larger. But scaling the upload capacity of a server becomes technologically and financially difficult as N becomes very large. So the alternative is to deploy more servers. Well, these N clients can also become servers themselves, and we call them peers. A *hybrid* peer and server deployment is what actually happens, but it is the P2P part of the network that scales itself as the number of peers increases.

Suppose each peer i has an upload capacity u_i, in addition to a download capacity d_i as before. These upload capacities may be much smaller than the download capacities, because the traditional design assumes that the Internet traffic is primarily unidirectional. With user-generated content on the rise and P2P protocols heavily used, this assumption is no longer valid. In some cases, upload capacities $\{u_i\}$ are quite large. Peers with larger upload capacities can help distribute the files by sitting closer to the root of the multicast distribution tree in Figure 15.3.

Suppose these distribution trees can be perfectly designed to fully utilize all the upload capacities. Then we can say that for the total number of bits to be shared, NF, the total upload bandwidth available to the whole network is $u_s + \sum_{i=1}^{N} u_i$. So the time it takes is $NF/(u_s + \sum_{i=1}^{N} u_i)$ seconds.

Of course, the server still needs to send out each bit at least once to some peer, taking F/u_s seconds, and the slowest peer still needs to receive each bit, taking F/d_{min} seconds. Therefore, the time it takes to distribute the file throughout the network is now

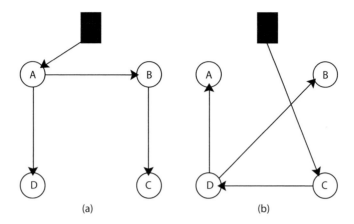

Figure 15.6 Two peering relationship trees. The squares are servers. The circles are peers. If only one of them is used, it is impossible to avoid wasting some nodes' uplink capacities: the tree in (a) wastes C's and D's uplink capacities, while the tree in (b) wastes A's and B's. This problem can be mitigated if we use both tree (a) and tree (b) for the same multicast session.

$$T = \max \left\{ \frac{F}{u_s}, \frac{F}{d_{min}}, \frac{NF}{u_s + \sum_{i=1}^{N} u_i} \right\}. \tag{15.2}$$

Let us compare (15.1) with (15.2). Among the terms in (15.2), only the third term has a numerator that scales with N, the number of peers. But that is divided by the summation of u_i over all N nodes. Therefore, T itself no longer scales with N. The network performance scales itself with the network size.

15.3.2 Constructing trees

The above back-of-the-envelope calculation assumes that all the peer upload capacities can be fully utilized. That is hard to do, and sometimes downright impossible, especially when you have only one distribution tree. As shown in Figure 15.6; in tree (a), peers C and D's upload capacities are not used. In tree (b), peers A and B's upload capacities are not used.

How about using both trees at the same time? This is called the **multi-tree** construction of peering relationships. That helps, but it is still not clear what is the *best* way to construct all the trees needed to utilize the upload capacities. The basic idea, however, is clear: those peers with a lot of leftover upload capacities should be placed higher up in the constructed trees. Determining exactly how to do that involves solving a difficult combinatorial optimization problem; embedding even one tree in a general graph is hard, let alone multiple trees. That is the subject in the Advanced Material.

But, here is a special case that is easy to solve. Assume that the download capacities of peers are not the bottlenecks, e.g., the $\{d_i\}$ are large enough. Now we want to prove

$$T = \max \left\{ \frac{F}{u_s}, \frac{NF}{u_s + \sum_{i=1}^{N} u_i} \right\}.$$

To show this, we need to construct a multi-tree, i.e., a set of multicast trees that collectively achieve the desired rates among N peers. Clearly, it suffices to show that the maximum broadcast rate of the multi-tree is

$$r_{max} = \min \left\{ u_s, \frac{u_s + \sum_{i=1}^{N} u_i}{N} \right\}.$$

To show this, we reason through two cases.

Case 1: If $u_s \leq (u_s + \sum_{i=1}^{N} u_i)/N$, then the maximum broadcast rate of $r_{max} = u_s$ should be supported. The server upload capacity is too small. So we consider a multi-tree that consists of N trees, such that each ith tree is two-hop, e.g., the server takes peer i as its child and peer i takes the other $N-1$ peers as its children. Collectively these trees should deplete the upload capacity of the server. Furthermore, trees with more capable peers near the root should stream at a higher rate. Let each tree i carry a rate proportional to u_i:

$$r_i = \left(\frac{u_i}{\sum_{j=1}^{N} u_j} \right) u_s, \quad i = 1, \dots, N,$$

as illustrated in Figure 15.7.

This rate assignment is possible because the total upload required for the server equals its capacity:

$$\sum_{i=1}^{N} r_i = u_s.$$

So is the total upload capacity required for peer i:

$$(N-1)r_i = (N-1) \frac{u_i}{\sum_{j=1}^{N} u_j} u_s \leq u_i,$$

since $N u_s \leq u_s + \sum_{j=1}^{N} u_j$ by the assumption of this case. This implies that

$$N u_s \left(\frac{u_i}{\sum_j u_j} \right) \leq \frac{u_s u_i}{\sum_j u_j} + u_i,$$

which further implies that

$$(N-1) \frac{u_i}{\sum_{j=1}^{N} u_j} u_s \leq u_i.$$

Now each peer receives a data stream directly from the server and also receives $N-1$ additional data streams from the other $N-1$ peers. So the aggregate broadcast rate at which *any* peer i receives is

$$r_{max} = r_i + \sum_{j \neq i} r_j = \sum_{i=1}^{N} r_i = u_s.$$

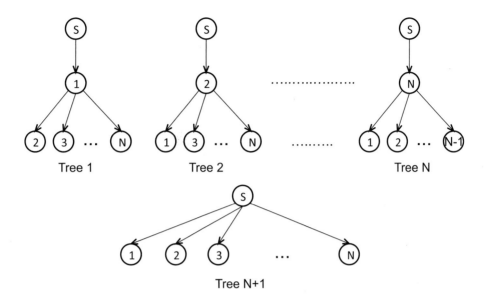

Tree 1   Tree 2   Tree N

Tree N+1

Figure 15.7 Multi-tree construction to maximize the multicast rate among N peers. In case 1, the server upload capacity is relatively small, and we use trees 1 to N. In case 2, the server upload capacity is sufficiently large, and we use trees 1 to $N + 1$, but with a different assignment of rates on the first N trees. In both cases, we need only shallow trees for this simple problem, and the multi-tree depth is 2.

Hence, it takes F/u_s seconds to transfer the whole file.

Case 2: If $u_s > (u_s + \sum_{i=1}^{N} u_i)/N$, then we need to show that the maximum broadcast rate of $r_{max} = (u_s + \sum_{i=1}^{N} u_i)/N$ can be supported. In this case, the server upload capacity is large enough for a different set of trees, including one tree where the server directly connects to all the peers (a server–client tree), so as to fully utilize its upload capacity.

Consider a multi-tree that consists of $N + 1$ trees, such that the ith tree is two-hop and carries a rate of

$$r_i = \frac{u_i}{N - 1},$$

i.e., equal distribution of each peer's uplink capacity among the other peers. And the $(N + 1)$th tree is one-hop directly from the server, which carries a rate of

$$r_{N+1} = \frac{u_s - \frac{\sum_{i=1}^{N} u_i}{N-1}}{N}.$$

This is the leftover uplink capacity from the server (after sustaining the first N trees) evenly distributed among all the N peers.

On the ith tree, for $i = 1, 2, \ldots, N$, the server has peer i as its child and peer i has the other $N - 1$ peers as its children. In contrast, on the $(N + 1)$th tree, the server has all peers as its direct children. The tree construction is shown in Figure 15.7. This is possible because the total upload capacity required for peer

i is exactly

$$(N-1)r_i = u_i,$$

and the total upload capacity required for the server is exactly u_s:

$$\sum_{i=1}^{N} r_i + N \cdot r_{N+1} = \sum_{i=1}^{N} \frac{u_i}{N-1} + N\frac{u_s - \frac{\sum_{i=1}^{N} u_i}{N-1}}{N} = u_s.$$

Of course, the above two equalities are true by the way we design the rates on these $N+1$ trees.

Now each peer receives two data streams directly from the server and also receives $N-1$ additional data streams from the other $N-1$ peers. So the aggregate broadcast rate at which any peer i receives is

$$r_i + r_{N+1} + \sum_{j\neq i} r_j = \frac{u_i}{N-1} + \frac{u_s - \frac{\sum_{i=1}^{N} u_i}{N-1}}{N} + \sum_{j\neq i} \frac{u_j}{N-1} = \frac{u_s + \sum_{i=1}^{N} u_i}{N}.$$

Hence, it takes $NF/(u_s + \sum_{i=1}^{N} u_i)$ seconds to transfer the whole file.

Combining the two cases above produces our desired results.

15.4   Advanced Material

15.4.1   Capacity of P2P

In the example above, we assumed many ideal conditions. Peering relationship construction in a general, large-scale network is much more challenging. Structured P2P overlay carefully designs the peering relationships on the basis of criteria such as throughput, latency, and robustness. Some of these topologies are inspired by what we saw in Chapters 9 and 10, e.g., the Watts–Strogatz graph. This leads us to a graph-theoretic optimization problem.

Consider the following problem: given a directed graph with a source node and a set of receiver nodes, how do we embed a set of trees spanning the receivers and determine the amount of flow in each tree, such that the sum of flows over these trees is maximized? The constraints of this problem include an upper bound on the amount of flow from each node to its children, the maximum degree of a node allowed in each tree, and other topological constraints on the given graph. This is the general problem of P2P capacity computation.

What is the P2P capacity and what is an optimal peering configuration to achieve the capacity? Here, "capacity" is defined as the largest rate that can be achieved for all receivers in a multicast session, with a given source, a set of receivers, and possibly a set of helper (non-receiver relay) nodes.

There are in fact at least sixteen formulations of this question, depending on whether there is a single P2P session or multiple concurrent sessions, whether the given topology is a full mesh graph or an arbitrary graph, whether the number of peers a node can have is bounded or not, and whether there are helper nodes or not. In each formulation, computing the P2P capacity requires determination of (1) how to embed an optimal set of multicast trees, and (2) what the rate should be in each tree.

We outline a family of algorithms that can compute or approximate the P2P capacity and the associated multicast trees. In general this problem is intractable; it is difficult to find efficient algorithms that solve the problem exactly. The algorithm we summarize below can solve, in polynomial time, seven of the sixteen formulations arbitrarily accurately, and eight other formulations to some constant-factor approximations.

We will be reformulating the optimization to turn the combinatorial problems into linear programs with an exponential number of variables. The algorithms combine a primal–dual update outer loop (similar to what we saw in Chapter 14) with an inner loop of "smallest-price-tree construction" (similar to what we just saw in the last section), driven by the update of Lagrange dual variables in the outer loop. Graph-theoretic solutions to various cases of the smallest-price-tree problem can then be leveraged, although that is beyond the scope here. Our focus will be on formulating this problem of embedding multiple trees in a graph, and the generalization of congestion control's primal–dual solution to a more complicated case where each inner loop is not just a simple rate- or price-update equation.

Consider a multicast file-sharing or video-streaming session. It originates from one source, and is distributed to a given set of receivers. For example, in video conferencing, there are multiple participants; each may initiate a session and distribute her video to others, and each participant can subscribe to others' videos. In an IPTV network, different channels may originate from different servers, with different sets of subscribers. Denote by s the original source, by R the set of receivers, and by H the set of helpers. We say that the session has rate r bps if all the receivers in this session receive the packets at a rate of r or above.

As illustrated in Figure 15.8, we consider a P2P network as a graph $G = (V, E)$, where each node $v \in V$ represents a peer, and each edge $e = (u, v) \in E$ represents a *neighboring* relationship between vertices (u, v). A peer may be the source, or a receiver, or a helper that serves only as a relay. A helper does not need to get all chunks, just the ones that it relays.

- This graph is an overlay on top of the given underlay graph representing the *physical connections* among users. The underlay graph may constrain the design of *peering* relationships: if two nodes u and v are not physically connected by some path of reasonable length, they cannot be neighbors in the overlay graph, and do not stand a chance of becoming peers either.
- Neighbors do not have to become peers. The neighboring relationship is given, while the peering relationship is to be designed as part of the P2P capacity computation.

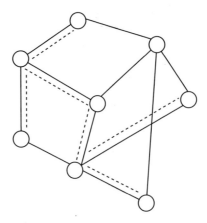

Figure 15.8 An overlay network where each node represents a peer, and each link (shown by a solid line) represents a neighboring relationship. The job of computing P2P capacity is to construct multiple trees so that their rates add up to the largest possible for a multicast session. A particular tree is shown by the dotted lines, which represent peering relationships designed under the constraint of the given neighboring relationships.

- The graph G may or may not be full mesh. Typically, full mesh is possible only in a small network with a small number of users, while a large network has a sparse topology.

Consider one chunk of a given stream. It starts from the source s, and traverses over all nodes in R, and some nodes in H. As in a homework problem in Chapter 13, the traversed paths form a Steiner tree in the overlay graph $G(V, E)$, a tree that spans all the nodes in only a given *subset* of V. In this case, the subset is R.

Different chunks may traverse different trees. We call the superposition of all the trees belonging to the same session a *multi-tree*. For each tree t, denote y_t as the sub-stream rate supported by this tree. Here, t is not a time index, but a complicated data structure: a tree.

The use of a P2P protocol imposes certain constraints on sub-trees. The most frequently encountered one is the node-degree constraint. For example, in Bit-Torrent, although a node may have fifty neighbors in G, it can upload to at most five of them as peers. This gives an outgoing degree bound for each node and constrains the construction of the trees. We will examine the case of the degree bound for each node *per* tree. Let $m_{v,t}$ be the number of outgoing edges of node v in tree t, and the bound be M_v. This gives an inequality constraint on allowed trees: $m_{v,t} \leq M_v, \forall t$. The more general case of a degree bound for each node across *all* the trees is even harder.

We denote by T the set of all allowed sub-trees. Obviously, the multicast rate r is the sum of all the rates on the trees:

$$r = \sum_{t \in T} y_t.$$

We will focus on the basic case with the following assumptions: there is a static set of stationary users; all desired chunks are available at each node; and data-rate bottlenecks appear only at user uplinks. The last assumption is widely adopted because, in today's Internet, access links are the bottlenecks rather than backbone links, and uplink capacity is several times smaller than downlink capacity in typical access networks. Denote by C_v the uplink capacity of node v. We have the following bound on the total uplink rate U_v for each node v:

$$U_v = \sum_{t \in T} m_{v,t} y_t \le C_v.$$

A rate is called *achievable* if there is a multi-tree in which all trees satisfy the topology constraint $(t \in T)$ and transmission rates satisfy the uplink capacity constraint $(U_v \le C_v)$. We define the **P2P capacity** as the largest achievable rate.

15.4.2   A combinatorial optimization

Now we can represent the (single-session) P2P capacity problem as the following optimization problem:

$$\begin{array}{ll} \text{maximize} & \sum_{t \in T} y_t \\ \text{subject to} & \sum_{t \in T} m_{v,t} y_t \le C_v \ \ \forall v \in V \\ \text{variables} & y_t \ge 0, \ \ \forall t \in T. \end{array} \tag{15.3}$$

This representation of the problem is deceptively simple: the difficulty lies in searching through all the combinations of trees t in the set of allowed trees T. For those trees not selected in the optimizer, their rates $\{y_t\}$ are simply 0. The problem has a large number of optimization variables. Compared with all the other optimization problems in previous chapters, this one is much more difficult. It has a combinatorial component and is not convex optimization. It has coupling across trees and cannot be readily decomposed.

Still, we can try to derive the Lagrange dual problem. From Lagrange duality, solving the above problem is equivalent to solving its dual problem, and an optimizer of the dual problem readily leads to an optimizer of the primal algorithm. The dual problem associates a non-negative variable p_v interpreted as the price with each node. It turns out to be as follows:

$$\begin{array}{ll} \text{minimize} & \sum_{v \in V} C_v p_v \\ \text{subject to} & \sum_{v \in V} m_{v,t} p_v \ge 1, \ \ \forall t \in T \\ \text{variables} & p_v \ge 0, \ \ \forall v \in V. \end{array} \tag{15.4}$$

We can interpret the dual problem similarly to the dual congestion control problem in Chapter 14: p_v is the price (per unit flow) for any edge outgoing from v. If node v uploads with full capacity, the incurred cost is $p_v C_v$. There are $m_{v,t}$

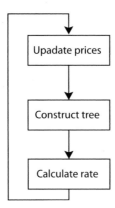

Figure 15.9 A summary of the key building blocks of the two-level iterations to approximately compute the P2P capacity and the corresponding set of sub-trees.

connections outgoing from node v in tree t, and thus the *total tree price* for tree t is simply $\sum_{v \in V} m_{v,t} p_v$. Therefore, the dual problem is to minimize the total tree cost provided that the tree price is at least 1, and the minimization is over all possible price vectors $\mathbf{p}$. This is a generalization of link (and path) prices used in solving the NUM problem for distributed capacity allocation in Chapter 14.

In general, the number of trees we need to search when computing the right multi-tree grows exponentially with the size of the network. This dimensionality increase is the consequence of turning a difficult graph-theoretic, discrete problem into a continuous optimization problem. The primal problem has too many variables and its dual problem has too many constraints, neither of which is suitable for a direct solution. However, the above representations turn out to be very useful by allowing a primal–dual update outer loop, which converts the combinatorial problem of multi-tree construction into a much simpler problem of "smallest-price-tree" construction.

15.4.3   Two-level iterations

We will use the two levels of iterations to solve the primal and dual problems, as summarized in Figure 15.9. In the outer loop, flows are augmented in the primal solution and dual variables are updated iteratively. In the inner loop, an easier tree-construction problem is solved under the guidance of pricing (dual) variables. The algorithm constructs peering multi-trees that achieve an objective function value within $(1 + \epsilon)$-factor of the optimum.

For a given tree t and prices $\mathbf{p}$, let $Q(t, \mathbf{p})$ denote the left-hand-side of constraint (15.4), which we call the price of tree t. A set of prices $\mathbf{p}$ is a feasible solution for the Lagrange dual problem if and only if

$$\min_{t \in T} Q(t, \mathbf{p}) \geq 1. \tag{15.5}$$

The algorithm works as follows. Start with initial weights $p_v = \delta/C_v$ for all $v \in V$. The parameter δ depends on the accuracy target ϵ. Repeat the following steps until (15.5) is satisfied.

---

1. Compute a tree t^* for which $Q(t, \mathbf{p})$ is minimum. We call t^* a **smallest-price tree**.

2. Send the maximum flow on this tree t^* such that the uplink capacity of at least one internal node (neither the root nor the leaf nodes of the tree) is saturated. Let $I(t)$ be the set of internal nodes in tree t. The flow sent on this tree t^* can only be as large as

$$y = \min_{v \in I(t^*)} \frac{C_v}{m_{v,t^*}}.$$

3. Update the prices p_v as

$$p_v \leftarrow p_v \left(1 + \frac{\beta m_{v,t^*} y}{C_v}\right), \quad \forall v \in I(t^*),$$

where the stepsize β depends on ϵ.

4. Increment the flow Y sent so far by y.

The optimality gap can be estimated by computing the ratio of the primal and dual objective function values in each step of the iteration above, which can be terminated when the desired proximity to optimality is achieved. (When the iteration terminates, primal capacity constraints on each uplink may be violated, because we were working with the original uplink capacities at each step in the outer iteration. We need to scale down the flows uniformly so that uplink capacity constraints are satisfied.)

To analyze both the accuracy and time-complexity of the algorithm, we use the method developed by **Garg–Konemann** for approximation algorithms in multi-commodity flow problems. It turns out that, for any given target accuracy $\epsilon > 0$, the algorithm we just discussed computes a solution with an objective function value within a $(1 + \epsilon)$-factor of the optimum, for appropriately chosen algorithmic parameters. Using the big-O notation from the theory of algorithms, it runs in polynomial time in the network size:

$$O\left(\frac{N \log N}{\beta^2} T_{spt}\right),$$

where N is the number of peers, and T_{spt} is the time taken to compute a smallest-price tree.

The core issue now lies with the inner loop of smallest price tree computation: can this be accomplished in polynomial time for a given price vector? This graph-theoretic problem is much more tractable than the original problem of searching for a multi-tree that maximizes the achievable rate. However, when the given graph G is not full-mesh, or when there are degree bounds on nodes in each tree, or when there are helper nodes, computing a smallest-price tree is still difficult.

Moreover, how to approach the P2P capacity with distributed and low-complexity algorithms is another challenge, even when the capacity can be efficiently computed with a centralized optimization of peering multi-trees. Part of this challenge is the control signaling, and part of the solution is spatial hierarchy. Unstructured P2P with smart peering mechanisms, like BitTorrent, may be able to get reasonably close to P2P capacity with much lower control overhead.

Control signaling in P2P can rely either on a centralized tracker or on broadcasting, the so-called query flooding, so that each peer has a local copy of the topology and the network states. In-between these two options is hierarchical overlay. It turns out a two-level hierarchy is often used both in theory and in practice, and both for control signaling and for the actual data sharing.

Summary

Box 15 P2P enables efficient multicast

The P2P law quantifies the positive network effect in application-layer multicast. A P2P network builds an overlay graph of logical connectivity. This graph can be viewed as a collection of multicast trees. Careful design of the trees can be carried out through a combinatorial optimization. Smart heuristics such as tit-for-tat enables large-scale P2P systems like BitTorrent to scale up, as the number of sources goes up together with the number of destinations.

Further Reading

There have been many measurement, modeling, and design papers written about P2P systems since the early 2000s.

1. The founder of BitTorrent wrote the following widely cited paper explaining some of the design choices in BitTorrent's incentive mechanisms:
B. Cohen, "Incentives build robustness in Bit Torrent," in *Proceedings of Workshop on Economics of Peer-to-Peer Systems*, 2003.

2. The following paper provided a comprehensive survey of P2P-based IP telephony systems, including reverse engineering some of the Skype details:
A. B. Salman and H. Schulzrinne, "An analysis of the Skype peer-to-peer Internet telephony protocol," in *Proceedings of IEEE Infocom*, 2006.

3. A large-scale peering-topology measurement project was reported in
C. Wu, B. Li, and S. Zhao, "Exploring large-scale peer-to-peer live streaming topologies," *ACM Transactions on Multimedia*, vol. 4, no. 3, pp. 1–23, August 2008.

4. The following is a seminal paper on modeling P2P streaming applications:
R. Kumar, Y. Liu, and K. Ross, "Stochastic fluid theory for P2P streaming systems," in *Proceedings of IEEE Infocom*, 2007.

5. The approach of P2P capacity computation was developed in the following recent paper:
S. Sengupta, S. Liu, M. Chen, M. Chiang, J. Li, and P. A. Chou, "Peer-to-peer streaming capacity," *IEEE Transactions on Information Theory*, vol. 57, no. 8, pp. 5072–5087, August 2011.

Problems

15.1   *Embedding trees* ⋆

Consider a network of one server and $N = 3$ peers, given that (1) $u_s = 2$, $u_1 = 3$, $u_2 = 2$, and $u_3 = 1$ (all in Mbps), (2) all nodes have unlimited download capacity, and (3) each peer in a multicast tree can upload to any number of peers.

(a) Find the maximum multicast rate r_{max}. Draw a multi-tree that achieves this maximum.

(b) Find the rate r_i of the ith multicast tree, for all $i = 1, \ldots, T$, where T is the number of multicast trees from part (a).

(c) Now consider a network with one server and $N = 3$ peers, with $u_s = 3$, $u_1 = 3$, $u_2 = 2$, and $u_3 = 1$ (in Mbps). But now we impose the constraint that each peer in a multicast tree can only upload to at most one peer, so as to limit the overhead in maintaining the states of the peers. Draw the resulting multi-tree that achieves the maximum multicast rate, and compute the per-tree rates r_i.

15.2   *Delay components in P2P streaming* ⋆⋆

Consider a P2P multicast tree that consists of N nodes. Suppose the tree is balanced (i.e., the depths of the leaf nodes never differ by more than 1), and the tree fanout is M (i.e., every internal node has M children). Every node has the same upload capacity C, and every link connecting two nodes has the same latency L. Let B be the chunk size.

(a) Suppose there is no congestion-induced queueing delay. So the streaming delay consists of two components: node (transmission) delay and link (propagation) delay. What is the maximum streaming delay over all the nodes in this tree?

(b) Let $N = 1000$, $C = 1$ Mbps, $B = 20$ kB, $L = 50$ ms, and $M = 5$. Which delay component is more significant?

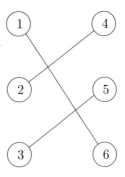

Figure 15.10 An example of a matching on a bipartite graph, where nodes 1, 2 and 3 are assigned to nodes 6, 4, and 5, respectively.

(c) Supposing that the less significant delay component can be ignored, what is the optimal fanout M to choose in order to minimize delay?

15.3　*Stable marriage matching* ★ ★ ★

The **stable marriage matching** problem is an extensively-studied problem and has many applications, from matching medical students to hospitals to analyzing voting systems and auctions. Figure 15.10 illustrates a matching in a bipartite graph. We saw a type of matching, the maximum weight matching, in VCG auctions in Chapter 2.

Suppose a set of partners can be split into two equal-sized subsets A and B. Each element $a \in A$ has a strict ranking of potential partners $b \in B$ (and vice versa). A stable matching assigns each $a \in A$ to some $b \in B$ and each $b \in B$ to some $a \in A$, such that there does not exist a pair $(a, b) \in A \times B$ with a preferring b to the $b' \in B$ that a is currently assigned to, and b preferring a to the $a' \in A$ that b is currently assigned to (for that would have been two unstable marriages).

A standard solution to the stable marriage matching problem is the **Gale–Shapley algorithm**. It follows a simple, iterative procedure. A man (a node in set A), or a woman (a node in set B), can be unengaged, engaged, or married. In each round, an unengaged man proposes to the most preferred woman among those to whom he has not yet proposed in previous rounds, and each woman chooses to become engaged to the suitor whom she prefers most and rejects the rest. As the iteration continues, engagements may be broken. When there is no longer an unengaged man, the iteration stops, and the engaged pairs at that point are married.

(a) Run the algorithm to find a stable matching between the two sets $A = \{a, b, c\}$ and $B = \{d, e, f\}$ with the following rankings:

$$a : d > e > f,$$
$$b : d > e > f,$$
$$c : d > f > e,$$
$$d : c > b > a,$$
$$e : c > a > b,$$
$$f : a > c > b.$$

(b) Argue that, at the conclusion of the algorithm, a stable marriage matching must have been found.

15.4   *BitTorrent as a game* ★★

BitTorrent's upload-incentive mechanism can be analyzed with game theory. Let there be N peers indexed $1, 2, \ldots, N$, and let c_i, u_i, and d_i be peer i's upload capacity, upload speed, and download speed, respectively (all in Mbps). The speeds u_i and d_i can vary with time. We assume peers have unlimited download capacities.

Each peer i can directly control its u_i (constrained by $u_i \leq c_i$) but not its d_i, and its aim is to maximize d_i and to minimize u_i. There is a tradeoff between the two objectives: if peer i makes u_i small, other peers will realize that peer i is selfish, and refuse to upload to it, resulting in a small d_i. BitTorrent's peer-selection mechanism aims to enforce this tradeoff so as to make u_i large and to encourage uploads.

Now consider the following set of rules, which are a simplified version of Bit-Torrent's peer-selection mechanism.

(1) Peers take turns to update u_i in the ascending order of i and then wrap around, e.g., $u_1, u_2, u_3, u_1, u_2, u_3, \ldots$ for three users.

(2) When peer i updates its u_i, all other peers see this change, and choose to upload to the top n_u (an integral parameter) peers in terms of the u_j values (and break ties by choosing randomly). The upload speeds are shared evenly among the n_u peers.

(3) Peer i chooses u_i by anticipating the d_i it receives according to rule (2): u_i is chosen to maximize the expected d_i. If multiple u_i values result in the same d_i, choose the smallest one. Then add a small constant ϵ.

Here come your two tasks in this homework problem. Let there be $N = 4$ peers with each peer uploading to $n_u = 2$ peers, and set $\epsilon = 0.1$.

(a) Suppose $c_1 = 1$ and $c_2 = c_3 = c_4 = 2$. Initially it is peer 1's turn to update u_1 with $u_2 = u_3 = u_4 = 1.1$. We have the following line of reasoning.

(i) Regardless of the value of u_1, no peer will upload to peer 1 ($d_1 = 0$) because $0 \leq u_1 \leq c_1 < u_2, u_3, u_4$, so peer 1 sets $u_1 = 0 + \epsilon = 0.1$ by rule (3).

(ii) In the next timeslot, it is peer 2's turn to update u_2, which becomes $u_1 + \epsilon$ because u_2 needs to be the third largest, i.e., greater than u_1, so that peers 3 and 4 will upload to it.

Continue this line of reasoning to show that the u_i values never converge to fixed values.

(b) Suppose now $c_1 = c_2 = c_3 = c_4 = 2$. Show that setting $u_1 = u_2 = u_3 = u_4 = 2$ constitutes a Nash equilibrium. You may first show that, if it is peer i's turn to set u_i, setting u_i to be any value other than $c_i = 2$ will not improve d_i.

(For more detail, see D. Qiu and R. Srikant, "Modeling and performance analysis of BitTorrent-like peer-to-peer networks," in *Proceedings of ACM Sigcomm, 2004*.)

15.5   *Private torrent games* ★★★

We have been discussing the public BitTorrent. There are also many private torrents that create their own rules of rewarding seeders and encouraging uploads beyond the simple tit-for-tat in BitTorrent. More than 800 such private communities were found in 2009. In this homework problem, we explore one possible rule of proportional reward, again through the modeling language of game theory and utility functions.

Let d_i and u_i be the actual download and upload volumes, now measured in bytes, for peer i in a fixed population of peers. A standard incentive mechanism in private BitTorrents is the following ratio incentive:

$$d_i \leq f(u_i),$$

i.e., the download volume cannot be bigger than some function of the upload volume. An affine parameterization of this function is

$$f(u_i) = \frac{u_i}{\theta} + \Delta.$$

Here, as in the leaky-bucket admission control that we will see in Chapter 17, θ is the upload–download ratio targeted, and Δ is the slack: the amount of data a peer can download outside of the ratio rule.

Each peer's net utility function V_i (we use V instead of U here, since U is already used to denote "upload"") can be parameterized as follows:

$$V_i(d_i, u_i) = B(d_i) - C(u_i) + \beta(f(u_i) - d_i),$$

as long as $f(u_i) - d_i \geq 0$. It becomes $-\infty$ (meaning that it will be evicted from the community) otherwise. Here, B and C are some utility and cost functions, and β is a positive weight.

Suppose the strategy for peer i is the two-tuple of target download and target upload (per unit time): (δ_i, σ_i), over the strategy spaces of $\delta_i \in [0, D]$ and $\sigma_i \in [0, U]$.

(a) How can we express the actual upload and download amounts $\{u_i, d_i\}$ as functions of the strategies $\{\delta_i, \sigma_i\}$ chosen by all the peers? This can be very difficult, but if we make an assumption that all the downloads add up to be exactly the same as the sum of all the uploads, then there is a closed-form answer. What is that answer?

(b) Now we want the Nash equilibrium to be efficient: each peer chooses the target download and upload to be just D and U. Prove that $(\delta_i, \sigma_i) = (D, U)$, $\forall i$, is indeed a Nash equilibrium if $f(u) > u$ and $f'(u) > C'(u)/\beta$. (Hint: You can assume that all users have upload and download capacities U and D except possibly user i. This will simplify the expression found in (a).)

(c) As a corollary to part (b), show that if the ratio incentive parameters (θ, Δ) are such that $u/\theta + \Delta > u$ and $\beta > \theta C'(u)$ for all $u \in [0, U]$, then using a ratio incentive implies that $(\delta_i, \sigma_i) = (D, U)$, $\forall i$, is the unique Nash equilibrium. (Hint: Assume that each user except user i follows the same strategy, and argue in three separate cases of that strategy.)

(For more details, see Z. Liu, P. Dhungel, D. Wu, C. Zhang, and K. W. Ross, "Understanding and improving incentives in private P2P communities," in *Proceedings of IEEE International Conference on Distributed Computing Systems*, 2010.)

16 What's inside the cloud of iCloud?

16.1 A Short Answer

In June 2011, Apple announced its iCloud service. One of the eye-catching features is its digital rights management of music content. The other part is its ability to let you essentially carry your entire computer hard drive with you anywhere and stream music to any device.

Cloud is more than just storage. For example, in the same month, Google introduced ChromeBook, a "cloud laptop" that is basically just a web browser with Internet connectivity, and all the processing, storage, and software are somewhere in Google servers that you access remotely.

These new services and electronics intensify the trends that started with web-based emails (e.g., Gmail), software (e.g., Microsoft Office 365), and documents (e.g., Google Docs and Dropbox), where consumers use the network as their computers, the ultimate version of online computing.

In the enterprise market, many application providers and corporations have also shifted to cloud services, running their applications and software in rented and shared resources in **data centers**, rather than building their own server farms. Data centers are facilities hosting many servers and connecting them via many switches. Large data centers today is typically over 300,000 square feet, house half a million servers, and cost hundreds of millions of dollars to build.

There are three major cloud providers as of 2012: Amazon's EC2, Microsoft's Azure, and Google's AppEngine. A pioneering player in cloud services is actually Amazon, even though to most consumers Amazon stands for an online retail store. In Amazon's S3 cloud service today, you can pay $0.115 per hour for a small standard instance, featuring 1 virtual core, 1 EC2 unit, and 160 GB of storage. Any amount of data coming into the EC2 is free, and for outgoing data, each month the first GB is free, and the next 10 TB is $0.120 per GB.

For many years, it has been a goal in the computing and networking industries that one day users could readily rent resources inside the network (the "cloud" on a typical network illustration), in a way that makes economic sense for all the parties involved. That day is today. Thanks to both technological advances and new business cases, cloud services are taking off and evolving fast.

Many features of cloud services are not new, some are in fact decades-old. Several related terms have been used in the media somewhat confusingly too:

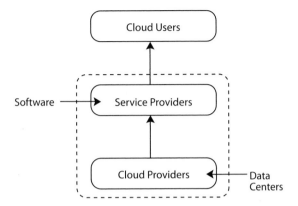

Figure 16.1 Three segments of the cloud service industry. Cloud providers operate data centers that house the hardware, including inter-connected processors, storage, and network capacity. Service providers run cloud services through their software. Cloud users include both consumer and enterprise customers for cloud services provided collectively by the service providers and cloud providers.

cloud computing, utility computing, clustered computing, software as a service, etc. To clarify the terminology, we refer to the graph in Figure 16.1. There are three key components of the "food chain."

Cloud providers build and manage the hardware platform, consisting of computing resources (servers), networking resources (switches), and storage resources (memory devices) organized inside data centers. There is a network within each data center where the nodes are servers and switches, and each data center in turn becomes a node in the Internet.

Service providers offer software and applications that run in data centers and interface with users. For example, an iPad application developer may use the computing and storage resources in Amazon's EC2 cloud to deliver its services. Sometimes, the service provider is the same as the cloud provider. For example, the iCloud music storage and streaming service from Apple runs in Apple's own data centers.

Cloud users are consumers and enterprises that use services running in data centers. Users can get the content they want (e.g., documents, books, music, video) or the software they need (e.g., Office software, an iPhone application, or in the cloud laptop's case, pretty much any software you need) from the cloud. And get them on demand, anytime, anywhere, and on any device with an Internet connection.

16.1.1 What features define cloud services?

To make the overall cloud service food chain work, we need all of the following ingredients:

1. *Large-scale computing and storage systems*, often leveraging virtualization techniques in sharing a given hardware resource among many processes as if they each had a slice of a dedicated and isolated resource.

2. *Networking* within a data center, across the data centers, and to the end-users (often with a wireless hop like WiFi or 4G). This networking dimension naturally will be the focus of our discussion in this book, especially networking within a data center.

3. *Software* that provides a graphical user interface, digital rights management, security and privacy, billing and charging, etc.

If I open up my home desktop's CPU and hard drive to renters, does that constitute a cloud service? Probably not. So, what are the defining characteristics of a cloud service? The keyword is *on demand*, along two dimensions: time and scale.

- On demand in timing: a cloud service allows its users to change their requests for resources at short notice, and possibly only for a short period of time.

- On demand in scale: a cloud service allows its users to start at a very small minimum level of resource request (e.g., 1.7 GB of RAM and 160 GB of memory on Amazon's EC2 today), and yet can go to really large scale (e.g., Target, the second-largest retail store chain in the USA, runs its web and inventory control in a rented cloud).

16.1.2   Why do some people hesitate to use cloud services?

Cloud services face many challenges, even though they are increasingly out-weighed by the benefits. Let us briefly bring them up before moving on.

Similarly to the pros–cons comparison between packet switching and circuit switching, once you are in a shared facility, the performance guarantee is com-promised and so are security and privacy. If you ride a bus instead of a taxi, you pay less but you might not have a seat and you will be seen by the fellow bus riders. That is the price you pay to enjoy the benefits of cloud services. There are many technologies that mitigate cloud's downsides for various market segments, but riding a bus will never be exactly the same as taking a taxi.

As illustrated by the Amazon cloud outage in April 2011 and more recently the storm-induced electricity outage to some of the Netflix service in June 2012, availability of service in the first place is the top performance concerns. The main root causes for unavailability include network misconfigurations, firmware bugs, and faulty components. The best way to enhance availability is *redundancy*: spread your traffic across multiple cloud providers (assuming it is easy enough to split and merge the traffic), and across different reliability zones in each of the providers.

16.1.3   Why do some people like cloud services?

Why does it make sense to provide and to use cloud services? The answers are similar to those relating to many other rental businesses, such as libraries and rental car companies. We summarize the arguments below and will go through an example in a homework problem.

To the cloud users, the key benefit is *resource pooling*. The cost of building and provisioning resources is now shared by many other users. This is called the "CapEx to OpEx conversion:" instead of spending money in capital expenditure to build out dedicated facilities, a user pays rent as part of its operational expenditure to share facilities. This is analogous to going from circuit switching to packet switching in the design of the Internet. The risk of miscalculating resource need shifts to cloud providers, a significant advantage if the resource demand varies a lot or is just hard to predict.

But why would cloud *providers* be interested? A main reason is the *economy-of-scale* advantages, both on the supply side and on the demand side of the business. On the supply side, a cloud provider can procure the servers, switches, labor, land, and electricity at significantly discounted prices because of its large-scale and bargaining power. Even when compared with a medium-sized data center with thousands of servers, a large scale data center with a hundred thousand servers can often achieve a factor of $5-7$ advantage in cost per GB of data stored or processed.

On the demand side, scale helps again, through *statistical multiplexing*. Fluctuations of demand for each user are absorbed into a large pool of users, as

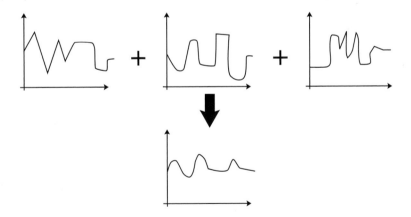

Figure 16.2 Statistical multiplexing smoothes out the burstiness of individual users. Suppose there are three users with their transmission rates over time as charted above. Their aggregate transmission rate, shown in the lower graph, is much smoother. Cloud providers leverage burstiness reduction as the scale goes up to reduce capacity provisioning, as long as the users' demands do not peak together. The scale of the lower graph's y-axis is about three times that of the upper graphs'.

illustrated in Figure 16.2. This is the same principle as that behind ISPs over-subscribing at each aggregation point of their access networks: aggregating many bursty users reduces the burstiness. Of course, the overall pool may still exhibit time-of-day peak–valley patterns. The *average* utilization of servers in a data center is often below 20% today. These peak–valley patterns can be further smoothed out by time-dependent pricing as discussed in Chapter 12.

Cloud is all about *scale*. Today's large-scale data centers are indeed huge, so big that electricity and cooling costs sometimes amount to more than half of the total cost. If an iPhone is one of the smallest computers we use, each data center is one of the largest. We have made an important assumption, that it is actually *feasible* to scale up a data center. Otherwise, we would have to truncate all the benefits associated with scaling up. But, as we saw in Chapter 10, scale can also be a *disadvantage* when each (reasonably priced) network element can have only a small number of high-performance ports. Unless you have the right network topology, building a 100,000-server data center can be much more expensive, in unit price of capacity (or, "bandwidth" in this field's common terminology), than building a 10,000-server data center. This echoes Chapter 10's theme: (high-throughput) connectivity per node does *not* scale up beyond a certain point in either technology or human networks. Yet we want (high-throughput) connectivity in order for the whole network to keep scaling up. That is the subject of the next section: how to achieve the advantages of scale for a network without suffering the limitation of scale of each node.

16.2   A Long Answer

16.2.1   Building a big network from small switches

We need a network within a data center. Many applications hosted in a data center require transfer of data and control packets across the servers at different locations in that big building. A natural, but inferior solution, is to build a tree like Figure 16.3(a), where the leaf nodes are the servers, and the other nodes are the routers. The low-level links are often 1 Gbps Ethernet, and upper level ones 10 Gbps Ethernet links. The top-of-the-tree switches are big ones, each with many 10 Gbps links. It is expensive to build these big switches. As the number of leaf nodes increases to 100,000 and more, it becomes technologically impossible to build the root switch. A high-end switch today can support only 1280 servers.

So we need to start *oversubscribing* as we climb up the tree. Sometimes the **oversubscription ratio** runs as high as 1:200 in a large data center. What if all the leaf-node servers want to fully utilize their port bandwidths to communicate with other servers at the *same* time? Then you have a 200-factor congestion. The whole point of resource pooling is defeated as we "fragment" the resources: idle servers cannot be utilized because the capacity between them cannot be used in

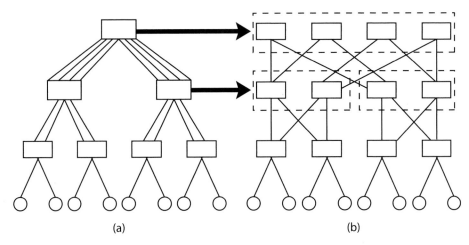

(a)   (b)

Figure 16.3 From tree to fat tree. Graph (a) shows a tree supporting eight servers (the leaf nodes represented by circles), with four switches with two inputs and two outputs, two switches with four inputs and four outputs, and one switch with eight inputs and eight outputs. It is expensive to build the larger switches. When the root switch is too big, it becomes impossible to build one, and oversubscription has to take place as we go up the tree. In contrast, in graph (b), two small switches (each with two inputs and two outputs) collectively serve the role of a 4×4 switch, and four small switches collectively serve the role of an 8×8 switch.

an oversubscribed tree. No matter how you use it, a tree is not scalable. Many trees would have been better, as in P2P in Chapter 15, but we cannot swap the leaf nodes upstream in multi-trees here, because the leaf nodes are servers and the upstream nodes are switches.

Is it still possible to build a large network with small switches, just like building a reliable network out of unreliable components?

The answer is yes, if you are smart enough with the network topology, and go from a tree to a **multi-stage switched network**. Instead of building a large scale network by scaling *up* the number of ports per router and hitting the economic and technology ceilings, we scale *out* the topology and can do that as much as we want. We use many small switches to make a large switch, with the same number of links per switch at each level.

This branch of networking has long been studied in the old days of circuit switching, when it became impossible to build a large enough single switch to handle all phone calls. Then **interconnection networks** were studied for multicore processors and parallel computation. Now the study of interconnection networks has been revived in the context of data center networking for cloud services. The key message here is that *connectivity itself is a resource* that we can build and need to build carefully.

There are many ways to quantify how good an interconnection topology is. We focus on throughput rather than latency.

- The worst-case pairwise end-to-end capacity (e.g., from one leaf node to another in a tree) is one possibility.
- **Bisection bandwidth** is another: as in the social graph partitioning in Chapter 8, the capacity on all the links between two equal-sized halves of the network is called a **bisection**, and the worst case of that over all possible combinations of halves of the network is called the bisection bandwidth.
- A third metric is a classic one often used in circuit switching: a network is called **non-blocking** if any pair of (unused) input and (unused) output can be connected as each traffic session (a switching request) arrives. It is called **rearrangeably non-blocking** if some existing pairs' connection needs to be rearranged in order to achieve the nonblocking property.

In 1953, the most famous interconnection network topology, called the **Clos network**, was invented. An example is shown in Figure 16.4. The general definition of a Clos network is as follows. Each 3-stage Clos network is specified by three integers: (n, m, r). Each input switch is $n \times m$, and there are r of them. Symmetrically, each output switch is $m \times n$, and there are also r of them. The middle-stage switches of course must be $r \times r$, and there are m of them. Each of the input and output switches is connected to each of the middle-stage switches. This is a rich connectivity, using only small switches to support rn input–output pairs of ports, as illustrated in Figure 16.4.

Assuming a centralized controller decides the switching patterns, it can be readily seen that if $m \geq 2n - 1$, then a Clos network is non-blocking. Consider a new switching request arriving at a particular input port of an input switch A, to be switched to an output port of an output switch B. In the worst case, each

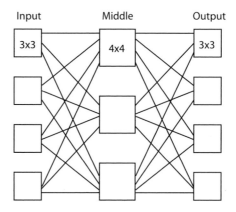

Figure 16.4 An example of a Clos network. This is a three-stage, $(3, 3, 4)$ Clos network, acting as a 12×12 switch using only 3×3 and 4×4 switches. There are four 3×3 switches on the input stage, four of them on the output stage, and three 4×4 switches in the middle stage. Each input switch is connected to each of the middle switches. Each output switch is also connected to each of the middle switches.

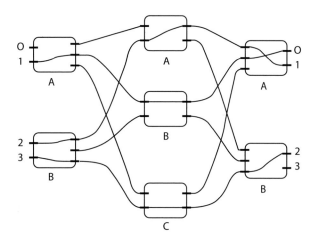

Figure 16.5 An example illustrating that, when $m \geq 2n - 1$, the Clos network is non-blocking. The key point is that, with a sufficiently large m, there are enough middle-stage switches even under the worst-case traffic pattern and wiring. In this example, $n = 2$, $m = 3$, and $r = 2$. Wiring within each switch determines the switching pattern. Wiring among the switches is fixed a priori by the design of the Clos network.

of the other $n-1$ ports on A is already occupied, each of the other $n-1$ ports on B is already occupied, and, most importantly, each of these $2n - 2$ connections goes through a *different* middle stage switch. This means we need to have an additional middle-stage switch in such a worst-case scenario. That means that if there are $m \geq 2n - 1$ middle switches, we have the non-blocking property.

This is illustrated in Figure 16.5 with a (2, 3, 2) Clos network. The traffic pattern is that input port 1 needs to be connected to output port 0, input port 2 to output port 1, input port 3 to output port 2, and input port 0 to output port 3 now. At this point, three input–output port pairs have already been connected as shown. We need to connect input port 0 to output port 3. The other input port sharing the same input-stage switch A connects through middle-stage switch B. The other output port sharing the same output-stage switch B connects through middle-stage switch C. Had there not been another middle-stage switch, A in this case, there would have been no way to connect input port 0 and output port 3 without rearranging existing connections. But with the presence of switch A, the Clos network is non-blocking.

It is interesting to see that r does not factor into the non-blocking condition $m \geq 2n-1$. But of course, if r is too big, the middle-stage switches will have large port counts. Bigger r means there are more input- and output-stage switches, and larger middle-stage switches, which can be recursively built from three-stage Clos networks using only small switches.

It takes a little longer to show that, if $m \geq n$, a Clos network is *rearrangeably* non-blocking. We will go through that argument in a homework problem.

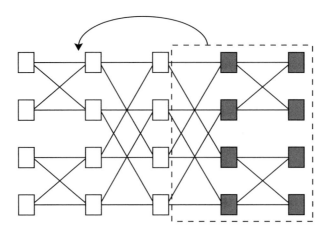

Figure 16.6 From Clos network to fat tree. Due to symmetry around the middle stage of this five-stage Clos network, the right two stages can be folded to the left two stages. We call a folded Clos network a fat tree. Now each link is bidirectional in the fat tree.

A Clos network can have its input part and output part folded. A folded Clos network is often called a **fat tree**. Do not be confused by this terminology, a fat tree enjoys scalability beyond what a tree can. It can achieve the maximum bisection bandwidth without having to oversubscribe traffic as we go up the tree levels. This is shown in Figure 16.6. Fat trees have been implemented in many data centers as the standard way to scale up.

There are several alternatives to Clos networks, such as the hybercube, mesh, bufferfly, etc. A variant called VL2 builds small trees while trees are still scalable, and then a Clos network among the roots of these trees, as illustrated in Figure 16.7.

Given a topology, we still need to run routing, congestion control, and scheduling of traffic on it. Some of these topics will be touched upon in the Advanced Material. When the connectivity in a Clos network is sufficiently rich, even simple, randomized routing can load balance the traffic well enough to achieve near optimal bisection bandwidth. In some proprietary systems such as Infiniband, deterministic routing, an even simpler approach, is used.

16.2.2   Comparing network topologies: the Internet, data centers, and P2P

Before we move on to illustrative examples of data-center topologies, we pause to reflect upon three ways of drawing and sizing topologies, one for each key type of wireline technology networks. We then explore the root causes behind these different design choices.

1. *The Internet backbone*: Overprovision link *capacities* and then carefully run IP routing, possibly multipath routing. Since routing is not responsive to the

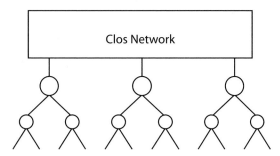

Figure 16.7 The "Virtual Layer 2" (VL2) design leverages spatial hierarchy in data centers. Many servers are connected by trees, and then the trees are connected to each other through a Clos network.

link load in real time; that job of congestion control is given to the transport layer, and end hosts react to varying loads on a fast timescale by TCP.

2. *Data-center networks*: Overprovision *connectivity* by increasing the number of paths available, and then run massive amounts of multipath routing, either carefully or randomly. Why not do this for the Internet backbone too? Because overprovision connectivity is even more expensive than overprovision capacity in the Internet's spatial scale, unless you overlay, like in P2P.

3. *P2P multicast overlay networks*: Overprovision *connectivity* rather than capacity, by increasing both the number of paths *and* the number of source nodes, and then run massive amounts of multi-tree construction by picking not just routing paths but *source–destination relationships*, either carefully or randomly. More than creating just a richly connected topology and then picking many paths, this creates many concurrent multicast trees, possibly one per chunk of bytes.

The progression of the above three designs can be summarized as follows. (1) Fix a topology, make pipes fatter and use the pipes intelligently. (2) Enrich the topology by increasing connectivity. (3) Create many topologies to choose from at the same time.

In the end, (2) and (3) can get close to their bisection-bandwidth limit and P2P capacity limit, respectively, but (1) cannot get to full utilization of backbone bandwidth. Furthermore, if there is enough overprovisioning of connectivity, you even get to choose among the connections randomly (like in VL2 and BitTorrent) and be pretty close to the limit. Overprovisioning connectivity pays off better than overprovisioning capacity, if you can *afford* it.

Why are there such differences among (1), (2), and (3)? The choice of network design also depends on the dominant cost drivers. In the Internet backbone, digging trenches is the dominant cost for the ISPs. And links are long haul, covering thousands of miles, constrained by the presence of fibers and population. It is very expensive to create connectivity. In a data center, the network inside a large building is a relatively small fraction of the cost, compared with the server, electricity, and cooling costs. So overprovisioning connectivity makes economic sense. P2P is an overlay network, so connectivity is a logical rather than physical concept and even cheaper to overprovision. P2P connectivity can be dynamically managed through control signals without digging any trenches.

In addition to cost structures, traffic demand's predictability and flexibility is another root cause for these fundamentally different choices in network design. In the Internet, traffic matrices are relatively predictable. Traffic-matrix fluctuation on the Internet is over time rather than space, and thus can be mitigated by either capacity overprovisioning or time-dependent pricing. In data centers, traffic demands are quite volatile and not yet well understood, another reason to overprovision connectivity. In P2P, you have the option of *changing* the traffic matrix, by picking different peers. So, leveraging that flexibility gives the biggest "bang for the buck."

16.3   Examples

16.3.1   Expanding and folding a Clos network

In this example, we demonstrate how to expand a Clos network from three stages to five stages, and then rearrange for symmetry before folding into a fat tree.

As illustrated in Figure 16.8, we follow a sequence of five steps.

- Step 1: We start with a particular three-stage Clos network where $n = 2, m = 2$ and $r = 4$. We would like to replace the middle-stage larger switches with small, 2×2 switches.
- Step 2: We construct a new three-stage Clos network where $n = 2, m = 2$ and $r = 2$. Each of these Clos networks can act as a 4×4 switch.
- Step 3: We now replace each center-stage switch in Step 1 with the new three-stage Clos network in Step 2. This recursive operation expands the original three-stage Clos network into a five-stage one. There are more switches, but they are all small ones (2×2) now.
- Step 4: We conform to the standard input-stage connection pattern by appropriately rearranging the positions of the switches in stage 2 and stage 4.
- Step 5: Finally, we can fold the five-stage Clos into a three-stage fat tree, each link being bidirectional now.

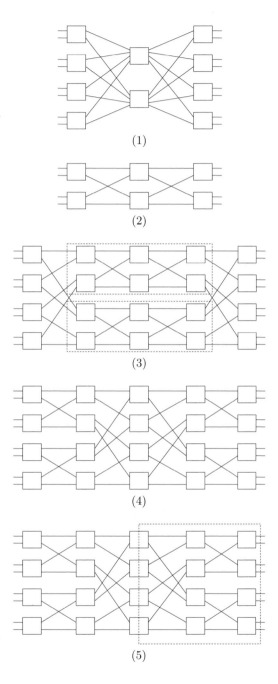

Figure 16.8 Convert a three-stage (2, 2, 4) Clos network into a three-stage fat tree with only 2×2 switches. Each of the middle-stage switches in the original Clos network is replaced by a three-stage (2, 2, 2) Clos network. Then the switches are rearranged in stages 2 and 4. Finally fold the network into a more compact one with bidirectional links.

16.3.2 Max flow and min cut

The bisection bandwidth is a special case of the cut size of a network. There is a celebrated result connecting the sizes of cuts in a graph with the maximum amount of flow that can be routed through the network.

Consider a directed graph with the edge capacities illustrated in Figure 16.9(a). We want to find out the maximum flow from source s to destination t. Figure 16.9(b) gives a solution of maximum flow computed using the Ford–Fulkerson algorithm that we went through in a homework problem in Chapter 13.

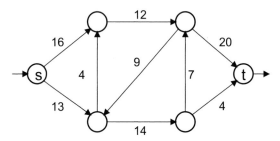

(a) Network topology and capacity.

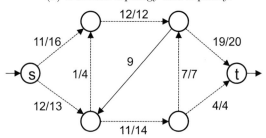

(b) The maximum flow from s to t is 23.

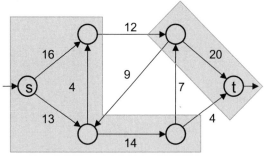

(c) The minimum cut $c(S, T)$, where $s \in S$ and $t \in T$, is $12 + 7 + 4 = 23$.

Figure 16.9 The maximum flow equals the minimum cut. In the middle graph, a/b means that a units of capacity, out of a total of b units available, is used on a link. Algorithms such as the Ford–Fulkerson algorithm can compute the maximum flow from source s to destination t. That must be equal to the minimum size of the cut where s belongs to one side of the cut and t the other.

In general, a **cut** (S, T) of the network $G = (V, E)$ is defined with respect to a given source–destination pair (s, t). It is a partition of the node set V into two subsets: S and $T = V - S$ (nodes in set V other than those in set S), such that $s \in S$ and $t \in T$. As in Chapter 8, the capacity of the cut (S, T), or its cut size, is the sum of the capacities on the links from S to T. A *minimum cut* of a network is a cut whose capacity is the minimum over all cuts of the network. Figure 16.9(c) shows a minimum cut in the network. The **max-flow min-cut** theorem states that the maximum amount of flow passing from the source to the destination is equal to the minimum cut of the network with respect to that source–destination pair. In this example, both numbers equal 23 units. The min cut divides the network into the left four nodes and the right two nodes as shown.

16.4   Advanced Material

Data centers host a multitude of services, which range from financial, security, web search, and data mining, to email, gaming, content distribution, and social networking. These services operate on varying timescales and belong to different traffic classes with diverse quality-of-service needs. For instance, interactive services such as gaming, video streaming, and voice-over-IP are delay-sensitive, whereas other services such as large file transfers and data-mining applications are throughput-sensitive. In addition, for efficient sharing of workload, components of a single application can be distributed, processed, and assembled on multiple servers that are located at different data centers. All these result in interesting traffic patterns, both within a data center and over the backbone network that interconnects multiple data centers over a large geographic span. We focus on traffic management within a data center here, especially on four degrees of freedom: (1) topology, (2) placement, (3) routing, and (4) scheduling.

(1) *Data-center topologies need to be scalable.* Topology of the connected servers is a key design issue as discussed in this chapter. Figure 16.10 shows some typical data-center topologies by inter-connecting switches. Most data centers, e.g., tree, VL2, and fat tree, follow a three-tier architecture. At the bottom is the access tier, where each server connects to one (or two) access switches. Each access switch connects to one (or two) switches at the aggregation tier, and, finally, each aggregation switch connects to multiple switches at the core tier. In BCube, servers are assumed to have multiple input and output ports, so that they can be part of the network infrastructure and forward packets on behalf of other servers.

(2) *VM placement can help localize traffic.* Cloud customers usually rent multiple machine instances with different capabilities as needed and pay at a per-machine-hour billing rate. Virtualization-based data centers are becoming the mainstream hosting platform for a wide spectrum of application mixtures. The virtualization technique provides flexible and cost-effective resource sharing in

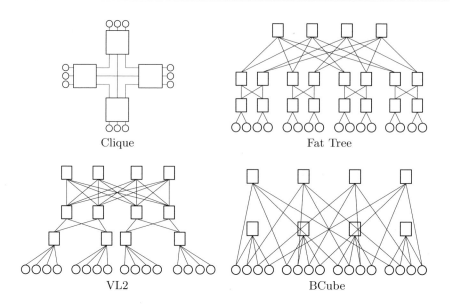

Figure 16.10 There are many topologies to scale up a data center. We have seen a fat tree as a folded Clos network and VL2 as trees combined with a Clos network. There are other options like cliques and cubes. In these graphs, circles are servers and squares are switches.

data centers. For example, Amazon EC2 uses Xen virtualization to support multiple **Virtual Machine** (VM) instances on a single physical server. An application job usually subscribes a handful of VMs placed at different hosts that communicate with each other, with different amounts of resource requirements by each VM for CPU and memory.

A number of proposals have been made to improve the agility inside a data center, i.e., any server can be dynamically assigned to any host anywhere in the data center, while maintaining proper security and performance isolation between services. Maximizing the network bisection bandwidth could be viewed as a global optimization problem – servers from all applications must be placed so as to ensure that the sum of their traffic does not saturate any link.

(3) *Route selection can exploit the multipath capability.* Data centers also rely on the path multiplicity to achieve scaling of host connectivity. Data-center topologies often take the form of multi-rooted spanning trees with one or multiple paths between hosts. Route selection can then be congestion-adaptive to the bandwidth availability between different parts of the network.

(4) *Scheduling* is yet another degree of freedom in managing traffic in a data center. The basic problem of job scheduling is essential to a wide variety of systems, from an operating system of a smart phone to an interconnection network on a factory floor. We assume for now that there is a centralized computer to collect all the inputs and compute the output. In Chapter 18, we will encounter the challenge of distributed scheduling in WiFi.

- The input to the scheduling problem is a list of jobs, each with some attributes: the job size (or more generally, the amount of resource required for each type of resources), the arrival time, a strict deadline or a cost of exceeding the deadline, and a quality of service expected (e.g., the minimum throughput). Sometimes there are also dependences among the jobs, which can be visualized as a graph where each node is a job and each directional link is a dependence between two jobs.
- The output is a schedule: which job occupies which parts of each resource during each timeslot.
- Some criteria to judge how good a schedule is, e.g., the distribution of response times (the time it takes to start serving a job), the distribution of job completion times, the efficiency of using all the resources, and the fairness of allocating the resources. Some of these metrics will be explored in homework problems and in Chapter 20.

Scheduling has been extensively researched in queuing theory, dynamic programming, algorithm theory, graph theory, etc. There are too many variants to list in this brief summary. Some major scheduling policies include the following, with self-explanatory names. We will see some of these again in Chapter 17.

- *First come, first serve*: Whenever there is an available resource, the first job to arrive claims it.
- *Smallest job first*: When there are multiple jobs waiting to be scheduled, the smaller jobs get served first. This helps reduce response times, if we just count the number of jobs without weighting them by the job sizes.
- *First to finish first*: This helps reduce completion times by the count of the number of jobs.
- *Longest queue first*: We group the jobs by the type of resource they request, and each group has its queue to hold jobs waiting to be scheduled. This scheduling policy helps avoid long queues building up as more jobs arrive at the queues.

Joint optimization: Topology, VM placement, routing, and job scheduling are four of the degrees of freedom in managing traffic inside a data center. Optimizing on any one of these alone can be quite inefficient.

- Ill-designed network topology limits the bisectional bandwidth and the path diversity between communicating nodes.
- Suboptimal VM placement introduces unnecessary cross traffic.
- Oblivious routing even in well-designed topologies can under-utilize the network resource.
- Inefficient scheduling may be constrained by routes available among the servers, which in turn reduces the efficiency of routing.

Having joint control over all the "knobs" provides an opportunity to fully utilize the data-center resources. For example, the operators have control over both where to place the VMs that meet the resource demand, and how to route the traffic between VMs.

Summary

> **Box 16** Cloud scales-up by scaling-out
>
> Cloud services provide the benefit of resource pooling to its users and that of the economy-of-scale to its providers. A large network can be built from small switches using multi-stage switched networks. Connectivity becomes a resource to provision just like capacity. A Clos network is a typical example illustrating how these connectivity patterns can be non-blocking and achieves a high efficiency of resource utilization.

Further Reading

The subject of building large networks from small switches is both old and new. Its root goes back to the telephone network design and its current driving application is one of the "hottest" buzz words in networking industry today.

1. The classic paper by Clos in 1953 initiated many further study in switched network design:
   C. Clos, "A study of non-blocking switching networks," *Bell System Technical Journal*, vol. 32, no. 2, pp. 406–424, 1953.

2. The following is a standard graduate-level textbook on interconnection networking:
   W. J. Dally and B. P. Towles, *Principles and Practices of Interconnection Networks*, Morgan Kaufmann, 2004.

3. The following book provides a down-to-earth introduction to key elements in cloud computing, with details on the practical side:
   B. Sosinsky, *Cloud Computing Bible*, Wiley, 2011.

4. The following is an accessible, general overview of cloud research problems:
   A. Greenberg, J. Hamilton, D. A. Maltz, and P. Patel, "The cost of a cloud: Research problems in data center networks," *ACM Sigcomm Computer Communication Review*, vol. 39, no. 1, pp. 68–73, January 2009.

5. Here is a comprehensive book on the mathematical foundation of distributed computation:

D. P. Bertsekas and J. N. Tsitsiklis, *Parallel and Distributed Computation: Numerical Methods*, Athena Scientific, 1997.

Problems

16.1   *To cloud or not to cloud* ⋆

The Bumbershoot Corporation's biology research center produces 600 GB of new data for every wet lab experiment. Assume the data generated can be easily parallelized, with a negligible overhead.

(a) Suppose Amazon Web Services (AWS) sells CPU hours at the price of $0.10 per hour per Elastic Compute Cloud (EC2) instance, where each instance takes 2 hours to process 1 GB of the experimental data. The data-transfer fee is $0.15/GB. What is the price Bumbershoot will need to pay for processing an experiment using the EC2 service?

(b) Suppose the data-transfer rate from the research center to AWS is 20 Mbps. What is the total time required to transmit and process the experimental data using the EC2 service?

(c) The Bumbershoot Corporation has 24 computers itself, each taking 2 hours to process 1 GB of data. Suppose the overall cost (including electricity, software, hardware, etc) is $15 per computer per experiment. What is the total amount of time and cost required to process the experiment? Will Bumbershoot Corporation be willing to use the EC2 service?

(d) We saw in (b) and (c) that the data transmission time can be a problem in cloud computing. Can you think of a way to overcome this obstacle, so that Bumbershoot can still process the experiment using the EC2 service within a day?

16.2   *Ideal throughput* ⋆⋆

Consider an interconnection network on a microprocessor chip, represented by a directed graph $G = (V, E)$, where $V = \{v_i : i = 1, \ldots, |V|\}$ is the set of nodes representing the terminals and routers on the chip, and $E = \{e_c : c = 1, \ldots, |E|\}$ is the set of links called "channels." Define the following symbols:

$\lambda_{s,d}$,   the traffic from input port s to destination port d;
$x_{d,c}$,   the traffic with destination d on channel c;
b_c,   the bandwidth of channel c;
γ_c,   the load of channel c;

A, the node-channel incidence matrix, where

$$A_{ic} = \begin{cases} +1 & \text{if } c \text{ is an outgoing channel from node } i \\ -1 & \text{if } c \text{ is an incoming channel to node } i \\ 0 & \text{otherwise.} \end{cases}$$

(a) What is the incidence matrix of the graph in Figure 16.11?

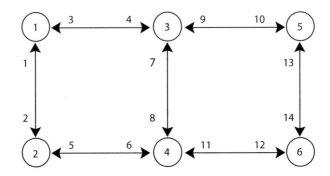

Figure 16.11 A simple network represented by a directed graph.

(b) Define

$$f_{d,i} = \begin{cases} \lambda_{i,d} & \text{if } i \neq d \\ -\sum_{j \neq d} \lambda_{j,d} & \text{if } i = d. \end{cases}$$

What is the relationship between $f_{d,i}$ and $x_{d,i}$?

(c) Express the load γ_c in terms of the traffic $x_{d,c}$.

(d) The ideal throughput Θ^* is the maximum throughput achievable in the network. What is the ideal throughput in terms of γ_c and the bandwidth b_c?

(e) Formulate the optimization problem of maximizing the ideal throughput via flow control (i.e., varying the traffic $x_{d,i}$), for a given traffic pattern $\lambda_{s,d}$ and incidence matrix **A**.

16.3 *Packaging optimization* ⋆

The nodes and links (channels) of on-chip interconnection networks are constructed on packaging modules. The network topology along with the packaging technology determines the constraints on the channel's bandwidth. In this question, we aim to derive an upper bound on the smallest channel width w_{min}.

Consider a network where channels are composed of unidirectional wires, each having a bandwidth of f units. For an arbitrary node v_n, suppose it has W_n pins available, along with δ_n^+ outgoing channels and δ_n^- ingoing channels. Since

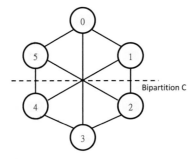

Figure 16.12 A bisection of a Cayley graph with six nodes. Each link represents two unidirectional channels going in opposite directions.

all $\delta_n = \delta_n^+ + \delta_n^-$ channels connecting to node v_n need to share the W_n pins, we have the following upper bound:

$$w_{min} \leq f \frac{W_n}{\delta_n}.$$

Furthermore, consider an arbitrary bisection C of the network, where there are B_C channels in between the two sets of nodes. In a practical packaging technology, because of the limited space inbetween the two sets of nodes, the number of wires inbetween is bounded by some number W_C as well. So we have the following upper bound:

$$w_{min} \leq f \frac{W_C}{B_C}.$$

Now consider a **Cayley graph**, along with a bisection, as shown in Figure 16.12. Suppose each wire has bandwidth $f = 1$ Gb, each node has $W_n = 140$ pins, and there can be at most $W_C = 200$ wires in between bipartition C. Give an upper bound of the minimum channel bandwidth w_{min} of this network.

16.4 *Alternatives to Clos networks* ★★★

(a) Consider the **butterfly network** as shown in Figure 16.13, where each channel has one unit of bandwidth, and the packets are sent from the input ports (denoted by the left circles) to the output ports (denoted by the right circles).

What is the ideal throughput assuming the random traffic pattern, i.e., each input port s sends $\frac{1}{8}$ unit of traffic to each output port d under unit throughput? You will need to calculate the channel load of all the links incident on each stage.

What is the ideal throughput assuming the "bit rotation permutation" traffic pattern? That is, the input port with the address (in binary) $a_2 a_1 a_0$ sends

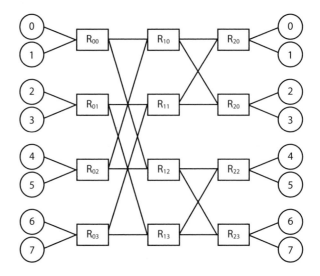

Figure 16.13 An example of butterfly networks.

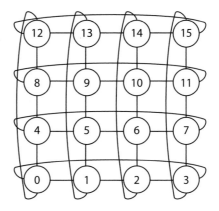

Figure 16.14 An example of cube networks.

packets to the output port with the address $a_1 a_0 a_2$. For example input port $5 = (101)_2$ sends packets only to output port $(011)_2 = 3$.

(b) Repeat (a) for the **cube network** as shown in Figure 16.14, where each channel has one unit of bandwidth, with each node acting as both an input port and an output port. (Hint: Write an inequality expressing the relationship between the aggregate channel load in the network and the aggregate traffic pattern across all source-destination pairs. What does this tell you about the maximum channel load?)

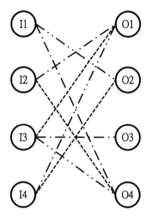

Figure 16.15 Consider the (3, 3, 4) Clos network we saw. In this graph, the left/right nodes represent input/output switches, and the links with three different types of dotted styles represent the assignment of the middle switches.

16.5 *Rearrangably non-blocking Clos networks* ★ ★ ★

We have seen how big m needs to be for a Clos network to be non-blocking. It turns out that m can be smaller if we ask for a weaker condition of rearrangably non-blocking.

(a) Design an algorithm for routing traffic on an (m, n, r) Clos network with $m \geq n$.

(b) Consider a $(3, 3, 4)$ Clos network along with its traffic illustrated in Figure 16.15. In Figure 16.15, each node represents an input/output switch, and links with three different dotted styles represent the middle switches assigned: sessions $(I2, O4), (I3, O1)$, and $(I4, O2)$ are routed through middle switch 1; sessions $(I1, O4), (I2, O1)$, and $(I3, O3)$ are routed through middle switch 2; and sessions $(I1, O2), (I3, O4)$, and $(I4, O1)$ are routed through middle switch 3.

Now, route a new call $(I4, O3)$ using your algorithm from (a).

17 IPTV and Netflix: How can the Internet support video?

We saw in Chapter 13 that the Internet provides a "best effort," i.e., "no effort" service. So, how can it support video distribution that often imposes stringent demands on throughput and delay?

17.1 A Short Answer

17.1.1 Viewing models

Watching video is a significant part of many people's daily life, and it is increasingly dependent on the Internet and wireless networks. Movies, TV shows, and home videos flow from the cloud through the IP network to mobile devices. This trend is changing both the networking and the entertainment industries. As of 2011, there were more than 100 million IPTV users in the USA, and Youtube and Netflix together take up about half of the Internet capacity usage. As the trend of decoupling among contents, content delivery channels, and content-consuming devices intensifies, IP has become the basis of almost all the content distribution systems.

This trend is bringing about a revolution in our viewing habits.

- *Content type*: Both user-generated and licensed content have become prevalent. Clearly, more user-generated content implies an increasing need for upload capacity, which is traditionally designed to be much smaller than download capacity.
- *When*: For many types of video content, we can watch them anytime we want, with the help of devices like a Digital Video Recorder (DVR) on IPTV or services like HBO Go.
- *Where*: We can watch video content almost anywhere, at least anywhere with a sufficiently fast Internet connection.
- *How*: Instead of just on the TV and desktop computers, we can watch video on our phones, tablets, and any device with a networking interface and a reasonable screen.
- *How much*: We are watching more video, thanks to applications like Netflix, Hulu, Deja, and embedded videos on many websites. For example, in February 2012, Hulu had roughly 31 million unique viewers, watching 951 million

videos. Comcast NBS Universal had 39 million unique viewers, watching 205 million videos (more than doubling the number from summer 2011). Some of these are free, some are free but with intrusive advertisements, some require a monthly subscription, some are pay-per-view, and some are part of a bundled service (e.g., the triple play of IPTV, Internet access, and VoIP). If the Internet connection charge is usage-based, there is also the "byte-transportation" cost per GB, as discussed in Chapter 11.

We can categorize viewing models along four dimensions. Each combination presents different implications to the network design in support of the specific viewing model.

- *Real time vs. precoded*: Some videos are watched as they are generated in real time, e.g., sports, news, weather videos. However, the vast majority are precoded: the content is already encoded and stored somewhere. In some cases, each video is stored with hundreds of different versions, each with a different playback format or bit rate. Real-time videos are more sensitive to delay, while precoded videos have more room to be properly prepared. Some other video-based services are not only real-time, but also two-way interactive, e.g., video calls, video conferencing, and online gaming. Clearly, interactive video has even more stringent requirements on delay and jitter (i.e., the variance of delay over time).

- *Streaming or download*: Some videos, like those on Netflix and YouTube, are streamed to you, meaning that your device does not keep a local copy of the video file (although Netflix movies sometimes can be stored in a local cache, and YouTube has started a movie-rental service). In other cases, e.g., iTunes, the entire video is downloaded first before played back at some later point. Of course, the content itself may be automatically erased from local storage if digital rights are properly managed, like in movie rentals. In-between these two modes there is the possibility of *partial* download and playback. As shown in the Advanced Material, this reduces the chance of jitter, and is followed in practice almost all the time except for interactive or extremely real-time content.

- *Channelized or on-demand*: Some contents are organized into channels, and you have to follow the schedule of each channel accordingly. This is the typical TV experience we have had for decades. Even with DVR, you still cannot jump the schedule in real time. In contrast, Video on Demand (VoD) allows you to get the content when you want it. Both YouTube and Netflix are VoD. There are also VoD services on TV, usually charging a premium. Sometimes the content owner changes the model, e.g., in 2011 HBO in the USA changed to a VoD model with its HBO Go services on computers and mobile devices. In-between the two extremes, there is *NVoD*, Near Video on Demand, which staggers the same channel every few minutes, so that, within a latency tolerance of that few minutes, you get the experience of VoD.

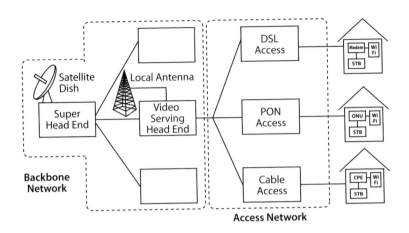

Figure 17.1 A typical architecture of IPTV. The content is collected at the super head end and distributed to different local video-serving head ends across the country, which also collect local content. Then it is further distributed to access networks running on copper, fiber, or cable, before reaching the homes. This is often carried out in private networks owned and managed by a single ISP.

- *Unicast or multicast*: Unicast means transmission from one source to one destination. Multicast means from one source to many destinations, possibly tens of millions for events like the Olympic Games, that belong to a multicast group. An extreme form of multicast is *broadcast*: everyone is in the multicast group. If you do not want certain content, you do not have to watch it, but it is sent to you anyway. TV is traditionally multicast, sometimes through physical media that are intrinsically multicast too, such as satellite. The Internet is traditionally unicast. Now the two ends are getting closer. We see unicast capabilities in IP-based video distribution, but also multicast in IP networks (carried out either in the network layer through IP multicast routing or in the application layer through P2P).

It seems that there are $2^4 = 16$ combinations using the above taxonomy of video viewing modes. Obviously some combinations do not make sense, for example, real-time video must be streaming-based and cannot be download-based. But precoded video can be either streaming- or download-based. Or, true VoD cannot be multicast since each individual asks for the content at different times, but channelized or NVoD content can be either unicast or multicast.

17.1.2 IP video: IPTV and VoI

The term "IP video" actually encompasses two styles: (1) IPTV and (2) VoI. IPTV turns TV channel delivery into IP-based, whereas VoI views the Internet as a pipe that can simply support any type of content delivery. Increasingly VoI

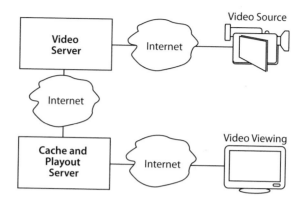

Figure 17.2 A typical architecture of video over the Internet. Video sources, ranging from iPhones to professional video cameras, upload content to video servers, which then distribute them through local caches to the viewers around the world who download the video to their devices. This is all carried out in the public Internet.

is becoming the more popular way for people to consume video content than IPTV. If all content is available on demand through VoI, what advantages does IPTV offer to the consumer experience?

(1) **IPTV** is often included as part of the triple- or quadruple-play service bundle provided by an ISP. It is delivered over a *private and managed network*, with a set-top box on the customer's premises. This private network uses IP as a control protocol, but many parts of it are deployed and operated by a single ISP, e.g., a telephone or cable company offering the Internet access. This makes it easier to control the quality of service. The content is often channelized, multicast, and streaming-based but with recording capability using DVR.

Before TV turned to the Internet access networks, it was delivered primarily through one of the following three modes: broadcast over the air, via satellites, or through cables. So why is the IPTV revolution happening now? There are a few key reasons.

- *Convergence*: almost everything else is converging on IP, including phone calls. Putting video on IP makes it a unified platform to manage.
- *Cost*: Having a uniform platform reduces the costs of maintaining separate networks.
- *Flexibility*: IP has demonstrated that one of its greatest strengths is the ability to support diverse applications arising in the future.
- *Compression* has got better and access network *capacity* has increased sufficiently that it has become possible to send HD TV channels.

(2) **Video over the Internet** (**VoI**) is delivered entirely over *public networks*, often via unicast, and to a variety of consumer devices. Given the current evolution of business models, VoI is increasingly taking over the IPTV business

as consumers access video content over the IP pipes without subscribing to TV services. There are three main types of VoI.

- The content owner sends videos through server–client architectures without a fee, e.g., YouTube, ABC, and the BBC.
- The content owner sends videos through server–client architectures with a fee, e.g., Netflix, Amazon Prime, Hulu Plus, and HBO Go.
- Free P2P sharing of movies, e.g., Bit Torrent and PPLive.

We touched upon the revenue models for IPTV and VoI. As to the cost models, they often consist of the following items.

- *Content*: The purchase of content-distribution rights. Popular and recent movies and TV series are naturally more expensive.
- *Servers*: The installation and maintenance of storage and computing devices.
- *Network capacity*: The deployment or rental of networking capacity to move content around and eventually deliver it to consumers.
- *Customer premises equipment*, such as set-top boxes and games consoles.
- *Software*: The software systems that manage all of the above and interface with consumers.

Whether it is IPTV or VoI, the quality measures depend on the bit rate, delay, and jitter. What kind of bit rates do we need for videos? It depends on a few factors, e.g., the amount of motion in the video, the efficiency of the compression methods, the screen resolution, and the ratio of viewing distance and screen size. But, generally speaking, the minimum requirement today is about 300 kbps. Below that, the visual quality is just too poor even on small screens. For standard-definition movies we need at least 1 Mbps, and a typical movie takes $1 - 2$ GB. For high-definition movies we need at least $6 - 8$ Mbps, and a typical movie takes $5 - 8$ GB. Truly HD video needs $20 - 25$ Mbps to be delivered. And the latest standard of UltraHD needs over 100 Mbps. As we will see in the next section, to make IP video work, we need technologies from both multimedia signal processing and communication networking.

17.2   A Long Answer

As shown in Figure 17.3, the overall protocol stack for IP video includes the following: MPEG over HTTP/SIP/IGMP/RTSP, over RTP/UDP/TCP, over IP, over ATM or Ethernet or WiFi, over wireless/fiber/DSL/cable. In this section, we go into some detail regarding the top three layers: compression, application, and transport, trying to highlight interesting networking principles beyond an "alphabet soup" of acronyms.

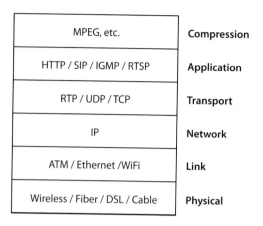

MPEG, etc.	**Compression**
HTTP / SIP / IGMP / RTSP	**Application**
RTP / UDP / TCP	**Transport**
IP	**Network**
ATM / Ethernet /WiFi	**Link**
Wireless / Fiber / DSL / Cable	**Physical**

Figure 17.3 A layered network architecture in support of video traffic. Compression standards such as MPEG use various application layer protocols, which in turn rely on combinations of transport and network layer protocols. The focus of this section is on a few key ideas in the top three layers.

17.2.1   Compression

A video is a sequence of frames moving at a particular speed. Each frame is a still picture consisting of **pixels**. Each pixel is described by its colors and luminance digitally encoded in bits. The number of bits per frame times the number of frames per second gives us the **bit rate** of a video file. Typical frame rates are 25 or 29.97 frames per second for standard definition, and 50 or 60 frames per second for high definition. Typical pixels per frame for high-definition video are $1280 \times 720 = 921600$ or $1920 \times 1080 = 20736004$. If we had to send all these bits, we would not have been able to deliver video of any reasonable quality to the vast majority of Internet-connected devices today.

In order to put more content into a given pipe, we need **compression**. This is the process of taking out redundancies in signals. If the resulting file can be later recovered, say at the consumer device, to be exactly the same as the original one, it is called **lossless compression**, e.g., the Lempel–Ziv compression used in zipping files. Otherwise, it is called **lossy compression**, and there is a tradeoff between the compression ratio (the size of the file after compression relative to that before compression) and the resulting fidelity. This is called the **rate–distortion** tradeoff, as shown in Figure 17.4.

In many VoI services, each video clip is precoded into many bitstreams, each at a different point on the rate–distortion curve. Then a specific one is chosen to be sent over the Internet, depending on any combination of the following three factors:

- the screen resolution of the end-user device,
- the throughput supported by the end-to-end path between the video server and the device, and

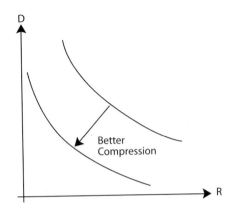

Figure 17.4 Rate–distortion curves for two different lossy compression schemes. Distortion can be measured by objective metrics or subjective tests. A higher rate leads to lower distortion. The closer to the origin the tradeoff curve, the better the tradeoff.

- as in the recent proposal of Quota-Aware-Video-Adaptation (QAVA), the data quota usage pattern of each user under usage-pricing plans in Chapters 11.

There are many techniques to help achieve the best shape and the largest range of such a tradeoff by taking out redundancies in the bits, e.g., transform coding (seek structures in the frequency domain of the signals), Huffman coding (reduce the expected length of the compressed signal by making frequently appearing codes shorter, as we saw in Chapter 10), and perceptual coding for video (let people's perceptual process guide the choice of which pixels and which frames to compress). Many frames also look alike. After all, that is how a perception of continuous motion can be registered in our brain. So video compressors often only need to keep track of the differences between the frames, leading to a significant saving in the number of bits needed to represent a group of pictures.

Video compression has made substantial progress since the early 1990s.

- **MPEG1**, a standard in 1992: this was used for VCD (which uses 1 Mbps bit rate).
- **MPEG2** (called H.262 by a different standardization body called the ITU-T), a standard in 1996: this was used for DVD (which uses about 10 Mbps bit rate).
- **MP3**, the layer 3 of the MPEG2 standard (there is no MPEG3, that track of standard was absorbed into MPEG2): this popular standard for the online music industry is for encoding just audio, and can achieve a 12:1 compression ratio.
- **MPEG4**, a standard in 2000: this is the current video compression standard family.

- MPEG4 Part 10 (also called AVC or H.264) in 2004: with 16 profiles, it offers substantial flexibility. It is also at least twice as good as MPEG2's compression capability. It is used for HDTV (with $20 - 25$ Mbps bit rate) and Blu-ray (with 40 Mbps bit rate). It can readily achieve a compression factor of 100.

- There are also other non-MPEG formats: H.261 was popular in IP video in the early days, and Quick Time by Apple is merging into MPEG4. There are also Windows Media Player by Microsoft, Flash by Adobe, and Real Media Viewer by Real Networks.

A key idea in MPEG compression is to exploit the redundancy across frames when the motion is not rich. This is called motion compensation with inter-frame prediction. There are three types of frames, and collectively a set of them form a **Group of Pictures** (GoP).

- **I frame (Intra-coded)**: This is an independent frame. Its encoding does not depend on the frames before or after it.
- **P frame (Predictive-coded)**: This type of frame depends on the previous I (or P) frame, but not the one after it.
- **B frame (Bidirectionally predictive-coded)**: This type of frame depends on both the I (or P) frames before and after it.

Each GoP must start with an I frame, followed by a sequence of P and B frames, as shown in Figure 17.5. The I frame is the most important one, while P and B frame losses are much more tolerable. At the same time, I frames are

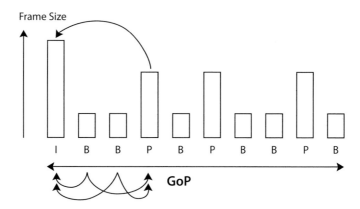

Figure 17.5 A typical structure of a Group of Pictures (GoP). Each GoP starts with an I frame, followed by a sequence of B and P frames. The I frame is independent and the most important one in each GoP. Each P frame depends on the previous I/P frame. Each B frame depends on both the previous and the following I/P frames. Some of these dependence relationships are indicated by arrows. Choosing the length and structure of a GoP affects bit rates, error resiliency, and delay in channel change.

harder to compress than P and B frames, which use motion prediction to assist in compression.

The length of a GoP influences several metrics.

- *Bitrate efficiency*: If the GoP is longer, there are more P and B frames (the more easily compressible ones). The bit rate becomes lower.
- *Error resilience*: If an I frame is lost and, consequently, the entire GoP needs to be retransmitted, a longer GoP means that more frames need to be retransmitted. There is a tradeoff between efficiency and resilience, as we will quantify in a homework problem.
- *Instant channel change*: For channelized video content, the ability to change channels fast is important if the traditional TV viewing experience is to be replicated on IPTV. Since GoP represents the logical basic unit for playback, a longer GoP means that the viewer needs to wait longer to change channels. There are also other factors at play for channel change, such as multicast group operations to be explained next.

17.2.2   Application layer

In addition to the ability to do multicast routing, we also need Internet Group Management Protocol (**IGMP**). It runs the multicast group management and tells the router that a client (an end-user device) wants to join a particular multicast group. There are only two essential message types: the router asks a `membership-query` to clients, and each client replies with a `membership-report` telling the router which groups it belongs to. There is an optional message `leave-group` from clients to routers. This message is optional because, by the principle of *soft state* (explained in Chapter 19), if the membership report information does not periodically refresh the groups, the client is assumed to leave the group.

Joining and leaving multicast groups incur propagation and processing delays. Instant channel change may become frozen. To accelerate a channel change, an IPTV service provider may send some unicast GoP to the end-user when she first requests a channel change. It goes into multicast mode once the "join group" request is processed. There is clearly a tradeoff between network efficiency and user experience. We will see more tradeoffs like this involved in optimizing the networked delivery of video.

For streaming applications, we often use Real Time Streaming Protocol (**RTSP**). It allows a media player to control the transmission of a media stream, e.g., fast forward, rewind, pause, play. It is independent of the compression standard or transport protocol.

A typical procedure is shown in Figure 17.6. The request for content first runs over HTTP from a web browser to a web server. Then, knowing the type of media and compression used (by reading the response received from the web server), the client can open the right media player, which then uses RTSP to carry out

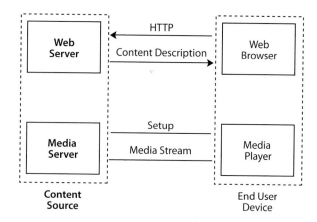

Figure 17.6 Real Time Streaming Protocol (RTSP) at work. An HTTP session first enables control information to be exchanged between the client web browser and the web server. This is followed by the actual media stream from the media server to the media-player software on the end-user device.

message passing between the client and the media server that actually holds the video file. Unlike HTTP, RTSP must keep track of the current state of the file at the client media player so that it can operate functionalities like pause and play. You must have realized there is a lot of overhead associated with managing video traffic, and we will see more of such overhead in Chapter 19.

Another protocol often used for IP multimedia transmission is Session Initiation Protocol (**SIP**) from the Internet Engineering Task Force (IETF). It establishes a call between a caller and callee over an IP network, determines the IP address, and manages the call, e.g., adds callers, transfers or holds calls, or changes voice encoding. Together with video standards, SIP can also provide a mechanism for video conferencing.

Yet another commonly used protocol is **H.323** from the International Telecommunication Union (ITU). It is actually a large suite of protocols involving many components already discussed in this subsection.

As you can see, there are many standardization bodies: the IETF standardizes many Internet related protocols, the ITU and the Institute of Electrical and Electronic Engineers (IEEE) have many standardization bodies, and some major standards have their own governing bodies, e.g., 3GPP and 3GPP2 for cellular, WiMax Forum for WiMax, DSL Forum for DSL, MPEG for video compression, etc. The way they operate is a mix of technology, business, and political factors, but they all share a common goal of inter-operability among devices so that the positive network effect of technology adoption can be achieved.

17.2.3   Transport layer

We have seen the *connection-oriented* transport protocol of TCP in Chapter 14. But much multimedia traffic, especially real-time or interactive types, runs instead over User Datagram Protocol (**UDP**) in the transport layer. UDP is *connectionless*, meaning that it does not try to maintain end-to-end reliability, not even sort packets in the right order. It works with IP to deliver packets to the destination, but if the packets do not get there, it will not try to solve that problem. For example, Skype uses UDP unless the client sits behind a firewall that allows only TCP flows to pass through. UDP also handles multicast well, since it does not require the receipt of a packet at all the destinations. IGMP that we just saw often runs on UDP.

UDP is fundamentally different from TCP in the end-to-end control of the Internet. Why would applications with tight deadlines prefer UDP? It boils down to the tradeoff between timeliness and reliability.

- TCP uses a three-way handshake (explained in Chapter 19) to establish a session, whereas UDP does not introduce that latency. TCP uses congestion control to regulate source rates, whereas UDP sends out the packet as soon as it is generated by the application layer.
- TCP ensures reliability through packet retransmission in the transportation layer, but many multimedia applications have their own built-in error resilience in the application layer. For example, losing a B frame in a GoP can often be concealed in a media player so that the viewers cannot tell. Moreover, a lost and retransmitted packet will likely be too late to be useful in playback by the time it arrives at the destination. It is instead more important to avoid holding back playback and just proceed.

In addition to real-time or interactive multimedia, many network management or signaling protocols, like SNMP in Chapter 19 and RIP in Chapter 13, also run on top of UDP. For these signaling protocols, there are two more reasons to prefer UDP.

- TCP maintains too many states for each session compared with UDP. So UDP can support many more parallel sessions at the same time.
- TCP header is 20 bytes and UDP is 8 bytes. For small control packets, this difference in overhead matters.

A protocol on top of UDP is Real-time Transport Protocol (**RTP**), which is heavily used in many IP multimedia applications including VoIP. It specifies a format to carry multimedia streams. The key challenge here is to support many types of media formats, including new ones coming up. And the key solution is to specify a range of profiles and payload formats, specific to each media type, without making the header dependent on the media type. RTP runs on the data plane that transmits the actual data, and its companion RTP Control Protocol (**RTCP**) runs on the control plane that sends control signals. RTCP keeps track of the RTP stream's information, such as the number of packets, the number of bytes, and the timestamp information. It monitors the statistics and synchronizes multiple streams.

Now that we have finished an overview of the top three layers in Figure 17.3, we will proceed to two examples of tradeoffs in IP video delivery.

17.3 Examples

17.3.1 Video quality and I/P/B frames

In this example, we will explore the effect of dropped I, P, and B frames on video quality. Consider each grayscale value of the pixel at position (x, y) in frame i. The transmitted frame is $\bar{p}_i$ and the received frame is p_i. For our metric of video quality, we use the L-1 norm, the sum of absolute differences across all the pixels and frames, i.e., error $= \sum_{x,y,i} |p_i(x, y) - \bar{p}_i(x, y)|$.

Consider a very small example with 2×2 pixels. For simplicity, assume all pixel values in a given frame are the same (this is a very boring video). A GoP consists of four frames, indexed by $i = 1, 2, 3, 4$. We will drop each frame of the GoP in turn, and quantify the effect on our error metric. Assume that frame 0 and frame 5 (belonging to the preceding and following GoPs, respectively) are always correctly received.

0 0	1 1	2 2	3 3	4 4
0 0	1 1	2 2	3 3	4 4
$i = 0$	$i = 1$	$i = 2$	$i = 3$	$i = 4$
Last GoP B frame	I frame	P frame	B frame	B frame

5   5
5   5
$i = 5$
Next GoP I frame

Recall that an I frame has no reference frame, a P frame uses the last I or P frame as the reference frame, and a B frame uses the last I or P frame and the next I or P frame as reference frames. An "error" in frame i means that either frame i is missing or it has to perform error concealment because frame i's reference is missing. We can set up a few error-concealment rules at the receiver.

- If the receiver misses any frame, it instead displays the last available frame.
- If the receiver detects an error in the reference frame of a P frame, it also displays the last available frame in place of the P frame.
- If the receiver detects an error in a reference frame of a B frame, it displays the other reference frame.

If the I frame of this GoP is dropped, the receiver displays what is summarized in Table 17.1.

	$\bar{p}(1,1)$	$\bar{p}(1,2)$	$\bar{p}(2,1)$	$\bar{p}(2,2)$
$i=0$	0	0	0	0
$i=1$	0	0	0	0
$i=2$	0	0	0	0
$i=3$	5	5	5	5
$i=4$	5	5	5	5
$i=5$	5	5	5	5

Table 17.1 Frame 1: Dropped, so repeat frame 0. Frame 2: Error in reference frame 1, so repeat frame 1. Frame 3: Error in reference frame 2, so display frame 5. Frame 4: Error in reference frame 2, so display frame 5.

Then the error between the transmitted and the received picture is

$$|p_1(1,1) - \bar{p}_1(1,1)| + |p_1(1,2) - \bar{p}_1(1,2)| + |p_1(2,1) - \bar{p}_1(2,1)| + |p_1(2,2) - \bar{p}_1(2,2)|$$
$$+|p_2(1,1) - \bar{p}_2(1,1)| + |p_2(1,2) - \bar{p}_2(1,2)| + |p_2(2,1) - \bar{p}_2(2,1)| + |p_2(2,2) - \bar{p}_2(2,2)|$$
$$+|p_3(1,1) - \bar{p}_3(1,1)| + |p_3(1,2) - \bar{p}_3(1,2)| + |p_3(2,1) - \bar{p}_3(2,1)| + |p_3(2,2) - \bar{p}_3(2,2)|$$
$$+|p_4(1,1) - \bar{p}_4(1,1)| + |p_4(1,2) - \bar{p}_4(1,2)| + |p_4(2,1) - \bar{p}_4(2,1)| + |p_4(2,2) - \bar{p}_4(2,2)|$$
$$= 4|p_1(1,1) - \bar{p}_1(1,1)| + 4|p_2(1,1) - \bar{p}_2(1,1)|$$
$$+4|p_3(1,1) - \bar{p}_3(1,1)| + 4|p_4(1,1) - \bar{p}_4(1,1)|$$
$$= 4|1 - 0| + 4|2 - 0| + 4|3 - 5| + 4|4 - 5|$$
$$= 24.$$

If instead the P frame is dropped, the receiver displays what is summarized in Table 17.2.

The error between the transmitted and the received picture is

$$4|p_1(1,1) - \bar{p}_1(1,1)| + 4|p_2(1,1) - \bar{p}_2(1,1)|$$
$$+4|p_3(1,1) - \bar{p}_3(1,1)| + 4|p_4(1,1) - \bar{p}_4(1,1)|$$
$$= 4|1 - 1| + 4|2 - 1| + 4|3 - 5| + 4|4 - 5|$$
$$= 16.$$

If the first B frame, frame 3, is dropped, the receiver displays what is summarized in Table 17.3.

	$\bar{p}(1,1)$	$\bar{p}(1,2)$	$\bar{p}(2,1)$	$\bar{p}(2,2)$
$i = 0$	0	0	0	0
$i = 1$	1	1	1	1
$i = 2$	1	1	1	1
$i = 3$	5	5	5	5
$i = 4$	5	5	5	5
$i = 5$	5	5	5	5

Table 17.2 Frame 2: Dropped, so repeat frame 1. Frame 3: Error in reference frame 2, so display frame 5. Frame 4: Error in reference frame 2, so display frame 5.

	$\bar{p}(1,1)$	$\bar{p}(1,2)$	$\bar{p}(2,1)$	$\bar{p}(2,2)$
$i = 0$	0	0	0	0
$i = 1$	1	1	1	1
$i = 2$	2	2	2	2
$i = 3$	2	2	2	2
$i = 4$	4	4	4	4
$i = 5$	5	5	5	5

Table 17.3 Frame 3: Dropped, so repeat frame 2.

The error between the transmitted and the received picture is

$$4|p_1(1,1) - \bar{p}_1(1,1)| + 4|p_2(1,1) - \bar{p}_2(1,1)|$$
$$+4|p_3(1,1) - \bar{p}_3(1,1)| + 4|p_4(1,1) - \bar{p}_4(1,1)|$$
$$= 4|1 - 1| + 4|2 - 2| + 4|3 - 2| + 4|4 - 4|$$
$$= 4.$$

If the second B frame, frame 4, is dropped, the receiver displays what is summarized in Table 17.4.

The error between the transmitted and the received picture is

$$4|p_1(1,1) - \bar{p}_1(1,1)| + 4|p_2(1,1) - \bar{p}_2(1,1)|$$
$$+4|p_3(1,1) - \bar{p}_3(1,1)| + 4|p_4(1,1) - \bar{p}_4(1,1)|$$
$$= 4|1 - 1| + 4|2 - 2| + 4|3 - 3| + 4|4 - 3|$$
$$= 4.$$

The greatest error resulted from dropping the I frame, followed by the P frame, and finally the B frames. More important frames cause more error in the GoP when dropped, leading to a bigger drop in the visual quality.

	$\bar{p}(1,1)$	$\bar{p}(1,2)$	$\bar{p}(2,1)$	$\bar{p}(2,2)$
$i = 0$	0	0	0	0
$i = 1$	1	1	1	1
$i = 2$	2	2	2	2
$i = 3$	3	3	3	3
$i = 4$	3	3	3	3
$i = 5$	5	5	5	5

Table 17.4 Frame 4: Dropped, so repeat frame 3.

17.3.2   Latency–jitter tradeoff

In this example, we look at the effect of initial buffering on streaming video's playback. We will see what the playback latency should be to provide a smooth viewing experience.

Assume there is one frame per packet. Figure 17.7 shows the transmitted, arrived (at the receiver), and played frames over time. We refer to these as the timing curves, abbreviated as the V, A, and P curves. The V and A curves are given by the source and the network, and our job is to design the best P curve.

The video source transmits at a constant rate, so V is a superposition of unit step functions. We abuse the notation a little to use vectors to represent the discrete jumps on the curves. Let V_i denote the time at which packet i is transmitted, A_i the time at which packet i is received, and P_i the time at which packet i is displayed to the user. The delay between the transmission and receipt of packet i is given by $d_i = A_i - V_i$. We have

$$\mathbf{V} = \begin{bmatrix} 1 \\ 2 \\ 3 \\ 4 \end{bmatrix}, \quad \mathbf{A} = \begin{bmatrix} 5.8 \\ 7.5 \\ 8 \\ 9.4 \end{bmatrix}, \quad \mathbf{d} = \mathbf{A} - \mathbf{V} = \begin{bmatrix} 4.8 \\ 5.5 \\ 5 \\ 5.4 \end{bmatrix}.$$

In reality, we cannot know $\mathbf{A}$ ahead of time and have to either estimate it or adapt in real time, like in a homework problem. For now, we assume it is known. P must be a unit step function since frames need to be displayed at a constant rate. When should the playback begin, i.e., what is the value of P_1? The goal is to minimize the total delay experienced by the user:

$$\text{minimize} \quad \sum_i (P_i - V_i)$$

$$\text{subject to} \quad P_{i+1} = P_i + 1, \ \forall i$$

$$P_i \geq A_i, \ \forall i$$

$$\text{variables} \quad \{P_i\}.$$

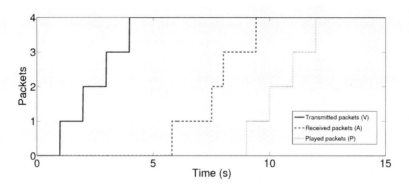

Figure 17.7 Playback buffer smoothes video playback. This graph shows curve V at the source: how packets are transmitted with a constant rate; curve A at the receiver: how the packets' arrival times vary as they traverse the network, and curve P at playback: how playback latency can smooth the arrival jitter. The V and A curves are given, and the P curve needs to be designed to make sure it is a superposition of unit step functions, lies below the A curve, and yet is as far to the left as possible.

The first constraint in the above optimization ensures that the playback curve P is a unit step function, and the second constraint says that playback of a packet can occur only after the packet has been received. The objective function can be further simplified; since P and V are both unit step curves, $P_i - V_i$ is the same for all i, so the objective function is equivalent to minimizing any single $P_i - V_i$.

This problem can be solved easily through visual inspection. Essentially, we shift a unit step function P to the left, until $P_k = A_k$ for some k, and $P_i \geq A_i$ $\forall i \neq k$. That is, we want to make curve P as close to curve A as possible but still remain below A.

From Figure 17.7, clearly $k = 2$. Since $P_2^* = A_2 = 7.5$ s, we have $P_1^* = 6.5$ s. This means playback should begin at 6.5 s, which in turn means delaying playback by $P_1 - A_1 = 6.5 - 5.8 = 0.7$ s. This playback buffering latency avoids frozen video due to the variation of packet arrivals through the network.

In general, for a constant-rate source, the playback should begin at $P_1^* = A_1 + D$, where $D = \max_i(d_i - d_1)$ and represents the maximum delay variation. This formula applies to the above example. Since $D = d_2 - d_1 = 5.5 - 4.8 = 0.7$ s, we have $P_1^* = A_1 + D = 5.8 + 0.7 = 6.5$ s.

17.4   Advanced Material

There are three general approaches in managing the quality of service on top of the simple connectivity services, which is offered by the thin waist of TCP/IP in the best effort style.

• Treat different sessions differently during resource allocation.

- Regulate which clients can be admitted.
- Distribute servers to strategic locations.

As an analogy, think of a highway's traffic control. Differentiating resource allocation is like reserving a lane for certain vehicles, like a car-pool lane. Regulating admission is like using the on-ramp traffic lights during the rush hours. Distributing servers is like constructing new exits for popular destinations such as grocery stores, which is clearly a much-longer-timescale operation than the other two.

17.4.1   Queueing policies

Different sessions can be treated differently, inside a node or along a link in a network. For example, there are several standard queueing disciplines in a router. As shown in Figure 17.8, the following three methods will create different sequences of service timing among the incoming sessions. But they are all work-conserving: they do not waste a timeslot if there is some packet to be served.

- *Round robin*: Each class of traffic takes turns.
- *Weighted fair queueing*: While taking turns, one class can receive a higher rate than another.
- *Priority queueing*: Higher-priority-class packets are processed before lower-priority ones, which have to wait until there are no more higher-priority packets in the queue.

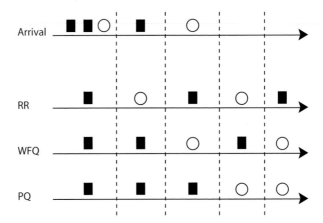

Figure 17.8 Three queueing disciplines give different orders of packet service. There are two classes arriving over the timeslots denoted by dotted lines. In Round Robin (RR) scheduling, the square packets and circle packets are served in turn, one packet in each timeslot. In Weighted Fair Queuing (WFQ), the two classes take turns, but square packets get a higher service rate. In Priority Queuing (PQ), square packets have strict priority, and circle packets get their chance only when all square packets have been sent.

What constitutes a *fair* allocation among competing sessions? This is yet another instance where we touch upon the notion of fairness. We will see a systematic treatment of the subject in Chapter 20.

Of course, differential treatment methods as above do not provide a guarantee on the quality of service. For that, we need methods for resource *reservation*. There are dynamic versions of establishing end-to-end circuits in the network layer and reserving adequate resources (such as capacity, timeslot, and processing power) so that the end-user experience is guaranteed to be good enough. Handoff in mobile networks in Chapter 19 will offer such an example.

17.4.2   Admission control

An alternative approach to managing resource competition is to regulate the demand. In some sense, TCP congestion control does that in a feedback loop on the timescale of RTT. Policing or throttling further shapes the rate of traffic injection into the network. Time-dependent pricing in Chapter 12 is another form of admission control on a longer timescale.

For an *open-loop* control, we can use admission control: deciding whether a session should be allowed to start transmitting at any given timeslot. Such admission control is usually carried out at the edge of the network, often at the first network element facing the end-user devices.

One possible admission control is to control the peak rate of traffic injection over some timescale. This is like the ramp light that regulates cars getting onto a highway and smoothes the traffic injection into the highway network during the rush hours. By reducing the rate of green lights at the ramp, we can lower the rate of adding traffic onto the highway (the backbone network) at the expense of causing congestion at the ramp (the access or edge network).

An example of admission control is **leaky bucket**, shown in Figure 17.9. In order for each packet to be admitted, there must be a token dripping from a (conceptual) leaky bucket. The bucket drips tokens at a rate of a, and has a volume of B tokens. In a homework problem, we will see that weighted fair queueing and leaky bucket can together shape any arrival patterns to a desirable one.

17.4.3   Content distribution

Yet another approach, in addition to differentiating resource allocation and controlling user admission, is to change the location of the source of content, so that the content is brought closer to the users, as shown in Figure 17.10. This leverages the continuous drop of storage cost and the widespread popularity of certain content to create *spatially pipelined* distribution of content.

The idea of changing the location of the source has long been practiced in web proxy servers since the mid-1990s. ISPs cache popular web content at local storage closer to end-users. A whole industry sector has also been generated,

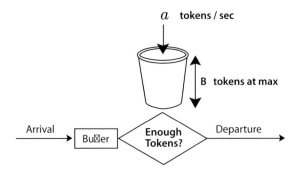

Figure 17.9 A leaky bucket for admission control. Each bucket can hold at most B tokens, and r tokens are added to the bucket per second. Each transmitted packet consumes one token. In order for a packet in the buffer to be transmitted on the egress link, there must be a token available in the bucket.

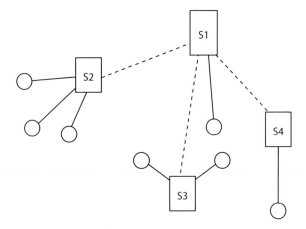

Figure 17.10 An illustration of a content distribution networks. The content originally placed in server S1 is replicated in three other locations, S2, S3, and S4, being sent through fiber links shown by dotted lines. When a client requests a piece of content, a particular server is selected as the source to serve that demand.

operators of **Content Distribution Networks** (CDN). They serve either ISPs or content owners, and manage the following processes.

- Deploy many servers, sometimes in private server farms and sometimes in shared data centers. These are often called mirror sites.
- Replicate content and place it in different servers. This involves optimization on the basis of the availability of high-speed links and high-capacity storage, as well as prediction of content popularity in different geographic areas.
- For each content request, select the right server to serve as the content source. This server-selection optimization tries to minimize the delay perceived

by a user. It is run by the CDN operator, under a given routing decided by the ISP. On the other hand, ISP runs routing optimization through traffic engineering like in Chapter 13, under a given traffic demand pattern between servers and user-devices. These two optimization problems are "mirror images" of each other, and the corresponding operations by the CDN operator and the ISP may have unintended interactions.

The convergence of content owner and content transporters is creating interesting dynamics. A proper operation of CDNs can create a win–win situation: consumers get better quality of service (as delay is reduced and throughput increased), while content owners or ISPs reduce the cost of deploying large-capacity servers and pipes (as the congestion in the network is reduced).

Summary

Box 17 Quality-of-service mechanisms

Contents are being decoupled from specific content delivery channels and content-consuming devices, and IP has become the basis of almost all the content distribution systems, including IPTV and VoI. Video compression, video network applications, and connectionless transport protocols all contribute to the proliferation of video on the Internet. Quality differentiation, admission control, and content distribution intelligence enable the best-effort Internet to support video applications.

Further Reading

The subject matter of this chapter spans both analytic models of quality of service and the systems design of multimedia protocols.

1. The fundamentals of video signal processing can be found in many graduate textbooks on the subject, including the following recent one:
   A. C. Bovik, *The Essential Guide to Video Processing*, Academic Press, 2009.

2. The following book provides a concise summary of all the major video-over-IP systems, including IPTV and VoI:
   W. Simpson, *Video over IP*, 2nd edn., Focal Press, 2008.

3. The following standard textbook of computer networking provides much more details about the application and transport layer protocols we mentioned, like IGMP, RTSP, SIP, UDP, and RTP:

J. Kurose and K. Ross, *Computer Networking: A Top-Down Approach*, 5th edn., Addison Wesley, 2009.

4. The following classic paper combines leaky-bucket admission control and generalized processor sharing to provide a delay guarantee:

A. Parekh and R. G. Gallager, "A generalized processor sharing approach to flow control in integrated services networks: The single-node case," *IEEE/ACM Transactions on Networking*, vol. 1, no. 3, pp. 344–357, June 1993.

5. The following book provides a concise survey both of the deterministic, algebraic approach and of the stochastic, "effective-bandwidth" approach to the design of quality guarantees in a network:

C. S. Chang, *Performance Guarantees in Communication Networks*, Springer Verlag, 2000.

Problems

17.1 *Video-viewing models* ⋆

Fill in the table indicating which video-watching models are infeasible. Provide examples of companies that follow each feasible model. Some rows have been filled out as an example.

Real-time or precoded	Streaming or download	Channelized or on-demand	Unicast or multicast	Companies
Real-time	Streaming	Channelized	Unicast	
Real-time	Streaming	Channelized	Multicast	
Real-time	Streaming	On-demand	Unicast	
Real-time	Streaming	On-demand	Multicast	
Real-time	Download	Channelized	Unicast	
Real-time	Download	Channelized	Multicast	
Real-time	Download	On-demand	Unicast	
Real-time	Download	On-demand	Multicast	
Precoded	Streaming	Channelized	Unicast	
Precoded	Streaming	Channelized	Multicast	

Precoded	Streaming	On-demand	Unicast	YouTube, Hulu, NBC, HBO Go
Precoded	Streaming	On-demand	Multicast	Infeasible (on-demand multicast)
Precoded	Download	Channelized	Unicast	
Precoded	Download	Channelized	Multicast	
Precoded	Download	On-demand	Unicast	
Precoded	Download	On-demand	Multicast	

17.2   *Compression–reliability tradeoff* ⋆

Let us examine the tradeoff between compression and error resilience through a back-of-the-envelope calculation. Suppose we have 15 frames to transmit and two possible GoP structures: (1) IPB and (2) IPBBB. Suppose an I frame costs 7 kB, a P frame costs 3 kB, and a B frame costs 1 kB.

If an entire GoP is not received correctly, we assume that the GoP must be sent again. As our metric of error resilience, consider the expected number of bits that must be retransmitted at least once. The probability of dropping a frame is 1% and assumed to be independent. (These assumptions are made to simplify this homework problem. In a realistic setting, if a P or B frame in a GoP is lost, the entire GoP does not need to be retransmitted. Loss is not independent and usually much less than 1%. And there should be many more frames in a video clip.)

(a) In case 1, the video frame structure is IPB/IPB/IPB/IPB/IPB. What is the total cost of the video in kB? What is the cost per GoP per kB?

(b) What is the probability that an entire GoP is transmitted successfully in case 1? What is the expected number of GoPs that are successful on the first attempt at the transmission of the entire video? What is the expected number of GoPs that must be retransmitted at least once? How much does the first retransmission cost in kB?

(c) Repeat (a) for case 2, where the video frame structure is now: IPBBB/IPBBB/IPBBB.

(d) Repeat (b) for case 2.

(e) Compare you results from (a), (b), (c), and (d) in terms of the tradeoff of compressibility vs. retransmission. What can you conclude?

17.3 *Playback buffer with random arrival time* ⋆⋆

We will look at a question similar to the example in Section 17.3.2 examining the latency–jitter tradeoff, but with a probabilistic packet arrival time. Suppose $V_1 = 0, V_2 = 1$, and $V_3 = 2$, i.e., a step function. Now the packets arrive independently at times A_1, A_2, and A_3, where A_i is drawn randomly between $\bar{A}_i - 1$ and $\bar{A}_i + 1$, and we set $\bar{A}_1 = 3, \bar{A}_2 = 4.2, \bar{A}_3 = 4.6$. What is the optimal playback time of the first packet p^* that minimizes latency but ensures that all packets are received with at least a 95% probability?

17.4 *Round robin, weighted fair queueing, and priority queueing* ⋆⋆

We compare three resource allocation policies. Recall the following.
- Round robin simply gives each queue a turn to transmit a packet.
- Priority queueing allows the queue with the higher priority to be continuously serviced until it is empty.
- A particular implementation of weighted fair queueing looks at the head of each queue and transmits the packet that would finish transmission quickest under the **Generalized Processor Sharing** (GPS) scheme. GPS is an ideal "fluid-flow" scheduler, and is defined as follows: if we have n queues with priority $p_1, p_2, \ldots, p_n$, then the bandwidth allocated to queue j per timeslot is $p_j / \sum_i p_i$.

Suppose we have queue A and queue B with packets arriving:

Time (s)	Queue A arrived packet size	Queue B arrived packet size
$t = 0$		3
$t = 1$	1	
$t = 2$	1	
$t = 3$		2
$t = 4$		
$t = 5$		4

Queue A has priority 1 and Queue B has priority 3 (higher number indicating higher priority). The outgoing link has bandwidth 1 Mbps. Once a packet has begun transmitting, it cannot be pre-empted by other packets. Fill in the following table for round-robin scheduling, priority queueing, and weighted fair queueing.

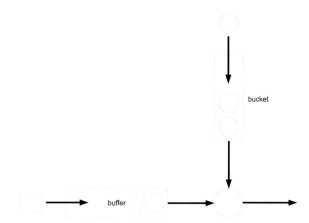

Figure 17.11 An illustration of a leaky bucket for admission control.

Time (s)	Queue A departed packet size	Queue B departed packet size
$t = 0$		
$t = 1$		
$t = 2$		
$t = 3$		
$t = 4$		
$t = 5$		

17.5 *Leaky bucket and GPS* ★ ★ ★

A link becomes congested when packets arrive at a faster rate than the link can support. The leaky bucket queueing system is one way to solve this problem. There is a bucket that contains tokens. For traffic class k, the bucket has size B_k and tokens refill the bucket at a rate a_k. Packets wait in the queue and can be released only with a token from the bucket. Therefore, the maximum number of packets that leave the queue during time interval $[u, t]$ is $B_k + a_k(t - u)$. This is illustrated in Figure 17.11.

Several leaky buckets drain into the same buffer. This buffer follows the Generalized Processor Sharing (GPS) service. Eeach traffic class k has weight w_k; C bps is the rate supported by the link out of the buffer; and $\rho_k = \frac{w_k}{\sum_j w_j} C$ is the instantaneous rate at which packets of traffic class k leave the GPS buffer. This is illustrated in Figure 17.12.

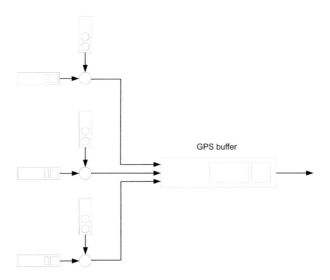

GPS buffer

Figure 17.12 An illustration of Generalized Processor Sharing.

(a) During the time interval $[u, t]$, at least $\rho_k(t - u)$ packets leave the GPS buffer. Let $B_{t,k}$ be the backlog of traffic class k in the GPS buffer at time t. Prove that the backlog of class-k traffic in the buffer cannot exceed some B_k if $\rho_k \geq a_k$. To start, assume that there is a time t when the backlog $B_{t,k} \geq B_k$. Consider the largest time $u < t$ such that $B_{u,k} = 0$, and write the inequality relating the change in backlog size from time u to time t.

(b) Prove that the delay experienced by a packet in class k in Figure 17.12 cannot exceed $\frac{B_k}{\rho_k}$ if $\rho_k \geq a_k$.

Let F_k denote the transmission time of packet k (the time when packet k leaves the buffer) under WFQ. Similarly define G_k for GPS. Let L_k denote the size (in bits) of packet k, and L_{max} is the largest L_k. We will show that

$$F_k \leq G_k + \frac{L_{max}}{C}, \quad \forall k. \tag{17.1}$$

GPS and WFQ both process packets at the same rate, so the total amount of packets in the system remains the same. Therefore, their busy and idle periods are the same, and we need only show that the result holds for a single busy period. Assume $F_1 < F_2 < \ldots < F_K$ correspond to K packets in one busy period of WFQ.

(c) Pick any $k \in \{1, 2, \ldots, K\}$ and find the maximum $m < k$ such that $G_m > G_k$. (If there is no such m, let $m = 0$.) This implies that $G_n \leq G_k < G_m$ for n in the set of indices $P = \{m+1, m+2, \ldots, k-1\}$. Let T_m denote the time when WFQ chose to transmit packet m. Now consider the time $S_m = F_m - T_m$. Show that the packets in set P must have arrived after S_m.

(d) Now consider the time interval $[S_m, G_k]$. Packets in set P arrived and departed during this interval. In addition, packet k was transmitted during this interval. Recall that the system is work conserving. Write an inequality relating the $[S_m, G_k]$ to the transmission times of packets in set P, and use this to show the main result in (17.1).

(e) Suppose that multiple-leaky bucket queues are multiplexed to a single WFQ buffer, similar to Figure 17.12. Combine (b) and (d) to show that the maximum delay experienced by a packet of class k is $\frac{B_k}{\rho_k} + \frac{L_{max}}{C}$.

18 Why is WiFi faster at home than at a hotspot?

A crude answer is that interference management in WiFi does not scale well beyond several devices sharing one access point. When the crowd is big, the "tragedy of the commons" effect, due to mutual interference in the unlicensed band, is not efficiently mitigated by WiFi. To see why, we have to go into the details of WiFi's medium access control in the link layer of the layered protocol stack.

18.1 A Short Answer

18.1.1 How WiFi is different from cellular

Since their first major deployment in the late 1990s, WiFi hotspots have become an essential feature of our wireless lifestyle. There were already more than a billion WiFi devices around the world by 2010, and hundreds of millions are added each year. We use WiFi at home, in the office, and around public hotspots like those at airports, in coffee shops, or even around street corners.

We all know WiFi is often faster than 3G cellular, but you cannot move around too fast on WiFi service or be more than 100 m away from an Access Point (**AP**). We have seen many letters attached to 802.11, like 802.11a, b, g, and n, shown on the WiFi AP boxes you can buy from electronic stores, but maybe do not appreciate why we are cooking an alphabet soup. We have all used hotspot services at airports, restaurants, hotels, and perhaps our neighbor's WiFi (if it does not require a password), and yet have all been frustrated by the little lock symbol next to many WiFi network names that our smartphones can see but not use.

When Steve Jobs presented iPhone 4 in a large auditorium, that demo iPhone could not get on the WiFi. Jobs had to ask all the attendants to get off the WiFi. Afterwards his iPhone managed to access the WiFi. Is there some kind of limit as to how many users a given WiFi hotspot can support?

In June 2012, five leading cable providers in the USA announced that they would join their fifty thousand hotspots into a single WiFi service. One year before then, the South Korean government announced a plan to cover the entire city of Seoul, including every corner of every street, with WiFi service by 2015. If many WiFi users aggregate around one popular street corner, how many hotspots need to be created to take care of that demand? And, more importantly, how can this traffic be backhauled from the WiFi air-interface to the rest of the Internet?

At home, a residential gateway of the Internet access is often connected to a WiFi AP, which provides the in-home wireless connectivities to desktops, laptops, games consoles, phones, tablets, and even TV's set-top boxes using the latest high speed WiFi variant. As each home adds more WiFi devices, will the quality of connection be degraded, especially in high-rise multi-tenant buildings?

The answers to these questions are continuously being updated by academia and industry, and we already know quite a bit about WiFi architecture and performance. Officially, WiFi should be called the **IEEE 802.11** standard. It is part of the 802 family of standards on Local Area Networks prescribed by the IEEE. The .11 part of the family focuses on wireless LAN using the **unlicensed spectrum**.

You must have a license from the government to transmit in the frequency bands used by all generations of cellular networks. For 3G and 4G, governments around the world sold these spectral resources for tens of billions of dollars, sometimes through auctions as we saw in a homework problem in Chapter 2. This avoids having too many transmitters crowding and jamming into those frequency bands.

In contrast, governments around the world also leave some bands unlicensed and free, as long as your transmit power is not too high. For example, the Industry, Science, and Medical (ISM) frequency ranges in the S-band around 2.4–2.5 GHz and in the C band around 5.8 GHz. They were originally allocated for use in the three fields, as suggested by the name ISM, but the most widely used appliance in the ISM S-band, other than WiFi, is actually the microwave oven. That band works well to excite water molecules. There are also other wireless communication devices running on bluetooth, zigbee, etc. sharing the same ISM band. Handling interference on an unlicensed spectrum is a major challenge.

In the mid 1990s, as the 2G cellular industry took off, people started wondering whether they could create an alternative in a wireless network: use the small amount of power allowed in ISM and short-range communication (around 100 m outdoors, a transmission range one order of magnitude smaller than that for cellular) for mostly stationary devices. Because this is not a single-provider network, an industry forum was needed to test the inter-operability of all the devices. It was established in 1999 as the Wi-Fi Alliance, where Wi-Fi stands for "Wireless Fidelity" and is a catchier name than "IEEE 802.11b."

Many versions of WiFi standards were created by the IEEE 802 organization, starting with 802.11b that uses the 2.4 GHz band and can transmit up to 11 Mbps. This was followed by two other main versions: 802.11g that uses a more advanced physical layer coding and modulation to get to 54 Mbps in the 2.4 GHz band, and 802.11a that can also achieve up to 54 Mbps in the 5.8 GHz band. Some of these standards divide the frequency band into smaller blocks in Orthogonal Frequency Division Multiplexing (**OFDM**). In contrast to anti-resource-pooling in Paris Metro Pricing in Chapter 12, this resource fragmentation is motivated

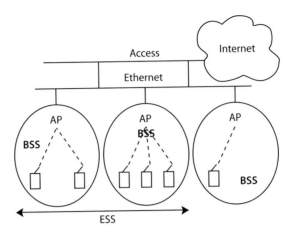

Figure 18.1 A typical topology of WiFi deployment. The air-interface provides bidirectional links between the APs and end user devices. Each BSS has an AP. A collection of BSSs that can readily support handoff is called an ESS. The air-interface is connected to a wireline backhaul, often an Ethernet, which is in turn connected to the rest of the access network and further to the rest of the Internet.

by better spectral efficiency because signals on smaller chunks of frequency bands can be more effectively processed. More recently, 802.11n uses multiple antennas on radios to push the transmission rate to over 100 Mbps. Augmenting the channel width to 40 MHz also helped increase the data rate. We will discuss OFDM and multiple antenna systems in the Advanced Material.

There have also been many supplements to improve specific areas of WiFi operation. For example, 802.11e improved the quality of service in its medium access control, a topic we will focus on in this chapter; 802.11h improved encryption and security, a major issue in the early days of WiFi; and 802.11r improved the roaming capability in WiFi to support, to some degree, the mobility of people holding WiFi devices.

Even though the nature of spectral operation is different, WiFi does share a similar topology (Figure 18.1) with cellular networks, except this time it is not called a cell (since there is often no detailed radio frequency planning before deployment), but a Basic Service Set (**BSS**). In each BSS there is an AP rather than a Base Station. A collection of neighboring BSSs may also form an Extended Service Set (ESS).

When your laptop or phone searches for WiFi connectivity, it sends probe messages to discover which APs are in its transmission range, and shows you the names of the BSSs. You might want to connect to a BSS, but, if it is password-protected, your device can associate with the AP only if you have the password to authenticate your status, e.g., as a resident in the building if the AP is run by the building owner, as an employee of the company if the AP is run by the corporation, or as a paying customer if the AP is part of the WiFi subscription service offered by a wireless provider. Increasingly you see more WiFi deployment, but *free* WiFi's availability may be on the decline.

These APs are tethered to a backhauling system, often a wireline **Ethernet** (another, and much older, IEEE 802 family member) that connects them to the rest of the Internet. This is conceptually similar to the core network behind the base stations in cellular networks, although the details of mobility support, billing, and inter-BSS coordination are often much simpler in WiFi.

If the channel conditions are good, e.g., you are sitting right next to the WiFi-enabled residential gateway at your home and no one else's signal interferes with yours, the data rate can be very impressive, especially if you are using 802.11n. It is faster than 3G, and probably even faster than the DSL or fiber access link that connects the residential gateway to the rest of the Internet. But if you sit outside the limited range of the AP or you start moving across the boundaries of an ESS, you can easily get disconnected. And if you share the air with ten or so other WiFi devices, the speed can drop substantially as you may have experienced in a crowded public WiFi hotspot.

There is actually also a peer-to-peer mode in 802.11 standards, the *infrastructureless*, ad hoc mode. WiFi devices can directly communicate with each other without passing through any fixed infrastructure like APs. You see this option when you configure the WiFi capability on your computers. But very few people use this mode today, and we will talk only about the infrastructure mode with APs.

18.1.2   Interference management in WiFi

Summarizing what we have talked about so far: WiFi is an evolving family of standards that enables short-range wireless communication over the ISM unlicensed bands for largely stationary devices. In contrast to cellular networks, WiFi networks are often deployed with very limited planning and managed only lightly, if at all.

WiFi is quite a different type of wireless networking from cellular, and its performance optimization requires some different approaches. Whether a WiFi hotspot works well or not really depends on how effectively such optimizations are carried out. We focus on the air-interface part between the AP and the devices (the terminology **station** covers both), even though the backhaul part could also become a bottleneck (e.g., when the DHCP server has a bug and cannot keep track of the IP addresses assigned to the devices, or simply because the backhauling capacity is limited.)

The first set of performance tuning involves the correct selection of the AP, the channel, and the physical layer transmission rate.

- *AP association*: A WiFi device has to regularly scan the air and then associate with the right AP, e.g., the one that offers the best SIR (and, of course, authenticates the device).

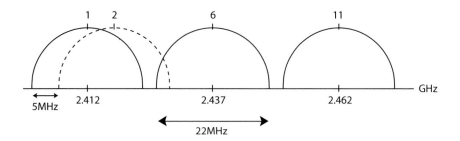

Figure 18.2 The 802.11b spectrum and channels. There are 11 channels in the USA. Each channel is 22 MHz wide, and 5 MHz apart from the neighboring channels. Therefore, only three channels, Channel 1, Channel 6, and Channel 11 are non-overlapping.

- *Channel selection*: The overall ISM frequency band is divided into channels. In 802.11b in the USA, for example, each channel is 22 MHz wide and 5 MHz apart from the neighboring channels. As shown in Figure 18.2, only those channels that are five channels apart are truly non-overlapping. So, if you want to have three devices on non-overlapping channels, the only configuration is for each of them to choose a different channel from among Channels 1, 6, and 11. Many WiFi deployments just use the default channel in each AP. If they are all on Channel 6, unnecessary interference is created right there.
- *Rate selection*: We mentioned that each of 802.11abgn can transmit *up to* a certain data rate. That is assuming a really good channel, with no interference and no mobility. In many WiFi hotspots, the channel condition fluctuates and interferers come and go. The maximum rate is rarely achieved, and the AP will tell the devices to backoff to one of the lower rates specified, e.g., all the way down to 1 Mbps in 802.11b, so that the decoding is accurate enough under the lower speed. A device knows it is time to reduce its rate to the next lower level if its receiver's SIR is too low for the current rate, or if there have been too many lost frames.

Suppose your WiFi device gets the above three parameters right. We need to take care of interference now. When two transmitters are within interference range of each other, and they both transmit a frame at similar time (t_1 and t_2, respectively), these two frames collide. There are three possible outcomes of a collision.

- Both frames are lost: neither receiver can correctly decode the intended frame.
- The stronger frame is properly received, but the weaker frame is lost: here, "strength" refers to SIR. This is called **capture**.
- Both frames are properly received. This is called **double capture**.

Now, which outcome will prevail? That depends, quite sensitively, on the following factors.

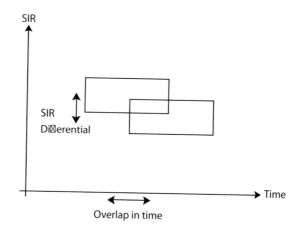

Figure 18.3 An energy-time diagram of two colliding frames. If collision is defined as two frames overlapping in their transmission time, the outcome of a collision depends quite sensitively on several factors: how long the overlap is, how big the differential in the received SIRs is, and how large an SIR is needed for proper decoding at each receiver.

- How long the frames overlap (based on their timing difference $t_1 - t_2$, frame sizes, and transmission rates).
- How big the differential in SIR between the two frames is (which depends on channel conditions and transmit powers). This is illustrated in Figure 18.3.
- How large an SIR is required for proper decoding at the receiver (which depends on transmission rates, coding and modulations, and receiver electronics).

It is an unpleasant fact that wireless transmissions may interfere with each other, since wireless transmission is energy propagating in the air. It is a particularly challenging set of physics to model, because collision is not just one type of event. In the rest of the chapter, we will simply assume that when a collision happens, both frames are lost.

Compared with power control in Chapter 1 for cellular networks, WiFi also uses a fundamentally different approach to manage interference, due to its much smaller cell size, the typical indoor propagation environment, a much smaller maximum transmit power allowed, and more uncontrolled interference in an unlicensed band. Instead of adjusting transmit powers to configure the right SIRs, WiFi tries to avoid collision altogether, through the mechanisms of **medium access control**.

Think of a cocktail party in Chapter 1 again, where guests' voices overlap in the air. With enough interference you cannot understand what your friend is trying to say. Cellular network power control is like asking each guest to adjust her volume without running into an arms race. WiFi medium access control is like arranging for the guests to talk at different times.

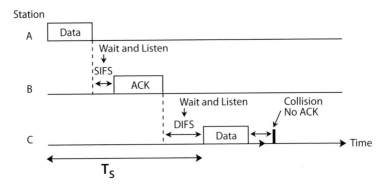

Figure 18.4 A timing diagram of basic WiFi transmissions. A station can be either a user device or an AP. First the transmitter of a session, station A, sends a data frame to its intended receiver, station B. Then, after a very short period of time with a predetermined length called the SIFS, B sends an acknowledgment frame back to A. After waiting for another slightly longer period of time called the DIFS, other nodes, like station C, can start sending new data frames. In the above example, node C's packet collides with some other packet transmitted by, say, station D.

You can either have a centralized coordinator to assign different timeslots for each guest to talk (scheduling), or you can ask each of them to obey a certain procedure for deciding locally when to talk and how long to talk (random access). We call these alternatives the Point Coordination Function (**PCF**) and the Distributed Coordination Function (**DCF**), respectively. The PCF, like a token ring in the wireline Ethernet protocol, represents centralized control (and dedicated resource allocation). It is complicated to operate and rarely used in practice. The DCF, in contrast, enables shared resource allocation. As you might suspect for any distributed algorithm, it can be less efficient, but easier to run. In practice, the DCF is used most of the time. We will discuss WiFi's DCF, which is a particular implementation of the Carrier Sensing Multiple Access (**CSMA**) random access protocol.

The basic operation of CSMA is quite simple and very intuitive. Suppose you are a transmitter. Before you send any frame, you regularly listen to the air (the part of the spectrum where your communication channel lies). This is called **carrier sensing**. As Figure 18.4 illustrates, every transmitter must observe a wait-and-listen period before it can attempt transmission. If the channel is sensed as busy (someone is using the medium to transmit her frames), you just stay silent. But if it is idle (no one is using it), you can go ahead and send a sequence of frames. You might want to send a lot of frames in a row, so you can send a control message declaring how long you intend to use the channel. Of course, the channel-holding time has an upper bound, just like in treadmill-sharing in a gym.

But if your frame collides with some other frames when you try to send it, your receiver will not get it (since we assumed a collision kills both frames).

So you will not get her acknowledgement. This is how you know you suffered a collision, and you need to *backoff*. This WiFi backoff is similar to the end-to-end TCP backoff by halving the congestion window in Chapter 14: you double the **contention window** in WiFi. And then you draw a random number between now and the end time of the contention window. That will be your next chance of sending the lost frame.

This protocol description might sound unmotivated at first. But there are actually two clever ideas of distributed coordination here: randomization and exponential backoff.

First, if stations A and B have their frames collide at one time, you do not want them to backoff to a common time in the future: there will be just another collision. They need to *randomly* backoff to minimize the chance of hitting each other again. This is exactly opposite to aiming at synchronization as in Chapter 8. Of course, it may so happen that they pick exactly the same timeslot again, but that is the price you pay for distributed coordination.

Second, if frames keep colliding, you know the interference condition is very bad, and you, as well as all those stations experiencing persistent collisions of their frames, should start backing off more upon receiving this implicit, negative feedback. Linearly increasing the contention window size is one option, but people thought that would not be aggressive enough. Instead, WiFi mandates *multiplicatively* backing off. Since the multiplicative factor is 2, we call it **binary exponential backoff**. This is similar to the multiplicative decrease of the congestion window size in TCP. As illustrated in Figure 18.5, when your contention window exceeds a *maximum value*, i.e., you have been backing off through too many stages, you should just discard that frame and report the loss to upper layer protocols so that they can try to fix it. The contention window may also have a *minimum value*, in which case a sender has to wait before its first attempt to send a frame.

So far so good. But it is another unpleasant fact of wireless transmission that sensing range is *not* the same as interference range: it might happen that stations A and B collide but they cannot hear each other, as shown in Figure 18.6. This is the famous **hidden node** problem, one of the performance bottlenecks of WiFi hotspots. This problem does not arise in TCP congestion control.

But there is a clever solution using a little explicit message passing this time, to help navigate through this challenging interference problem. It is called **RTS/CTS**. When station A wants to send a frame, it first sends a short control message called Request To Send (RTS). All stations within the sensing range of A receive that message, and each of them in turn sends a short control message called Clear To Send (CTS). All stations within sensing range of them receive the CTS and refrain from transmitting any frames in the near future. Of course station A itself also gets the CTS message back, and when it sees that CTS, it knows all those hidden nodes have also received the CTS and thus will not send any frames now. At that point, station A sends the actual frames.

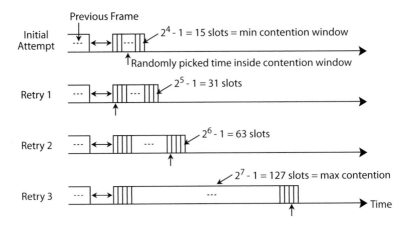

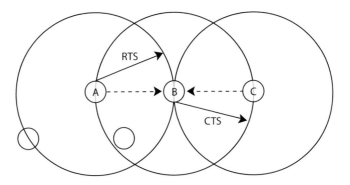

Figure 18.5 The exponential backoff in DCF. There are two key ideas. First, when two frames collide, both need to back off. In order to avoid both picking the same time to retransmit, each picks a random point over the contention window to retransmit. Second, if collisions continue, each sender needs to back off more. Doubling the contention window is a reasonable way to increase the degree of backing off. A homework problem will explore this further. The minimum window size is W_{min}, and the maximum number of backoffs allowed (before the frame is discarded) is B.

Figure 18.6 The hidden node problem: Stations A and C's transmissions to station B interfere with each other, but cannot sense each other. Dotted lines denote sensing/transmission range. RTS/CTS is a message passing protocol to help resolve the hidden node problem. Node A first sends an RTS. Upon hearing the RTS, all nodes (including node B) send a CTS. Upon hearing the CTS, all nodes (including node C) remain silent for a period of time, except node A itself, which initiated the RTS in the first place. Node A now knows it is safe to send the actual data frames without worrying about hidden nodes.

As Figure 18.7 illustrates, the brief period of idle time in between an RTS and the CTS is shorter than the wait-and-listen time between data transmissions. This is yet another clever idea in distributed coordination in wireless networks. By creating multiple types of wait-and-listen intervals, those transmissions that

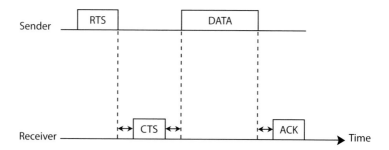

Figure 18.7 A timing diagram of RTS/CTS in WiFi DCF to help mitigate the hidden node problem. The durations of time between RTS and CTS, and between CTS and Data frames, are smaller than the period of time that other nodes need to wait for a clear channel before transmitting. This timing difference effectively provides the priority of CTS and the following data traffic over competing traffic.

only need to observe a shorter wait-and-listen interval are essentially given higher *priority*. They will be allowed to send before those who must observe a longer wait-and-listen period.

RTS/CTS is not a perfect solution either, e.g., RTS and CTS frames themselves may collide with other frames. Still, with the RTS/CTS message passing protocol, together with prioritization through different wait-and-listen intervals, distributed transmission through randomized transmit timing, and contention resolution through an exponentially backed-off content window, we have a quite distributed MAC protocol that enables the operation of WiFi hotspots as they scale up.

We will see several other wireless peculiarities and their effects on both efficiency and fairness in the Advanced Material. But first let us work out the throughput performance of WiFi devices in a hotspot running DCF.

18.2 A Long Answer

Random access offers a complementary approach to power control as an interference management method. To be exact, there is a power control functionality in WiFi too, but it is mostly for conforming to energy limits in the unlicensed band and for saving battery energy, rather than to manage interference. While power control can be analyzed through linear algebra (and some game theory and optimization theory) as presented in Chapter 1, random access involves probabilistic actions by the radios and its performance analysis requires some basic probability theory.

CSMA in WiFi DCF is not particularly easy to model either, because frame collisions depend on the actions of each radio, and the history of binary exponential backoff couples with the transmission decision at each timeslot. A well-known performance analysis model uses a two-dimensional Markov chain, which exceeds our prerequisite of basic probability concepts and becomes too complicated for this chapter. There is a simplified version that uses very simple arguments in basic probability and a little bit of handwaving to get the gist of the complicated derivation. And that is the approach we will follow. The main idea is to find two different expressions relating the probability of transmission at a given timeslot with the probability of collision, and then solve for both probabilities through these two equations. Then computing throughput becomes easy.

18.2.1   Expressing S as a function of τ

The expected throughput S bps for CSMA random access in WiFi DCF is defined as

$$S = \frac{\text{average number of bits transmitted successfully in a timeslot}}{\text{average length of a timeslot}}. \qquad (18.1)$$

(1) First, we examine the average number of bits transmitted successfully in a timeslot. It can be expressed as the product of three numbers:

$$P_t P_s L,$$

where P_t is the probability that there is at least one (could be more) transmission going on, P_s is the probability that a transmission is successful, and L is the average payload length (measured in bytes or bits).

Let τ be the probability that a station transmits at a given timeslot. Then we know

$$P_t = 1 - (1 - \tau)^N, \qquad (18.2)$$

since P_t equals 1 minus the probability that no station transmits, which is in turn the product of the probabilities that each station does not transmit: $1 - \tau$, over all the N stations. This assumes that all stations make transmission decisions independently. This is an example of the *diversity gain* of the network effect, except now it is about the probability of a good event (a transmission gets through without collision) rather than a bad event (some link fails).

We also know that

$$P_s P_t = N\tau(1 - \tau)^{N-1}, \qquad (18.3)$$

since the left hand side is the probability that there is a successful transmission at a given timeslot. For each station, that should be the probability that it transmits (τ) but none of the other stations does $(1 - \tau)^{N-1}$. For the whole network, it is $N\tau(1 - \tau)^{N-1}$.

(2) Now we examine the average length of a timeslot. That depends on what happens at that timeslot. There are three possibilities, as illustrated in Figure 18.4.

- No transmission at all: the probability is $1 - P_t$ and the timeslot is a backoff slot with length T_b.
- There is a transmission but it is not successful: the probability is $P_t(1 - P_s)$ and the timeslot is a collision slot with length T_c.
- There is a transmission and it is successful: the probability is $P_t P_s$ and the timeslot is a successful slot with length T_s.

In summary, if we know how to compute τ, we can compute P_t as well as P_s, and thus the expected throughput S:

$$S = \frac{P_t P_s L}{(1 - P_t)T_b + P_t(1 - P_s)T_c + P_t P_s T_s}. \tag{18.4}$$

Among the quantities shown above, N, L, T_b, T_c, and T_s are constants. So we just need to compute τ.

18.2.2   Computing τ

First, we can express c, the probability that a frame (transmitted by a particular station, say station A) collides with frames from other stations, as a function of τ. Assuming that the collision probability is independent of the backoff stage, we have

$$c = 1 - (1 - \tau)^{N-1}, \tag{18.5}$$

since c is simply the probability that at least one (could be more) of the other $N - 1$ stations transmits in addition to station A. So (18.5) follows the same argument as that behind the P_t expression in (18.2).

Suppose we can also do the reverse and express τ as a function of c. In that case, we can substitute c as a function of τ (18.5) and numerically solve for τ. So now everything boils down to the following: Find τ (the probability a station transmits) in terms of c (the probability that a transmitted frame collides).

Since there are many backoff stages indexed by i, each with a contention window doubling the previous stage's, we look at the joint probability that a station transmits while at backoff stage i. We can express this joint probability in two ways:

$$\text{Prob(transmit)Prob(in backoff stage } i|\text{transmit)}$$

and

$$\text{Prob(in backoff stage } i)\text{Prob(transmit}|\text{in backoff stage } i),$$

and the two expressions above must be the same. We give shorthand notation to the above expressions: $\tau P(i|T)$ and $P(i)P(T|i)$. So we have the following *Bayesian* expression, also used in Chapter 7 to analyze information cascades:

$$\tau \frac{P(i|T)}{P(T|i)} = P(i).$$

Summing over all the i from 0 (no backoff) to B (the maximum number of backoffs allowed) on both sides, we have

$$\tau \sum_{i=0}^{B} \frac{P(i|T)}{P(T|i)} = \sum_{i=0}^{B} P(i) = 1. \tag{18.6}$$

If we can express $P(i|T)$ and $P(T|i)$ in terms of c, we have an expression for τ in terms of c, which completes our derivation for S.

Computing $P(i|T)$ is easy: if a station transmits at the backoff stage i, it must have suffered i collisions in the past and one non-collision now. We can write down that probability, making sure it is normalized:

$$P(i|T) = \frac{c^i(1-c)}{1 - c^{B+1}}.$$

Computing $P(T|i)$ is also easy (with a little handwaving): the transmit slot is one slot on its own, so the lifetime of backoff stage i, on average, is $1 + T_i$ slots. Here, T_i is the average value of the backoff counter at stage i and can be expressed as

$$T_i = \frac{1}{2}(0 + 2^i \mathrm{W}_{min}),$$

where W_{min} is the minimum contention window size. As is obvious from the above, we have assumed that the timeslot to transmit is picked randomly between "right now" and "the upper limit of binary exponential backoff." The actual contention window size W is 2 raised to some integral power then minus 1. We ignore the "minus 1" part for simplicity. Now we have

$$P(T|i) = \frac{1}{1 + T_i}.$$

Finally, we can put everything back together into (18.6), and have the following expression of τ in terms of c:

$$\tau = \frac{1}{1 + \frac{1-c}{1-c^{B+1}} \sum_i c^i T_i}. \tag{18.7}$$

Therefore, just plug (18.5) into (18.7). We can solve for τ numerically, as a function of the number of backoff stages B and the minimum contention window size W_{min}.

18.2.3   Putting everything together

Once τ has been found, by (18.2, 18.3) we have P_t and P_s as well, and can in turn compute S (18.4) in terms of the WiFi DCF protocol parameters: L, B, W_{min}, the lengths of the three types of timeslots T_b, T_c, T_s, and the number of stations N.

18.3   Examples

18.3.1   Parameters and timeslots

Before doing any calculations, we need to specify the DCF protocol parameters L, B, W_{min} and the timeslot lengths T_b, T_s, T_c.

We consider DCF being used in 802.11g at 54 Mbps. From the protocol specifications and Figure 18.4, the relevant timing parameters are

$$\text{slot time} = 9 \ \mu s,$$
$$\text{SIFS} = 10 \ \mu s,$$
$$\text{DIFS} = \text{SIFS} + (2 \times \text{ slot time}) = 28 \ \mu s.$$

According to the specifications, $W_{min} = 15$. Also, we set $L = 8192$ bits and $B = 3$ as the default values. We will later sweep their values to explore the impact.

(1) *Duration of T_b.* It is simply the length of DIFS, i.e., $T_b = 28 \ \mu s$.

(2) *Duration of T_s.* A successful transmission consists of both the transmission of a data frame from the sender and the transmission of an ACK frame from the receiver, together with appropriate spacing: [data frame] + SIFS + [ACK frame] + DIFS.

A data frame consists of a 16 μs PHY layer preamble, a 40-bit PHY header, a 240-bit MAC header, the L-bit payload, and a 32-bit CRC code. The protocol specifications state that the PHY header is further split and sent at two different rates: the first 24 bits at 6 Mbps to be more robust against channel errors at the expense of a lower rate, and the remaining 16 bits at 54 Mbps. Hence the time taken to send a data frame is

$$16 + \frac{24}{6} + \frac{16 + 240 + 32}{54} + \frac{L}{54} = 25.33 + \frac{L}{54} \ \mu s.$$

Similarly, an ACK frame consists of a 16 μs PHY layer preamble, a 40-bit PHY header (again split and sent at different rates), and a 112-bit MAC layer frame (header and CRC). Therefore, the time taken to send an ACK frame is

$$16 + \frac{24}{6} + \frac{16 + 112}{54} = 22.37 \ \mu s,$$

and we have $T_s = 25.33 + L/54 + 10 + 22.37 + 28 = 85.70 + L/54 \ \mu s$.

(3) *Duration of T_c.* When there is a collision, the sender has to wait for the full duration of the ACK frame before deciding that there has been a collision (by noting the absence of an ACK), so $T_c = T_s = 85.70 + L/54 \ \mu s$.

18.3.2   Throughput

First, by plugging in $T_i = (0 + 2^i W_{min})/2$ into (18.7), we solve numerically for τ in

$$\tau = \frac{1}{1 + \frac{1-c}{1-c^{B+1}} \sum_{i=0}^{B} c^i 2^{i-1} W_{min}},$$

where $c = 1 - (1 - \tau)^{N-1}$.

Then we plug the solution of τ into the formula for S:

$$S = \frac{N\tau(1-\tau)^{N-1}L}{(1-\tau)^N T_b + [(1-(1-\tau)^N) - N\tau(1-\tau)^{N-1}]T_c + N\tau(1-\tau)^{N-1}T_s},$$

while varying the values of N, B, W_{min} and L. Unless specified, their default values are $N = 5$, $B = 3$, $W_{min} = 15$, and $L = 8192$. With these values, $\tau = 0.0765$. Now, what is the impact of the number of stations, the medium access aggressiveness, and the payload size?

(1) *Varying N* (Figure 18.8). This is the key graph we have been looking for in this chapter, quantifying the impact of the crowd size in the "tragedy of the WiFi commons."

- As N increases, the per-station throughput $S(N)/N$ decreases because more stations are competing for the same channel. The drop is quite sharp from $N = 2$ to $N = 15$. The throughput value becomes quite low, below 2 Mbps, once N becomes 10. This highlights the inscalability of CSMA.

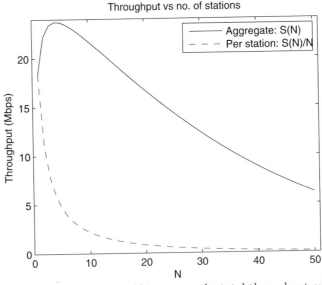

Figure 18.8 As the number of users N increases, the total throughput eventually drops. Per-user average throughput drops to small values around 2 Mbps when there are about ten users. The rapid drop of the per-user average throughput, even for a small N, highlights the unscalability of CSMA.

- The aggregate throughput $S(N)$ initially increases for small N because stations are fundamentally limited by their exponential backoff mechanism. Even when a station has no competitors for the channel, it still picks a transmit slot uniformly at random between 0 and W_{min}, and this leads to inefficiency. Adding stations helps in utilizing the channel, if there is not much contention (when N is small).
- Despite the advertised throughput of 54 Mbps, the actual maximum throughput is around 25 Mbps. This is in part because only the payload is sent at 54 Mbps, and there is a significant overhead in the PHY layer (e.g., preamble and header being sent at 6 Mbps).

(2) *Varying* W_{min} (Figure 18.9). A smaller W_{min} leads to higher aggressiveness of a station in using the channel.

- When the channel is not congested ($N = 5$), it helps to be more aggressive.
- When the channel is congested ($N = 20$), being aggressive leads to more collisions, so it is better to choose a larger W_{min} and be less aggressive.

(3) *Varing* B (Figure 18.10). A larger B tends to increase the average contention window size (i.e., become less aggressive). Hence the observation is similar to that of W_{min}.

- When $N = 5$, increasing B does not help at all.
- When $N = 20$, being less aggressive (increasing B) helps a lot.

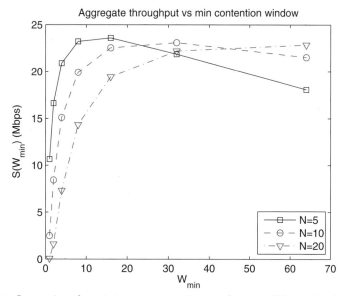

Figure 18.9 Increasing the minimum contention window size W_{min} initially has a positive effect on average throughput, since the contention is less aggressive. Past some threshold value, making the minimum window size larger hurts throughput (via its overhead in waiting) more than it helps. The more users there are, the higher this threshold value becomes.

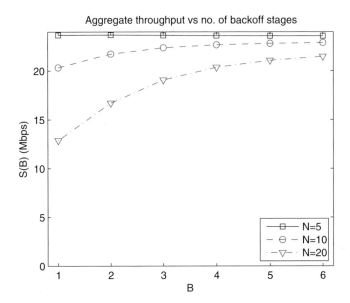

Figure 18.10 As B increases, more backoff stages are allowed, and the average backoff rises. More conservative contention increases the average throughput (but also increases latency).

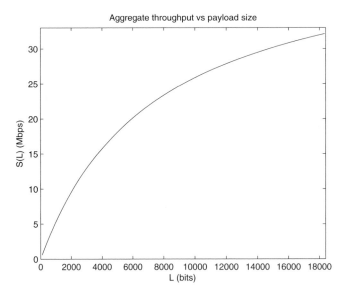

Figure 18.11 A larger payload L means a smaller percentage of overhead. The throughput naturally increases. If the model had incorporated the fact that a larger payload also increases the chance of collision, the curve would have bent over as L becomes too big.

(4) *Varying L* (Figure 18.11). The payload size relative to the overhead size also matters.

- Increasing the payload size helps because less overhead is incurred.
- But this model does not capture the effect of a frame with a larger payload having a larger collision probability. In reality we expect the throughput to increase until it reaches an optimum value, and then decrease.

18.4   Advanced Material

18.4.1   Impact of interference topology

We saw that the DCF deploys several smart ideas to enable a reasonable medium access control with a small amount of message passing overhead. But it is also well-known that the DCF can sometimes be inefficient and is often unfair, as will be shown in a homework problem. We have encountered quantification of fairness many times by now, and in Chapter 20 we will focus on fairness metrics.

Part of the reason behind DCF's inefficiency and unfairness stems from the tough reality of sensing and interference in wireless networks. The hidden node problem arises from the *difference* between sensing range and interference range, as we saw. Sensing range is also *asymmetric*: one (transmitter, receiver) pair might be able to sense and interfere with another, but not the other way around.

An example is the "flow in the middle" topology in Figure 18.12(a). The session in the middle can sense the other two sessions on each side, and has to be polite in waiting and listening to both. In contrast, either of the side sessions can only sense the middle session, and has less waiting and listening to do. A

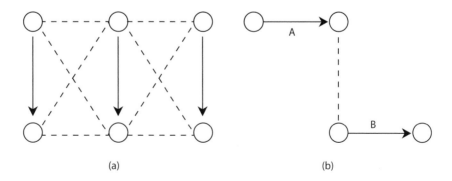

(a)   (b)

Figure 18.12 Two more important topologies, where solid lines represent intended transmission and dotted lines represent interference as well as sensing. (a) flow in the middle and (b) asymmetric sensing. In (a), the middle session has to wait for both the left and the right flows to be silent, and is often starved to near-zero throughput in practice. In (b), session A is interfered with by session B, but not the other way around. Session A sees much less throughput as a result.

proportionally fair solution should give the two side sessions the same throughput and the middle session half of that. But WiFi DCF almost certainly will starve the middle session to near-zero throughput.

Another example is the asymmetric sensing and interference topology in Figure 18.12(b). Session A cannot sense session B, and consequently they often collide. In a proportionally fair allocation, both sessions should have the same throughput. But in WiFi DCF, session A's throughput will be substantially less than that of session B.

There have been many papers on improving WiFi DCF, mostly because it is much more feasible to experiment in unlicensed band than in licensed band in university labs. For example, one of the more recent improvements of DCF comes from an optimization model of CSMA parameter adaptation: each transmitter changing its aggressiveness of contention and channel-holding duration by observing its recent throughput, balancing between the demand for capacity and the supply of allocated capacity (over the recent timeslots), like similar balancing acts in TCP congestion control or in cellular power control. Researchers then study whether the distributed action without any message passing (other than RTS/CTS) can converge to the optimal throughput for all devices. Hidden node, flow in the middle, and asymmetric sensing are three of the topologies that one needs to test for any of these alternatives to CSMA used in today's WiFi. The complicated nature of sensing (imperfect and asymmetric) and of interference (Figure 18.3) also need to be carefully incorporated in the evaluation.

18.4.2   WiFi management

There are many protocol parameters in WiFi and they are not easy to tune. Given some noisy reporting of passive measurement data or active probing data, we need to carry out performance analysis through correlation over different timescales and physical locations, and even across different protocol layers, in order to discover the root cause of the observed inefficiency or unfairness. Then some of these WiFi parameters need to be adjusted, either in real time right on the spot, or on a much slower timescale remotely. Among these parameter adjustments are the following.

- Dynamic adjustment of the rate–reliability tradeoff by using different modulation and codes, and thus different transmission speeds and decoding error probabilities for a given network condition.
- Dynamic adjustment of channel assignments (e.g., out of the 11 channels available in 802.11b) and transmit power levels to load-balance the stations among neighboring APs.
- Dynamic adjustment of the scan frequency, which determines how frequently a WiFi device can have an opportunity to reassociate itself with a different AP. Too frequent a scanning causes unnecessary interruption of ongoing

transmissions. Too infrequent a scanning creates undesirable configurations of transmitter–receiver pairs.

- Dynamic adjustment of the upper bound B on the number of retries a frame's transmission can go through. When the retry timer runs out, the frame is declared lost and the recovery is handed over to upper layers. Too early a retry timeout means the loss of an opportunity of recovering the frame, and too late a retry timeout means that upper layers, such as TCP and its congestion control, might invoke their own timeout timers.
- Dynamic adjustment of the frame size above which RTS/CTS is used. RTS/CTS is a smart idea but it introduces overhead. Only when the payload L is large enough will using RTS/CTS make sense. We will see an example in a homework problem.

18.4.3 OFDM and MIMO

Given the limited basic resource (the spectrum "crunch") and the interference-limited nature (negative externality) of wireless communications, people have given a lot of thought to designing both cellular and WiFi networks. We want to avoid the tragedy of the commons. There are several dimensions along which we can orient our thinking.

Time: e.g., CSMA is a distributed protocol that enables time sharing among WiFi stations and APs without either dedicated resource allocation or centralized control.

Frequency: We have seen dividing frequency for duplex communication: dividing the frequency band into two parts, one for uplink and another for downlink. Another idea, illustrated in Figure 18.13, is to divide each of these parts further into many small frequency chunks, sometimes called frequency carriers. In

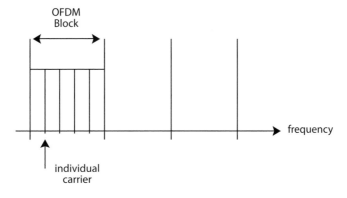

Figure 18.13 A conceptual illustration of OFDM, where the frequency band is chopped up into blocks and each block is further divided into a set of carriers. Signals are modulated onto each carrier for transmission, and signal processing within and across the blocks helps reduce signal distortion.

addition to Paris metro pricing, this is the second time we have seen the use of anti-resource-pooling. This time the justification behind anti-resource-pooling is signal processing rather than revenue maximization. There are efficient ways to process the signals transmitted on narrow frequency bands so that a higher spectral efficiency, measured in bits per second per Hz, can be maintained. Naturally, those carriers with a better channel quality will be used more, e.g., by spending more of the total power budget there. A common name for this idea is **OFDM**, and it is used in WiFi 802.11a and 802.11g to enable a physical layer speed of up to 54 Mbps. Similar methods are also used in cellular 4G and in DSL technologies.

Space: In wireless networks, as in social networks and P2P networks, the very concept of a "link" is tricky. There are no pipes in the air, just an electromagnetic field with energy propagating in it.

- Since the early days of wireless communications, a first step in utilizing the spatial dimension is to install multiple antennas at the *receiver*, so that energies bouncing off from different paths can all be collected. These signals are then processed, e.g., by quantifying how similar two signals are, using correlation coefficients as we saw in Netflix recommendation in Chapter 4, and by combining them with a weighted sum similar to AdaBoost in Chapter 6. This realizes the diversity gain (Figure 18.14 upper graph) in the wisdom of crowds, which is symbolically captured by $1 - (1 - p)^N$. Here, the "crowd" is the set of channels in-between the transmitter and the receiver, and p is the probability of one of these channels in a very noisy state.
- For over two decades, people have also studied how to install and then leverage multiple antennas on the *transmitter* side, sending different versions of the

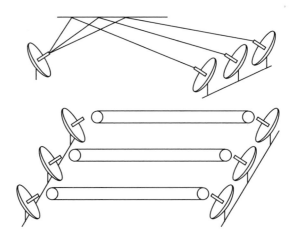

Figure 18.14 Multiple antennas can be used to deliver diversity gain (upper graph) or multiplexing gain (lower graph) in wireless networks. These two types of gains in the wisdom of crowds are also found in social networks.

signal from each of these antennas. These are called antenna arrays or smart antennas. Some cellular standards also used smart antennas to create Space Division Multiple Access.

- Since the late 1990s, Multiple Input Multiple Output (**MIMO**) systems, with multiple antennas at *both* the transmitter and the receiver, have gone from academic research to widespread usage. For example, the new 802.11n WiFi systems are MIMO based, with up to four transmit antennas and four receive antennas, a 4×4 configuration. In theory, if there is enough diversity in the channels and sufficient knowledge about the channel conditions at the transmitter or the receiver, we have a factor of N increase in speed for an $N \times N$ MIMO system. This is thanks to the multiplexing gain in the wisdom of crowds, since these signals sent along different channels created by MIMO collectively "add up" to create a high speed pipe. It realizes the multiplexing gain (Figure 18.14 lower graph). It also reminds us of the factor of N reduction in averaging of independent guesses in Chapter 5.

We now take OFDM, MIMO, and physical layer speeds to the next chapter and examine the throughput as observed by the end users of a wireless cellular network.

Summary

Box 18 Interference management through random access

WiFi uses the unlicensed band and relies on random access protocols to resolve the contention of signals and mitigate the negative network effect. Its DCF uses a version of CSMA that uses randomization, wait-and-listen periods, and limited explicit message passing for interference management. But its efficiency degrades rapidly as the number of users goes up and the contention becomes too intense for resolution by distributed scheduling.

Further Reading

There are literally thousands of research papers on all aspects of WiFi since the late 1990s, from performance to security, from new protocols to hardware experimentation. This is in part due to the availability of simulation and experimentation platforms for unlicensed band transmissions. Almost every research group on wireless communications and networking in the world has published some papers on WiFi.

1. The primary source of the past and ongoing standards can be found on the IEEE website:

IEEE 802.11 Wireless Local Area Networks Working Group, `www.ieee802.org/11`

A lot of whitepapers and educational material about the practical operations of WiFi can be found at

WiFi Consortium, `www.wi-fi.org`

2. The following is a well-written and comprehensive book on WiFi protocol details:

M. S. Gast, *802.11 Wireless Networks*, 2nd edn., O'Reilly, 2005.

3. The second edition of the following textbook on wireless communications has an extensive coverage of the physical layer technologies, including OFDM and MIMO, in the latest 802.11n standards:

A. F. Molish, *Wireless Communications*, 2nd edn., Wiley, 2011.

4. A short derivation of a widely-cited model on DCF performance can be found in the following paper, which we have followed in the example in this chapter:

G. Bianchi and I. Tinnirello, "Remarks on IEEE 802.11 DCF performance evaluation," *IEEE Communication Letters*, vol. 9, no. 8, pp. 765–767, August 2005.

5. CSMA in WiFi is a wireless variation of the original random access protocol in Ethernet, invented by Metcalfe at Xerox PARC in 1976:

R. Metcalfe and D. R. Boggs, "Ethernet: Distributed packet switching for local computer networks ," *Communications of the ACM*, vol. 19, no. 7, pp. 395–404, July 1976.

Problems

18.1   *Hidden nodes* ⋆

Consider the network in Figure 18.15.

(a) Suppose station 1 is transmitting to station 2. Which station(s) can cause the hidden node problem?

(b) What about station 1 transmitting to station 4?

(c) What about station 1 transmitting to station 5?

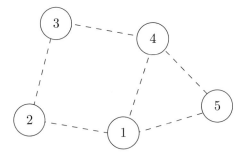

Figure 18.15 A simple network to illustrate hidden nodes. A dashed edge between two stations indicate that the stations can transmit to and interfere with each other.

18.2 *Flow in the middle* ⋆

Consider the "flow in the middle" topology in Figure 18.12(a), which has three sessions A, B, and C.

(a) Figure 18.16 shows the activity of sessions A and C in time. Draw in the figure the range of time when session B can transmit without colliding with the other sessions.

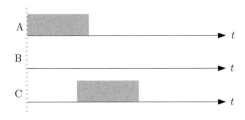

Figure 18.16 A session activity diagram. Gray areas indicate the time when a session is transmitting a frame.

(b) Figure 18.17 shows the activity of sessions B and C in time. Draw in the diagram the range of time within which session A can transmit without colliding with the other sessions.

(c) Explain why session B is disadvantaged.

18.3 *Sensing asymmetry* ⋆
Consider the "asymmetric sensing" topology in Figure 18.12(b), assuming RTS/CTS is enabled. Session A starts from station 1 to station 2, and session B from station 3 to station 4.

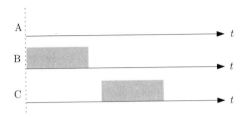

Figure 18.17 A session activity diagram. Gray areas indicate the time when a session is transmitting a frame. Note that the transmissions of B and C do not overlap in time.

(a) Figure 18.18 shows the activity of session B. At time t_1 station 1 senses the channel to be idle and sends an RTS frame to station 2 in an attempt to transmit data. What will happen?

(b) Suppose station 1 sends RTS frames at times t_2, t_3 and t_4. What will happen? Roughly speaking, what is the relationship between the time differences $t_2 - t_1$, $t_3 - t_2$ and $t_4 - t_3$?

(c) Explain why it is difficult for session A to transmit successfully.

(d) Suppose we reverse the statuses of sessions A and B, i.e., session A is transmitting and station 3 sends an RTS frame to initiate data transfer in session B. Explain why the problem in part (c) disappears.

Figure 18.18 A session activity diagram. Gray areas indicate the time when a session is transmitting a frame.

18.4   *Aloha* ★★

There is a simpler random access protocol than CSMA that is just as famous. It is called **Aloha**, as it was invented in Hawaii in the early 1970s, and further led to the development of packet radio technologies. The operation of (the slotted time version of) Aloha is easy to describe. During each timeslot, each of a given set of N users chooses to transmit a packet with probability p. We assume that if two or more users transmit at the same timeslot, all packets are lost. This is the only feedback available at each transmitter. Each lost packet is retransmitted with probability p too. We assume this process continues until a packet is eventually transmitted successfully.

(a) Assume the channel supports 1 unit of capacity (say, 1 Mbps) when there is a successful transmission. What is the throughput S as a function of N and p?

(b) What is the optimal p, as a function of N, to maximize the throughput?

(c) As the network becomes large and $N \to \infty$, what is the maximized throughput? You will see it is not a big number, which is intuitive since slotted Aloha described above has neither the listen-and-wait nor the exponential backoff features. Aloha takes the least amount of communication and coordination and it does not even use carrier sensing. It also has a low throughput. CSMA leverages implicit message passing through carrier sensing but requires no further explicit coordination. Its throughput can be high for a very small number of users but drops as the crowd gets larger. A centralized scheduler would have incurred even more coordination overhead, and would in turn provide the best performance. But in many networks, it is infeasible to afford a centralized scheduler.

18.5   *Alternative backoff rules* $\star \star \star$

Suppose there are two stations in a CSMA network attempting to transmit a data frame. The two stations start at stage 1 with some contention window size w_1, and each station chooses a timeslot within the contention window, uniformly at random. If the chosen timeslots collide, then the stations proceed to stage 2 with an updated contention window size w_2, and so on. Transmission completes at some stage i, if during this stage the two stations choose different timeslots. We are interested in the expected number of timeslots that will have elapsed before the completion of transmission. This expected number is a measure of how efficient the transmission is (the smaller the better).

To simplify the upcoming analysis, we assume there is no limit to the number of stages, i.e., the contention window size is unbounded.

(a) Suppose the two stations are in stage i with contention window size w_i. What is the probability that the stations choose the same timeslot? Conditioning on the two stations having chosen the same timeslot, what is the expected value of the timeslot chosen, given that they are indexed from 1 to w_i?

(b) What is the probability that the transmission completes at stage i?

(c) Given that the transmission completes at stage i, what is the expected number of timeslots elapsed?

(Hint: It is the sum of the expected values of timeslots chosen in previous stages (see part (a)), plus the expected value of the maximum of the two (different) timeslots chosen at stage i, which is $2(w_i + 1)/3$.)

(d) What is the expected number of timeslots elapsed before the completion of transmission?

(Hint: Apply the law of total expectation.)

(e) Now we plug in different contention window update rules. Consider the following three rules:

(1) binary exponential backoff; $w_i = 2^i$;

(2) additive backoff; $w_i = i$;

(3) super-binary exponential backoff; $w_i = 2^{2i}$.

Compute the expected number of timeslots in part (d) for the three cases. What do you observe? How does that match the intuition that the best backoff policy should be neither too conservative nor too aggressive?

19 Why am I getting only a few % of the advertised 4G speed?

By the end of this chapter, you will count yourself lucky to get as much as a few percent of the advertised speed. Where did the rest go?

19.1 A Short Answer

First of all, the terms 3G and 4G can be confusing. There is one track following the standardization body 3GPP called **UMTS** or **WCDMA**, and another track in 3GPP2 called **CDMA2000**. Each also has several versions inbetween 2G and 3G, often called 2.5G, such as EDGE, EVDO, etc. For 4G, the main track is called Long Term Evolution (**LTE**), with variants such as LTE light and LTE advanced. Another competing track is called WiMAX. Some refer to evolved versions of 3G, such as HSPA+, as 4G too. All these have created quite a bit of confusion in a consumer's mind as to what really is a 3G technology and what really is a 4G technology.

You might have read that the 3G downlink speed for stationary users should be 7.2 Mbps. But when you try to download an email attachment of 3 MB, it often takes as long as one and half minutes. You get around 267 kbps, 3.7% of what you might expect. Who took away the 96%?

Many countries are moving towards LTE. They use a range of techniques to increase the **spectral efficiency**, defined as the number of bits per second that each Hz of bandwidth can support. These include methods like OFDM and MIMO mentioned at the end of the last chapter and splitting a large cell into smaller ones. But the user observed throughput in 4G, while much higher than that for 3G, still falls short of the advertised numbers we often hear in the neighborhood of 300 Mbps. Why is that?

There are two main reasons: non-ideal network conditions and overheads. Many parts of the wireless network exhibit non-ideal conditions, including both the air-interface and the backhaul network. Furthermore, networks, just like our lives, are dominated by overheads, such as the overhead of network management in the form of control bits in packets or control sequences in protocols.

This chapter is in some sense the "overhead" of this book: there are no further "deep" messages other than the importance of overhead: networking is not just about maximizing performance metrics like throughput, but also involves the inevitable cost of managing the network.

Let us go into a little bit of detail on three major sources of "speed reduction," or more accurately, reduction in **useful throughput** in a session from a sender to the receiver. Useful throughput is defined as the number of bits of actual application data received, divided by the time it takes to get the data through. This is what you "feel" you are getting in your service, but might not be what advertisements talk about or what speed tests measure.

19.1.1   Air-interface

1. *Propagation channel*: Wireless channels suffer from various types of degradation, including **path loss** (the signal strength drops as the distance of propagation increases), **shadowing** (obstruction by objects), and multipath **fading** (each signal bounces off of many objects and is collected at the receiver from multiple paths). A user standing at the cell edge, far away from the base station and blocked by many buildings, will receive a lower rate than will another user standing right under a base station. These factors come into play even if there is only one user in the whole world.

2. *Interference*: There are also many users, and they interfere with each other. As mentioned in Chapter 1, if there are few strong interferers, or if the interferers are weak but there are many of them, the received SIR will be low. At some point, it will be so low that the order of modulation needs to be toned down and the transmission rate reduced so that the receiver can accurately decode. As we saw in Chapter 1, a typical instance of the problem in CDMA networks is the *near far problem*. Even power control cannot completely resolve this problem.

19.1.2   Backhaul network

There can be more than ten links traversed from the base station to the actual destination on the other side of a wireless session of, say, YouTube streaming. The session first goes through the radio access network, then the cellular core network also owned by the cellular provider, then possibly a long distance providers' links, then possibly multiple other ISPs composing the rest of the Internet, and, finally, to Google's data center network.

1. *Links*: Users' traffic competes with the traffic of other users on the links behind the air-interface in the cellular network. As explained in more detail in the next section, many wireless networks actually have most of their links in wireline networks. Congestion happens on these links and the resulting queuing delay reduces throughput. Plus there is also propagation delay simply due to the distance traversed. An increase in delay reduces throughput, since throughput is defined as the number of bits that can be communicated from the source to the destination per second.

2. *Nodes*: These links are connected through nodes of various kinds: gateways, switches, routers, servers, etc. Some of these, such as routers, store packets while waiting for the egress links to be ready, thus increasing packet delay. Others, such as servers, have processing-power limitations, and can become heavily congested when they are in popular demand. For example, a popular web server or video server may become so congested that it cannot process all the requests. This has nothing to do with the rest of the network, just a server that cannot handle the demand. Yet it does reduce the throughput for the session.

19.1.3 Protocols

1. *Protocol semantics*: Many functionalities require sequences of message passing. For example, in TCP, each session needs to be set up and torn down, through a three-way handshake and a four-way tear-down, respectively. This process is illustrated in Figure 19.1. Why does the network protocol designer bother to create such a complicated procedure just for setting up and terminating a session? Well, because in this way, for session establishment, both the sender and the receiver know that there is a session and that the other knows it too. And for session tear-down, four-way handshake ensures there is no dangling state of connection in either direction of a full-duplex connection (i.e., a bidirectional path where both ways can be carried out at the same

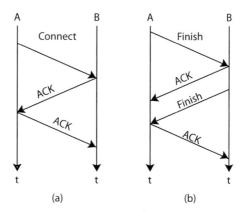

Figure 19.1 (a) Three-way session establishment and (b) four-way session tear-down in TCP. (a) When A initiates a connection with B, B sends an acknowledgement, and A acknowledges the acknowledgement, so that B knows that A knows there is now a connection established. (b) When A initiates a session tear-down, B first acknowledges that. Then B sends a tear-down message right afterwards, since TCP connections are bidirectional: A having no more messages for B does not mean B has no more messages for A. A has to hear that from B.

time). Obviously, for shorter sessions, these overheads occupy a larger fraction of the capacity used, leaving less for the application data.

2. *Packet headers*: As explained in Chapter 13, each layer adds a header to carry control information, such as address, protocol version number, quality of service, error check, etc. These headers also leave some space for flexible future use too. These headers add up, especially if the packet payload is small and the fraction of header becomes correspondingly larger. Some protocols also specify a packet-fragmentation threshold, so bigger packets are divided into smaller ones, adding to the fraction of header overhead.

3. *Control plane signaling*: Think about an air transportation network. The actual traffic of people and cargo is carried by airplanes flying between airports following particular routes. But the routing decision and many other control signals traverse entirely different networks, possibly the Internet or the telephone network. The data plane is separated from the control plane. On the Internet, the actual data traffic flows on **data channels** (a logical concept, rather than physical channels), while control signals travel on **control channels**. Control signals may have to travel half of the world even when the source and destination nodes are right next to each other. These signaling channels take portions of the available data rate and reserve them for control purposes. In 3G and 4G standards, a great deal of effort is put into designing control channels. Sometimes they are sized too small, causing extra delay and reducing throughput. Other times they are sized too big, taking up unnecessary amounts of the overall capacity and hurting throughput too.

In general, there are five main functionalities of network management.

- *Performance*: Monitor, collect, and analyze performance metrics.
- *Configuration*: Update configuration of the control knobs in different protocols.
- *Charging*: Maintain the data needed to identify how to charge each user, e.g., when a user uses the network in time-dependent pricing.
- *Fault-management*: Monitor to see whether any link or node is down, and then contain, repair, and root-cause diagnose the fault.
- *Security*: Run authentication, maintain integrity, and check confidentiality.

The messages of these functionalities sometimes run on channels shared with the actual data (in-band control), and sometimes run on dedicated control channels (out-of-band control). Collectively, they form the control plane. Protocols running network management include examples like Simple Network Management Protocol (**SNMP**) for the Internet.

19.2   A Long Answer

The speed of your wireless (or wireline) Internet connection is not one number, but *many* numbers depending on the answers to the following four questions.

First, speed as measured in *which layer*? This is often a primary reason for the confusion on speed-test results. For example, wireless standardization bodies often quote physical layer speeds, but users only experience application layer speed directly. Depending on which layer we are talking about, it also changes which parts of backhaul and protocols are involved. The more links traversed in the backhaul network, the more chances to run into congestion. The more protocols involved, the more overheads.

In Chapter 17, we saw a few key protocols often used in each of the layers. As another example, in LTE, just the link layer (layer 2) consists of three sublayers, each with its own overhead.

- *MAC layer*: Medium access control. It controls the multiplexing of data in different logical channels, and decides the scheduling of the packets.
- *RLC layer*: Radio link control. It controls the segmentation and reassembly of packets to fit the size that can be readily transmitted on the radio link. It also locally handles the retransmission of the lost packets.
- *PDCP layer*: Packet data convergence protocol. It processes header compression, security, and handover.

Second, speed as measured *where*? Depending on the locations of the two end points of the speed measurement, different parts of the backhaul networks and different protocols are involved. It is the weakest link that matters when it comes to a speed test. For example, speed is often measured between a mobile device and the base station in the cellular industry. But, as shown in Figure 19.2, the base station is followed by several gateways before reaching the IP Internet backbone and eventually reaching the other end, e.g., another mobile device or a media server. The speed measured will be different as we put the destination at each of these possible points along the path. If the speed test is between the smartphone and the base station, only the air-interface and physical and link layers' protocols are involved. If the speed test is between the smartphone and the media server, then the cellular backhaul, the backbone IP network, and even the content server's congestion conditions will all be relevant. So will network layer, transport layer, and application layer protocols.

Third, speed as measured *when*? At different times of the day, we see different amounts of congestion in the air-interface and in the backhaul. Traffic intensity during different hours of the day often exhibits a repetitive pattern, especially at higher aggregation levels. More user activity translates into more bits to be transmitted, causing

- *interference* (a multiplicative behavior) as we saw in Chapter 1: $x/y \geq s$ where s is some target SIR;
- *congestion* (an additive behavior) as we saw in Chapter 14: $x + y \leq c$ where c is some capacity limit; and
- *collision* (a binary behavior) as we saw in Chapter 18: $x, y \in \{0, 1\}$ if sessions x and y cannot transmit at the same time without colliding.

Fourth, speed as measured for *what application*? This matters for two reasons. (1) Different traffic runs different sets of protocols, some with more overhead than others. For example, texting may require only a little capacity in terms of the data channel traffic but it does require a lot of control channel traffic. Email download and web traffic can be less overhead-heavy than voice over IP or video conferencing. (2) User utility and expectation also differ a lot for different applications. Interactive gaming has very stringent demands on delay and jitter, while file download does not. Some applications can tolerate a longer initialization latency if the throughput is consistently high once it starts. Other applications are more sensitive to time shifts but can accommodate throughput fluctuations gracefully. Any objective measure of speed eventually must translate into subjective user experience.

19.3   Examples

We will now walk through a few numerical examples of non-ideal network conditions, before turning to a further discussion of protocol overhead in the Advanced Material.

19.3.1   An air-interface example

Here is an example for 4G LTE. The throughput, at the physical layer and over the air-interface only, can be estimated as follows.

For each subframe of data (one unit of time lasting 1 ms), we will count how many bits are sent. The LTE physical layer uses OFDM, dividing the spectrum into blocks and running signal processing on each of them. The number of bits transmitted equals (a) the number of symbols per frequency block multiplied by (b) the number of frequency blocks, then multiplied by (c) the number of bits per symbol per frequency block, then multiplied by (d) the coding and multi-antenna gain.

For part (a) above, the number of symbols is (a1) the number of symbols per frequency carrier, times (a2) the number of carriers per frequency block, with (a3) the control overhead deducted.

Therefore, we have the following formula to count bits sent in a subframe:

[(symbols/carrier – control overhead) × number of carriers per frequency block – channel estimation overhead] × bits/symbol × number of frequency blocks × coding and multi-antenna gain.

So, what are these factors' numerical values?

- Symbols/carrier: This is typically 12–14 in LTE.
- Control overhead/carrier: At least 1 but sometimes 2 or 3 symbols per carrier in a subframe will be for control signaling.
- Carriers/frequency block: Usually 12 carriers per frequency block of 180 kHz.

- Channel estimation overhead per frequency block: The overhead spent on sending pilot symbols to estimate the channel is ideally around 10 symbols, but often 20 symbols in practice, for 4×4 multi-antenna systems.
- Bits/symbol: This depends on the modulation, the received SIR, and the decoder capability. Ideally it can be 6 for LTE, using $2^6 = 64$ QAM modulation. But often it is 4, using the lower order of $2^4 = 16$ QAM modulation, when the channel condition is not good enough (e.g., when there is too much shadowing, fading, or the distance to the base station is long). If the channel is even worse, it can become 2.
- Number of frequency blocks: Suppose we can use the entire 20 MHz frequency band for an LTE channel, with a 1 MHz guard band on each side. Since each frequency block in OFDM is 180 kHz, we have 100 blocks. But in reality the two-way communication is carried out by either frequency division duplex (FDD), and we only get 10 MHz, i.e., 50 blocks, or time division duplex (TDD), and we get 40% of the time using the frequency for uplink and 60% of the time for downlink.
- Coding rate: The coding rate is the overhead due to channel coding, which adds redundancy to protect the transmitted bits against channel distortion. The higher the coding rate, the less redundancy is added. Ideally we want it to be close to 1, but it can be lower for certain codes. The higher the level of protection required and the less efficient a code, the lower this factor becomes.
- Multi-antenna gain: Ideally for 4×4 multi-antenna systems, it should be a factor of 4. But, due to the limitations of devices (sometimes only 2 antennas can be installed) and the channel correlation in space (when antennas are too close to each other, their received signals do not travel very different channels), it is often more like a factor of 2.

Therefore, the number of bits transmitted over a 1 ms subframe timeslot, in the *best case*, will be

$$[(14 - 1) \times 12 - 10] \times 6 \times 100 \times 0.9 \times 4 = 315360 \text{ bits,}$$

which translates into $315360/0.001 = 315$ Mbps. It would be really impressive if we could get that as the useful throughput all the time.

But in reality it is more likely going to be

$$[(12 - 2) \times 12 - 20] \times 4 \times 50 \times 0.7 \times 2 = 28000 \text{ bits,}$$

which translates into $28000/0.001 = 28$ Mbps, i.e., 8.8% of the original expectation.

If the MAC layer retransmission and overhead is counted: PDCP has 3.5 bytes of header plus the frame sequence number, RLC has 5.5 bytes of header plus sequence number, MAC has 1 byte and an error check (called Cyclic Redundancy Check, CRC) has 3 bytes, and there is at least another factor of 10% loss in useful throughput, lowering it to about 25 Mbps.

This is only for the PHY layer channel degradation and the MAC layer over-head. It is already less than 8% of the ideal number. If we count interference among users, the upper layer protocols, and the backhaul network congestion, the throughput number can easily drop by another factor of 2–5.

Now, 5–10 Mbps on a mobile device is still quite impressive and can enable a lot of applications including medium-quality video streaming. User experience is not bad, especially when most people's home WiFi throughput is only about 5–20 Mbps today, as constrained by either the 802.11b speed or the residential gateway backhaul speed. But you should not be surprised if you do not "feel" 300 Mbps on your LTE smartphone.

19.3.2   Backhaul examples

The cellular backhaul consists of some links (e.g., microwave links, free-space optical links, satellite links) that connect the air-interface with the rest of the end-to-end path in Figure 19.2, and then the cellular core network, and, finally, the public IP network. So many factors can reduce the useful throughput. We will just take a look at three TCP-related examples.

(1) As we saw in Chapter 14, the TCP throughput can be estimated as follows:

$$\text{TCP throughput} = \text{TCP Window Size}/\text{RTT},$$

where RTT denotes the Round Trip Time. The maximum window size is $2^{16}-1 = 65535$ bytes, due to the receiver window's 16-bit field in the TCP header. The purpose of the receiver window is to prevent a fast sender from overwhelming a slow receiver.

Suppose you are connected to a 1 Gbps Ethernet link and transmit a file to a destination with a long round-trip latency of 100 ms, using just one TCP session. In this case, the maximum TCP throughput that can be achieved is

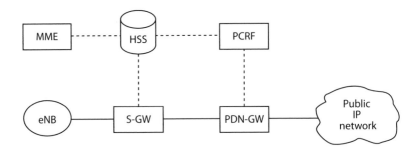

Figure 19.2 Main components of the Evolvable Packet Core in LTE: the air-interface (between an end-user device and the BS, also called eNB in LTE), the public IP network, and the rest of the wireline network in between called the cellular core network. The solid lines represent physical connections, with the link between the phone and the BS being the only wireless link. The dotted lines represent control plane channels. The BS passes through stages of gateways in the cellular core network before reaching the public IP network.

$$65535 \times 8/0.1 \approx 5.24 \text{ Mbps}.$$

Therefore, even if you are connected to a 1 Gbps Ethernet link, you should not expect any more than 5.24 Mbps when transferring the file, given the TCP window size and the round-trip time.

(2) In practice, TCP might not even attain the maximum window size because of the congestion control mechanism that prevents the sender from overloading the network. The TCP window is min{congestion window, receiver window}.

The congestion window is reduced when congestion is detected inside the network, e.g., a packet loss. The TCP throughput for long-distance links can be estimated below, as shown in a homework problem in Chapter 14:

$$\text{TCP throughput} \leq \text{MSS}/(\text{RTT} \times \sqrt{p}),$$

where MSS is the Maximum Segment Size (fixed for TCP/IP protocol, typically 1460 bytes), and p is the packet-loss rate. Consider a high packet-loss rate of 0.1%. This gives the following throughput upper bound:

$$1460 \times 8/(0.1 \times \sqrt{0.001}) \approx 3.69 \text{ Mbps}.$$

(3) Finally, consider the problem of *flash crowds*, i.e., a large number of visits to a popular web server, all within a short amount of time. The server's application might not generate the data fast enough because of server bottlenecks (on CPU, memory, or network bandwidth). This may trigger **Nagle's algorithm** that delays sending the data: the server combines small amounts of data together into large packets for better network utilization, but the delay goes up as a result. Suppose it takes 4 RTTs for the server to send out 1 MSS of data. The effective throughput now becomes extremely small:

$$1460 \times 8/(0.1 \times 4) \approx 29.2 \text{ kbps}.$$

This may explain why you often have to wait a long time for a simple webpage to show up on your browser. If the bottleneck is at the server, it does not matter whether your smartphone is on 4G LTE or on the old 2G GSM.

This is a typical tradeoff in managing overhead by aggregating frames: you might be able to reduce the amount of overhead, but at the expense of an increase in latency. A similar tradeoff shows up in aggregating acknowledgements: if an aggregated acknowledgement is lost, the sender would have to retransmit all the packets which have not been acknowledged, an effect that is similar to losing an I frame in a GOP in Chapter 17.

19.4 Advanced Material

We delve deeper into control protocol overhead in this section, with three cases on cellular core network management, mobility support, and local switching. In a homework problem, we will also explore the overhead associated with network security.

19.4.1   Cellular core network

A wireless network, cellular or WiFi, actually consists mostly of *wired* links. Usually only the link between the end-user device and the base station (or access point in WiFi's case) is wireless. If the destination is also a wireless device, there will be another wireless leg there, but otherwise it is all wired links now. So how does traffic go from the base station through the rest of the Internet and reach its destination?

A packet traverses through two segments of wireline networks.

- First is the cellular core network, backhauling the traffic from the base station through a set of servers and links carefully designed and managed by the cellular service provider.
- Then there is the public IP network, run by either one ISP or a federation of ISPs.

We have seen some features of the public IP network in Chapters 13 and 14, and now we sketch the cellular core network. We discuss only a tiny fraction of the cellular core that allows us to understand where the overheads might arise, and leads into the discussion of mobility management and network management.

Why do we even need a core network to support the radio air-interface of a wireless network? That is because there are many tasks beyond sending and receiving signals over the air. For example, some systems are needed to take care of billing, to support mobility of users, to monitor traffic, to provision the quality of service, and to ensure inter-operability between different types and generations of networks.

Each generation of cellular standards has a different design of the core network. For 4G LTE, it is called the Evolvable Packet Core (**EPC**) network. The starting point of the EPC is the base station, (called the **eNB**, the evolvable Node B, in LTE) and the associated controller (such as the Mobile Switching Center, **MSC**, in 3G and in the current standard of voice traffic in LTE). The end point is an IP router. Inbetween there are a few things going on.

- *Hardware*: There are multiple gateways filled with servers and switches. The two main ones are the Serving Gateway, abbreviated as S-GW, and the Packet Data Network (PDN) Gateway, abbreviated as P-GW.
- *Software*: There is a suite of software in charge of accounting and billing, monitoring and configuration, multimedia signal processing, mobility support, security and privacy, quality-of-service control, IP address allocation, etc.
- *Control channels*: The control signaling is sometimes carried out in-band: together with the actual data, and sometimes out-of-band: over special channels dedicated to control signals.

As shown in Figure 19.2, the main logical components of the EPC include the following.

- *PCRF*: Policy Control and charging Rules Function. It is responsible for policy-control decision making, e.g., authorization to differentiate the quality of service.
- *MME*: Mobility Management Entity. It controls the signaling between user devices and the EPC.
- *HSS*: Home Subscriber Server. Similar to a Home Location Register in earlier generations of cellular standards, this is a database to support mobility as we will explain next.
- *S-GW*: Server GateWay. It counts the number of bytes for charging purposes. It serves as the local mobility anchor, buffering data while the MME initiates a handoff. It also interfaces with 3G cellular standards, such as UMTS and GPRS, for a smooth integration. It processes at the link layer.
- *P-GW*: PDN GateWay. It allocates IP addresses, controls quality of service, and enforces other rules from the PCRF. It also interfaces with other 3G or 4G standards like CDMA2000 and WiMax. It processes at the network layer running IP.

19.4.2   Mobility management: Mobile IP and cellular handoff

Your device's IP address, or your phone number, is a type of unique ID. But for mobile devices, the problem is that the ID must be *decoupled* from a fixed location. When your laptop moves in a building, you may be communicating with a different AP. When you drive around, the base station serving you changes whenever the channel between a new base station and you is better than that between the current one and you. When your airplane lands at the airport and you turn your iPhone back on, all your emails and voicemails must locate where you are now.

The three cases above have different velocities of mobility and different spatial spans of change in location, but they all need to take care of mobility management. The key point in all solutions is to have an *anchor*, a point where others can reach you no matter where you are. That anchor keeps track of you, or who can find you. This is similar to the scenario when an elementary schoolmate wants to find you after you go to a different city for college. She may contact your parents' home, the "home agent," trusting that your parents will know how to contact your college dorm, the "foreign agent," and in turn reach you.

Consider a device A with a fixed IP address and a home network. There is a **home agent** in the home network. As the device moves into a foreign network as a visitor, it contacts an agent there in charge of taking care of visitors. It goes through an authentication process and tells that foreign agent its home agent's ID. Then the **foreign agent** will inform the home agent that device A is now residing in its network.

Now when someone, say, Bob, wants to send a message to device A, the message goes to the home agent. Bob does not know where device A might actually be, but he knows that the home agent would know. The home agent looks up a

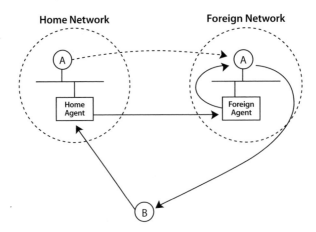

Figure 19.3 Indirect forwarding for mobility support. Device A moved from the location on the left in the home network to a new location on the right in a foreign network. The foreign agent informs the home agent. So, when another device B calls A, it first contacts the home agent, who can forward the call to the foreign agent, who in turn forwards the call to A. After this initialization, it becomes simpler for A to communicate back to B.

table that matches this device's IP address with the current foreign agent's ID, contacts that agent, and forwards the. message to it. The foreign agent then forwards the message to the device. This process of **indirect forwarding** is shown in Figure 19.3. If the device keeps moving, the first foreign agent becomes the **anchor foreign agent** and keeps forwarding the packets onwards.

Adding the foreign agent's address to the message makes it clear how the message has been handled. If you want to keep it completely transparent to device A, the home agent can *encapsulate* the actual message and *tunnel* it through to the foreign agent, who can then strip the encapsulation.

Having a home agent as the permanent address anchor and a continuously updated foreign agent is the key idea behind both mobile IP protocols and cellular handoff procedures. For example, Figure 19.4 shows the handoff procedure in 3G networks. Each cell phone has an entry in the permanent address database, often called **Home Location Register** (HLR). As it moves to foreign networks, a care-of-address is entered into the dynamic **Visitor Location Register** (VLR). The calling procedure then follows the above indirect forwarding method in Figure 19.3. In 4G LTE standards as of summer 2012, mobility support for voice and for data follow two different procedures, but the core concept of mobility support remains the same.

As to the actual resource allocation for handoff, say again in a 3G network, it depends on whether the phone is moving across boundaries of an MSC or not. Each MSC controls multiple BSs (or Node Bs). If it is moving across to a different BS but still with the same MSC, that MSC can manage the handoff. If it is also crossing to a new MSC, as shown in Figure 19.5, then the moment the

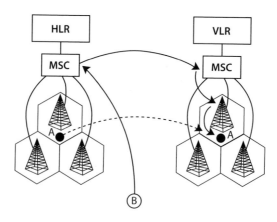

Figure 19.4 Handoff in wireless cellular networks with indirect forwarding. The solid undirected lines indicate the connections between the MSC and the BSs. The dotted line indicates the movement by phone A from one cell to another, controlled by a different MSC. The solid arrowed lines represent the communication path from a caller B, first to the MSC in A's home network, then to the MSC in A's visiting network, and, finally, to A.

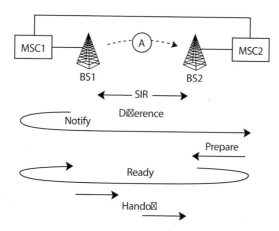

Figure 19.5 Handoff in wireless cellular networks with resource reservation. As phone A moves to some point between BS1 and BS2, and detects that the received SIR from BS2 is starting to be better than that from BS1, it initiates a handoff. MSC1 notifies MSC2, which asks a right BS to prepare the necessary radio resources, then notifies back to MSC1 that BS2 is ready to receive A. Finally, MSC1 tells BS1 to let A be handed off to BS2.

current base station detects that the SIR is low and a handoff is needed, it asks its MSC to notify a nearby MSC, which in turn asks the right BS to prepare for a handoff and to allocate the necessary radio resources. Then a "ready to go" signal is sent all the way back to the current base station, which then tells the phone to shift to the new BS.

19.4.3   Protocol overhead in switching and routing

In Chapter 13, we discussed routing using IP addresses. But each device's network adaptor actually recognizes only MAC addresses. Unlike the 32-bit, sometimes dynamically assigned, IPv4 address (e.g., 64.125.8.15), each MAC address is a 48-bit, hard-coded number, often expressed in the format of six segments of 8 bits each (e.g., 00-09-8D-32-B2-21). You can readily check the MAC address of each of your network gears.

When we discussed the DHCP's dynamic assignment of IP addresses in Chapter 13, we skipped the detail of how that is done when a device has neither the source IP address (that is precisely what needs to be done via DHCP) nor the destination address (it does not know where the local DHCP server is). Now we briefly go through these to highlight that distributed protocols carry a price of overhead.

First, the DHCP takes in a MAC address and returns an IP address, together with other local network information including the DNS server and gateway router. The dynamic IP address is leased for a period of time before it must be renewed. This is an example of a **soft state**.

In a soft state, a configuration needs to be periodically refreshed. If not, it is erased, i.e., the state disappears unless instructed otherwise. In contrast, in a **hard state**, a configuration stays there forever until an explicit "tear-down" command is issued, i.e., the state remains the same unless instructed otherwise. Many Internet protocols use soft state because of the risk of leaving dangling states when control packets get lost in the network. This is similar to mobile devices using soft state for keeping the screen on: unless you use the phone before the timer is up, it will go to some type of sleep mode to conserve energy even if you forget to put it in a sleep mode.

Now back to DHCP. How does a new device on a local network know how to contact the DHCP server? It does not. So it sends a **broadcast** message of DHCP server discovery to bootstrap. Some communication media, such as wireless, are also intrinsically broadcast in nature. Then the DHCP servers, hearing this message, each send a DCHP offer (which includes an IP address, a DNS server ID, a gateway router ID, and the lease time). The device then broadcasts an echo of the parameters from the DHCP server it chooses. Then the chosen DHCP server realizes it has been chosen and sends an acknowledgement. This four-way handshake is illustrated in Figure 19.6.

We are not done yet. Our second challenge is to determine how to deliver a message to the network adaptor of the destination device. This requires the help of another translator called the Address Resolution Protocol (**ARP**). It is the functional mirror-image of DHCP: given an IP address, the ARP provides the MAC address.

Suppose A (an iPhone) wants to communicate with B (e.g., a www.bbc.com server). To simplify the flow, we assume B is connected to A via just the gateway router.

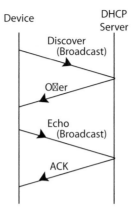

Figure 19.6 The basic protocol overhead associated with a device getting connected with a DHCP server so that a dynamic IP address can be obtained from the server. It illustrates the principle of "when in doubt, shout," as the device cold-starts with a broadcast message to discover the DHCP servers it can reach and then pick one.

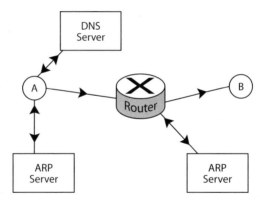

Figure 19.7 The source device with a network adaptor A wants to reach the destination device with a network adaptor B. It first needs to contact the DNS server to get the IP address of B. Then it needs to use an ARP server translate the IP address of the gateway router to the corresponding MAC address that the device driver can understand. The frame is sent from A to the router. Then, the router reads B's IP address from A's packet, gets B's MAC address from another ARP server, and finally finds a way to communicate with B.

First, A gets B's IP address via the DNS server as explained in Chapter 13. Now, A needs to translate the destination IP address to the corresponding MAC address, since that is what device drivers understand. To accomplish this, first A gets the gateway router's MAC address from the ARP. (It already knows the gateway router's IP address from the DHCP server, as shown in Figure 19.6.)

This establishes a link layer communication path between A and the gateway router. At this point, A encapsulates the IP packet into a MAC frame. Upon receiving it, the router extracts the IP packet from the MAC frame, and reads the destination IP address. Now the router gets B's MAC address from this IP address via another ARP, and can send it to B. This process is illustrated in Figure 19.7.

Across Figures 19.6 and 19.7, one DHCP server, one DNS server, and two ARP servers are used, and many rounds of control messages are signaled. All these overheads take up capacity and introduce delay, in turn slashing the useful throughput.

The protocols above may sound excessively complicated. But if you want to have distributed coordination in a layered architecture, some message passing among the network elements is a price you have to pay.

Summary

Box 19 Speed comes in many different flavors

A cellular network transmission traverses not just the air-interface, but also the core network and the public Internet. Useful throughput at the application layer is often limited by non-ideal conditions of the many nodes and links along the end-to-end paths and the inevitable overhead in packet formats, control messages, and protocol sequences. Speed tests' results depend on factors summarized as "which layer, where, when, and for what."

Further Reading

Non-ideal network conditions' impact on the application layer's throughput and the overhead of network protocols are relatively under-explored subjects in academic publications.

1. A famous debate in network management is that on the end-to-end principle, which was started with the following classic paper:

J. H. Saltzer, D. P. Reed, and D. D. Clark, "End to end arguments in system design," in *Proceedings of IEEE International Conference on Distributed Computing Systems*, 1981.

2. On the wireless side, there are very few well-written texts focusing on the cellular core networks, relative to the extensively-studied air-interface. The following recent book is one of the few:

M. Olsson, S. Sultanan, S. Rommer, L. Frid, and C. Mulligan, *SAE and the Evolved Packet Core*, Academic Press, 2009.

3. A classic graduate textbook on data networks from twenty years ago still contains some key messages relevant in today's study of networks:
D. P. Bertsekas and R. Gallager, *Data Networks*, 2nd edn., Prentice Hall, 1992.

4. Here is an interesting book discussing the underlying reasons and historical anecdotes behind why the Internet protocols operate the way they do today:
J. Day, *Patterns in Network Architecture: A Return to Fundamentals*, Prentice Hall, 2008.

5. Here is a practice-driven book recently written, focusing on the actual practice of designing and managing a global network:
G. K. Cambron, *Global Networks: Their Design, Engineering, and Operations*, John Wiley and Sons, 2013.

Problems

19.1 *RTS/CTS overhead* ⋆

In the Examples section of Chapter 18, we estimated T_s of 802.11g at 54 Mbps for the case in which RTS/CTS is disabled. Here we estimate T_s for the other case. Given that RTS/CTS is enabled, a successful transmission consists of the sequence [RTS frame] + SIFS + [CTS frame] + SIFS + [data frame] + SIFS + [ACK frame] + DIFS. The only addition is the RTS/CTS handshake that occurs before data transmission.

You are also given that (1) the time taken to transmit an RTS frame is 23.25 us, and (2) the time taken to transmit a CTS frame is 22.37 us.

(a) If $L = 8192$ bits, calculate T_s both for the case of RTS/CTS being enabled and for the case of RTS/CTS being disabled. Calculate the effective throughput as L/T_s.

(b) If $L = 320$ bits, calculate T_s and L/T_s for both cases again.

(c) In most home networks, RTS/CTS is disabled. Can you see why from (a) and (b)?

19.2 *Header overhead* ⋆

A typical IEEE 802.3 Ethernet packet structure is illustrated in Figure 19.8, with the terminology explained below.

- Preamble: Uses 64 bits to synchronize with the signal's frequency before transmitting the real data.

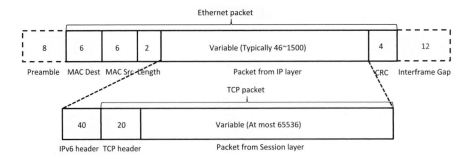

Figure 19.8 IEEE 802.3 packet structure. (lengths in bytes)

- MAC Dest/Src: Records the destination and source MAC addresses of the packet, each with 6 bytes.
- Length: Specifies the length of the IP packet with 2 bytes.
- Frame-check sequence: Contains a 32-bit cyclic redundancy check that enables the detection of corrupted data within the packet.
- Inter-frame gap: After a packet has been sent, transmitters are required to transmit a total of 96 bits of "idle line" state before transmitting the next packet.
- IPv6 header: 40 bytes in total.
- TCP header: 20 bytes in total.

What is the percentage of payload data rate if we are to send a 250-byte packet?

19.3   *The slow start phase's throughput* ★★

As mentioned in Chapter 14, TCP starts with a small congestion window, which is initially set to 1 MSS, and go through the slow start phase. The congestion window increases multiplicatively, e.g., 2 MSSs, 4 MSSs, 8 MSSs, ..., for every round trip time, until the slow start threshold is reached and the congestion avoidance phase is entered. Suppose you open a web browser and try to download a webpage of 70 MSSs.

(a) Assuming there is no packet lost, how many RTTs are required in order to download the webpage? Remember to add up one RTT of handshake to set up the TCP connection.

(b) If RTT = 100 ms, what is the average throughput in kbps? Can you see the impact of slow start on a short-duration session's throughput?

19.4   *Alternatives to indirect forwarding* ★★

Section 19.4.2 mentioned the process of indirect forwarding in mobility management. Can you think of an alternative forwarding method?

19.5   *Overhead associated with security* ★ ★ ★

We have not had a chance to talk about **network security** so far, a significant subject with many books written about it. There are several meanings to the word "security," and many famous methods to ensure or to break security in a network. We will simply walk through the main steps involved in ensuring confidentiality in Secure SHell (**SSH**), an application-layer security protocol heavily used in remote login to ensure that the server is the right one, the client is who it claims to be, and the communication between this client and the server is confidential. Along the way, we will see the amount of overhead involved in providing SSH's secure service.

There are numerous books and papers about encrypting texts. We will need only the following notion in this homework problem: **public key cryptography**. This is based on mathematical operations that are easy to run one way but very difficult the other way around, e.g., multiplying two large prime numbers is easy, but factoring a large number into two large prime numbers is very difficult. This enables the creation of a pair of keys: a public encryption key (known to anyone who wants to send a message to, say, Alice), and a private decryption key (known only to those who are allowed to decrypt and read the original message).

Now consider a client trying to remotely login to a server. If you are the inventor of a secure remote login protocol using public and private keys, what are the steps you would design? What would be the biggest vulnerability of your design?

20 Is it fair that my neighbor's iPad downloads faster?

We have come to the last chapter, on a sensitive subject that we touched upon many times in the previous chapters and forms an essential part of both social choice theory and technology network design: quantifying fairness of resource allocation. This may sound obvious, but it does not hurt to highlight: the scope of our discussion will be only on performance metrics, not on liberty and rights.

20.1 A Short Answer

20.1.1 Thinking about fairness

The naive view of "equality is fairness" is problematic in examining performance metrics of a group of users stemming from some allocation of resources. If you have to choose from an allocation of (1, 1) Mbps between two iPad users, and an allocation of (100, 101) Mbps, many people would choose (100, 101) Mbps even though it deviates from an equal allocation. Magnitude matters. Part of Rawls' theory of justice is the difference principle that we will discuss in the Advanced Material, which prefers a less equal allocation if that means everyone gets more. Of course, a more challenging choice would have been between (1, 1) Mbps and (1, 2) Mbps.

Another objection to marking equal allocations as the most fair stems from the differences in the contributions by, and the needs of, different users. If a user in a social network glues the entire network together, her contribution is higher than that of a "leaf node" user. If one works twice as hard or twice as effectively as another, these two people should not receive identical salaries. If instead of assigning one A+ and some D grades, a professor assigns a B grade to all students no matter their performance, that will neither be providing the right incentive for learning nor be deemed fair by many students.

And yet most people would also agree that a more lazy or less capable worker does not deserve to starve to death simply because she works slower. There are some basic allocations that should be provided to everyone. The debate surrounds the definition of "basic:" bread and water, or an annual vacation to the Bahamas (assuming that the latter is at all feasible)? Different notions of fairness define what is "basic" differently.

Throughout this chapter, we will examine approaches for discussing these views of fairness using less ambiguous languages.

20.1.2   Fairness measures from axioms

Given a vector $\mathbf{x} \in \mathcal{R}_+^n$, where x_i is the resource allocated to user i, *how fair is it?* This question is a special case of the general questions on fairness we saw.

Consider two feasible allocations, $\mathbf{x}$ and $\mathbf{y}$, of iPad download speeds among three users: $\mathbf{x} = [1\ 2\ 3]$ Mbps and $\mathbf{y} = [1\ 10\ 100]$ Mbps. (Since we will not be multiplying these vectors by matrices in this chapter, we skip the transpose notation here.) Among the large variety of choices we have in quantifying fairness, we can get many different fairness values, such as 0.33 for $\mathbf{x}$ and 0.01 for $\mathbf{y}$, or 0.86 for $\mathbf{x}$ and 0.41 for $\mathbf{y}$. That means $\mathbf{x}$ is viewed as 33 times more fair than $\mathbf{y}$, or just twice as fair as $\mathbf{y}$.

How many such "viewpoints" are there? What would *disqualify* a quantitative metric of fairness? Can they all be constructed from a set of axioms: simple statements taken as true for the sake of subsequent inference?

One existing approach to quantifying fairness of $\mathbf{x}$ is through a function f that maps $\mathbf{x}$ into a real number. These fairness measures are sometimes referred to as **diversity indices** in statistics. They range from simple ones, e.g., the ratio between the smallest and the largest entries of $\mathbf{x}$, to more sophisticated functions, e.g., Jain's index and the entropy function. Some of these fairness measures map $\mathbf{x}$ to a normalized range between 0 and 1, where 0 denotes the minimum fairness, 1 denotes the maximum fairness, and a larger value indicates more fairness. How are these fairness measures related? Is one measure "better" than any other? What other measures of fairness may be useful?

An alternative approach is the optimization-theoretic approach of α-fairness and the associated utility maximization problem. Given a set of feasible allocations, a maximizer of the α-fair (or isoelastic) utility function satisfies the definition of α-fairness. We have seen two well-known examples in the previous chapters: a maximizer of the log utility function ($\alpha = 1$) is proportionally fair, and a maximizer of the α-fair utility function as $\alpha \to \infty$ is max-min fair. It is often believed that $\alpha \to \infty$ is more fair than $\alpha = 1$, which is in turn more fair than $\alpha = 0$. But it remains unclear what it means to say, for example, that $\alpha = 2.5$ is more fair than $\alpha = 2.4$.

Clearly, these two approaches for quantifying fairness are different. One difference is the treatment of the efficiency, or magnitude, of resources. On the one hand, α-fair utility maximization results in Pareto optimal resource allocations. On the other hand, scale-invariant fairness measures (ones that map $\mathbf{x}$ to the same fairness value as a normalized $\mathbf{x}$) are unaffected by the magnitude of $\mathbf{x}$, and $[1,\ 1]$ is as fair as $[100,\ 100]$. Can the two approaches be unified?

To address the above questions, we discuss an *axiomatic* approach to fairness measures. There is a set of five axioms, each of which is simple and intuitive, thus being accepted as true for the sake of subsequent inference, like what we saw in Chapter 6. They lead to a useful family of fairness measures. As explained in the Advanced Material, we have the axioms of continuity, of homogeneity, of saturation, of starvation, and of partition. Starting with these five axioms, we can *generate* fairness measures. We derive a unique family of **fairness functions** f_β that includes many known ones as special cases and reveals new fairness measures corresponding to other ranges of β. Then we will remove one of the axioms and discover a more general class of fairness functions $F_{\beta,\lambda}$, with a new parameter λ capturing the relative weight put on fairness vs. efficiency.

While we start with the approach of the fairness measure rather than the optimization objective function, it turns out that the latter approach can also be recovered from f_β. For $\beta \geq 0$, α-fair utility functions can be factorized as the product of two components: (1) our fairness measure with $\beta = \alpha$, and (2) a function of the total throughput that captures the scale, or efficiency, of $\mathbf{x}$. Such a factorization quantifies a tradeoff between fairness and efficiency, addressing questions like "what is the maximum weight that can be given to fairness while still maintaining Pareto efficiency?" It also facilitates an unambiguous understanding of what it means to say that a larger α is "more fair" for general $\alpha \in [0, \infty)$.

20.2   A Long Answer

20.2.1   Constructing the fairness function

What does the fairness function look like? We are again condensing a vector into a scalar, a task we faced in rating averages in Chapter 5 and in opinion aggregation in Chapter 6. We first present a unified representation of the fairness measures constructed from five axioms (in the Advanced Material). It is also provably the *only* family of fairness measures that can satisfy all the axioms. It is a family of functions parameterized by a real number β:

$$f_\beta(\mathbf{x}) = \text{sign}(1 - \beta) \cdot \left[\sum_{i=1}^{n} \left(\frac{x_i}{\sum_j x_j} \right)^{1-\beta} \right]^{1/\beta}. \tag{20.1}$$

With the normalization term $\sum_j x_j$, we see that only the distribution matters, not the magnitude of $\mathbf{x}$. This is due to one of the axioms, the Axiom of Homogeneity that says the fairness function f should be a "homogeneous function" where scaling of the arguments does not matter. In the rest of this section, we will show that this unified representation leads to many implications.

We first summarize the special cases in Table 20.1, where β sweeps from $-\infty$ to ∞, and $H(\cdot)$ denotes the entropy function:

$$H(\mathbf{x}) = -\sum_i x_i \log x_i.$$

For some values of β, known approaches to measure fairness are recovered, e.g., Jain's index frequently used in networking research:

Value of β	Fairness measure	Known names
$\beta \to \infty$	$-\max_i \left\{ \frac{\sum_i x_i}{x_i} \right\}$	Max ratio
$\beta \in (1, \infty)$	$-\left[(1-\beta) U_{\alpha=\beta} \left(\frac{\mathbf{x}}{w(\mathbf{x})} \right) \right]^{1/\beta}$	α-fair utility
$\beta \in (0, 1)$	$\left[(1-\beta) U_{\alpha=\beta} \left(\frac{\mathbf{x}}{w(\mathbf{x})} \right) \right]^{1/\beta}$	α-fair utility
$\beta \to 0$	$e^{H\left(\frac{\mathbf{x}}{w(\mathbf{x})} \right)}$	Entropy
$\beta \in (0, -1)$	$\left[\sum_{i=1}^{n} \left(\frac{x_i}{w(\mathbf{x})} \right)^{1-\beta r} \right]^{1/\beta}$	No name
$\beta = -1$	$\frac{(\sum_i x_i)^2}{\sum_i x_i^2} = n \cdot J(\mathbf{x})$	Jain's index
$\beta \in (-1, -\infty)$	$\left[\sum_{i=1}^{n} \left(\frac{x_i}{w(\mathbf{x})} \right)^{1-\beta r} \right]^{1/\beta}$	No name
$\beta \to -\infty$	$\min_i \left\{ \frac{\sum_i x_i}{x_i} \right\}$	Min ratio

Table 20.1 Some known fairness metrics are recovered as special cases of an axiomatic construction in (20.1). For $\beta \in [0, \infty]$ the fairness component of an α-fair utility function is recovered. In particular, proportional fairness at $\alpha = \beta = 1$ is obtained from $\lim_{\beta \to 1} \frac{|f_\beta(\mathbf{x})|^\beta - n}{|1 - \beta|}$, as we will verify in a homework problem.

$$J(\mathbf{x}) = \frac{(\sum_i x_i)^2}{n \sum_i x_i^2}.$$

For $\beta \in (0, -1)$ and $\beta \in (-1, -\infty)$, new fairness measures are also revealed.

As an illustration, for two resource allocation vectors $\mathbf{x} = [1\ 2\ 3\ 5]$ and $\mathbf{y} = [1\ 1\ 2.5\ 5]$, we plot fairness $f_\beta(\mathbf{x})$ and $f_\beta(\mathbf{y})$ for different values of β in Figure 20.1. It is not trivial to decide which of these two vectors is more fair. Different values of β clearly change the fairness comparison ratio, and may even result in different fairness orderings: $f_\beta(\mathbf{x}) \geq f_\beta(\mathbf{y})$ for $\beta \in (-\infty, 4.6]$ but $f_\beta(\mathbf{x}) \leq f_\beta(\mathbf{y})$ for $\beta \in [4.6, \infty)$.

In a homework problem, we will see that a person's fairness parameter β can be reverse-engineered, through a series of questions asking the person to pick what she perceives as a fairer allocation between two choices. Different people will have different values of β in their minds, but we would hope that the same person will be self-consistent and keep the same β whether she is evaluating fairness of allocation to friends or foes.

You may have heard of **majorization**. It is a partial order over vectors to study whether the elements of vector $\mathbf{x}$ are less spread out than the elements of vector $\mathbf{y}$. Vector $\mathbf{x}$ is majorized by $\mathbf{y}$, and we write $\mathbf{x} \preceq \mathbf{y}$, if $\sum_{i=1}^{n} x_i = \sum_{i=1}^{n} y_i$, and

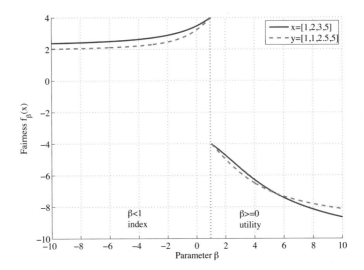

Figure 20.1 An example of fairness evaluation for two vectors of resource allocation over different β. One allocation is $\mathbf{x} = [1\ 2\ 3\ 5]$ and the other $\mathbf{y} = [1\ 1\ 2.5\ 5]$. It is not obvious which of these two allocations is more fair than the other. Indeed, the order is different depending on the value of β that defines the exact shape of the fairness measure.

$$\sum_{i=1}^{d} x_i^{\uparrow} \leq \sum_{i=1}^{d} y_i^{\uparrow}, \text{ for } d = 1, \ldots, n, \tag{20.2}$$

where $x_i^{\uparrow}$ and $y_i^{\uparrow}$ are the ith elements of $\mathbf{x}^{\uparrow}$ and $\mathbf{y}^{\uparrow}$, sorted in ascending order. According to this definition, among the vectors with the same sum of elements, the one with the equal elements is the most majorizing vector. For example, $[1\ 2\ 3\ 4] \succeq [1\ 1\ 2\ 6]$.

Intuitively, $\mathbf{x} \preceq \mathbf{y}$ can be interpreted as $\mathbf{y}$ being a fairer allocation than $\mathbf{x}$. However, majorization alone cannot be used to define a fairness measure since it is only a *partial* order and may fail to compare vectors. Still, if resource allocation $\mathbf{x}$ is majorized by $\mathbf{y}$, it is desirable to have a fairness measure f such that $f(\mathbf{x}) \leq f(\mathbf{y})$. A function satisfying this property is known as **Schur-concave**. In statistics and economics, many measures of statistical dispersion or diversity are known to be Schur-concave, e.g., the Gini coefficient. Fairness measure (20.1) also is Schur-concave.

There are many other properties that can be proved about this unique family of axiomatically constructed fairness measures. One good use of axiomatic construction is that if a conclusion is undesirable, we often have a guess as to which axioms need to be perturbed. For example, it is obvious that equal allocation maximizes our fairness measure. There are two ways to avoid this problematic view of fairness.

- Add user weights $\{q_1, q_2, \ldots, q_n\}$ in the axiomatic construction, which leads to another fairness measure that depends on both $\mathbf{x}$ and $\mathbf{q}$. We will largely skip this presentation, except in discussing Rawls' theory in the Advanced Material.

- Incorporate efficiency into the picture, which can be carried out by deleting the axiom that states fairness does not depend on magnitude of the resource allocation vector. We will follow this path now.

20.2.2 What does "larger α is more fair" mean?

To answer this question, we first show a factorization of the α-fair utility function U_α that we are so familiar with by now. Rearranging the terms, we have

$$U_{\alpha=\beta}(\mathbf{x}) = \frac{1}{1-\beta} |f_\beta(\mathbf{x})|^\beta \left(\sum_i x_i\right)^{1-\beta}$$

$$= |f_\beta(\mathbf{x})|^\beta \cdot U_\beta \left(\sum_i x_i\right), \tag{20.3}$$

where $U_\beta \left(\sum_i x_i\right)$ is the univariate version of the α-fair utility function with $\alpha = \beta \in [0, \infty)$.

Equation (20.3) demonstrates that the α-fair utility functions can be factorized as the product of two components:

- a fairness measure, $|f_\beta(\mathbf{x})|^\beta$; and
- an efficiency measure, $U_\beta \left(\sum_i x_i\right)$.

The fairness measure $|f_\beta(\mathbf{x})|^\beta$ depends only on the normalized distribution, $\mathbf{x}/(\sum_i x_i)$, of resources, while the efficiency measure is a function solely of the sum resource $\sum_i x_i$.

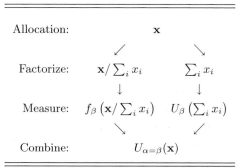

Allocation: $\mathbf{x}$

Factorize: $\mathbf{x}/\sum_i x_i$ $\sum_i x_i$

Measure: $f_\beta\left(\mathbf{x}/\sum_i x_i\right)$ $U_\beta\left(\sum_i x_i\right)$

Combine: $U_{\alpha=\beta}(\mathbf{x})$

Table 20.2 An illustration of the factorization of the α-fair utility functions into a fairness component of the normalized resource distribution and an efficiency component of the sum resource.

The factorization of α-fair utility functions is illustrated in Table 20.2. Decoupling into these two components helps us understand issues such as fairness–efficiency tradeoff and the feasibility of $\mathbf{x}$ under a given constraint set. For example, an allocation vector that maximizes the α-fair utility with a larger α need not be less efficient, because the α-fair utility incorporates both fairness and efficiency at the same time.

Guided by the product form of (20.3), we consider the following **welfare function**: a scalarization of the maximization of two objectives: fairness *and* efficiency:

$$\Phi_\lambda(\mathbf{x}) = \lambda \ell \left(f_\beta (\mathbf{x}) \right) + \ell \left(\sum_i x_i \right), \tag{20.4}$$

where $\beta \in (-\infty, \infty)$, the parameter prescribing the notion of fairness, is fixed, and $\lambda \in [0, \infty)$ is a weight specifying the emphasis placed on fairness relative to efficiency. And ℓ is just a shorthand notation for signed log:

$$\ell(y) = \text{sign}(y) \log(|y|).$$

The use of the log function later recovers the product in the factorization of (20.3) from the sum in (20.4).

An allocation vector $\mathbf{x}$ is said to be **Pareto dominated** by $\mathbf{y}$ if $x_i \leq y_i$ for all i and $x_i < y_i$ for at least some i. As mentioned in Chapter 1 in the case of power control, an allocation is called Pareto optimal if it is not Pareto dominated by any other feasible allocations. All the Pareto optimal points form a Pareto optimal tradeoff curve, a concept we mentioned many times before and have now defined in a rigorous way. To preserve Pareto optimality when using the welfare function Φ, we require that if $\mathbf{y}$ Pareto dominates $\mathbf{x}$, then $\Phi_\lambda(\mathbf{y}) > \Phi_\lambda(\mathbf{x})$.

Intuitively, if the relative emphasis on efficiency is sufficiently high, Pareto optimality of the solution can be maintained. The necessary and sufficient condition on λ, such that $\Phi_\lambda(\mathbf{y}) > \Phi_\lambda(\mathbf{x})$ if $\mathbf{y}$ Pareto dominates $\mathbf{x}$, is that λ must be no larger than a threshold $\bar{\lambda}$. That threshold turns out to be

$$\bar{\lambda} = \left| \frac{\beta}{1 - \beta} \right|. \tag{20.5}$$

A different notion of fairness, *i.e.*, a different β, leads to a different threshold $\bar{\lambda}$.

When weight $\lambda = \left| \frac{\beta}{1-\beta} \right|$, it turns out that function (20.4) becomes (20.3). That means α-fairness corresponds to the solution of an optimization that places the *maximum emphasis* on the fairness measure parameterized by $\beta = \alpha$ while preserving Pareto optimality. With this property, it indeed does make sense to say that a larger α is more fair.

20.2.3   Fairness–efficiency unification

Let us return to the Axiom of Homogeneity that says f needs to be a homogeneous function, e.g., $f_\beta([1, \ 1]) = f_\beta([a, \ a])$ for any $a > 0$. This axiom clearly takes efficiency out of the picture altogether. If fairness $F(\mathbf{x})$ satisfies a new set of axioms that removes the Axiom of Homogeneity, we can show that it must be of the form

$$F_{\beta,\lambda}(\mathbf{x}) = f_\beta(\mathbf{x}) \cdot \left(\sum_i x_i \right)^{1/\lambda}, \tag{20.6}$$

where $1/\lambda$ is called the "degree of homogeneity" (this is the same as weighting the fairness component $f_\beta(\mathbf{x})$ by λ), and $f_\beta(\mathbf{x})$ is the fairness function (20.1). There are now two parameters for fairness measures: a real number β and a positive number λ.

The new fairness measure F may not be Schur-concave, and equal allocation may not be fairness-maximizing any more. For example, you can easily verify that $F_{\beta,\lambda}(1,1) < F_{\beta,\lambda}(0.2,10)$ for $\beta = 2$ and $\lambda = 0.1$.

This family of fairness measures unifies a wide range of existing fairness indices and utilities, from diverse fields such as computer science, economics, sociology, psychology, and philosophy, including

- Jain's index,
- Jasso index,
- Theil index,
- Atkinson index,
- Shannon entropy,
- Renyi entropy,
- α-fairness,
- Foster-Sen welfare function, and
- p-norm.

These are all fairness measures that are global (mapping a given allocation vector in a system to a single scalar) and decomposable (subsystems' fairness values can be somehow collectively mapped into the overall system's fairness value). They can all be unified by (20.6).

20.3   Examples

Since fairness is evaluated in both technology networks and social networks, we will walk through two typical examples in both types of networks.

20.3.1   Capacity allocation

As discussed in Chapter 14, the typical optimization problem modeling capacity allocation is as follows:

$$\begin{array}{ll} \text{maximize} & U(\mathbf{x}) = \sum_{i=1}^n U(x_i) \\ \text{subject to} & \mathbf{x} \in \mathcal{C} \\ \text{variables} & \mathbf{x}, \end{array}$$

where U is the utility function for each session, and $\mathcal{C}$ is the constraint set of all feasible resource allocation vectors.

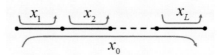

Figure 20.2 A linear network with L links and $n = L + 1$ sessions. All links have the same capacity of 1 unit. The long session is indexed as session 0. The short sessions are indexed by $i = 1, 2, \ldots, L$. Allocating capacity between the long session and the short sessions can strike different tradeoffs between fairness and efficiency, as driven by the choice of objective function.

We now consider the classic example of a linear network with L links, indexed by $l = 1, \ldots, L$, and $n = L + 1$ sessions, indexed by $i = 0, 1, \ldots, L$, shown in Figure 20.2. Session $i = 0$ goes through all the links and sources $i \geq 1$ go through links $l = i$. All links have the same capacity of 1 unit. We denote by x_i the rate of session i.

We will illustrate two points: how a given $\mathbf{x}$ can be evaluated by different fairness measures, and how $F_{\beta,\lambda}(\mathbf{x})$ acting as the objective function broadens the range of tradeoff between efficiency and fairness compared with $U(\mathbf{x})$.

Let us formulate a generalized NUM problem in this linear network: maximization of $F_{\beta,\lambda}(\mathbf{x})$, a generalization of the α-fair utility function, under link capacity constraints.

$$
\begin{aligned}
&\text{maximize} && F_{\beta,\lambda}(\mathbf{x}) = f_\beta(\mathbf{x}) \cdot \left(\textstyle\sum_i x_i\right)^{1/\lambda} \\
&\text{subject to} && x_0 + x_i \leq 1, \ \forall i && (20.7) \\
&\text{variables} && x_i \geq 0, \ \forall i.
\end{aligned}
$$

For $\beta \geq 0$ and $\lambda = \bar{\lambda} = (1-\beta)/\beta$, the optimal rate allocation maximizing (20.7) achieves α-fairness, if we just take parameter β in our fairness function to be parameter α in the isoelastic utility function. For $\lambda \leq (1 - \beta)/\beta$, problem (20.7) is still convex optimization after a logarithm change of the objective function, i.e., $\log F_{\beta,\lambda}(\mathbf{x})$.

Let us fix $\beta = 1/2$ and solve (20.7) for different values of $1/\lambda$. Figure 20.3 plots the optimal rate allocations $\mathbf{x}^*$ (showing only x_0^* and x_i^* for $i \geq 1$, since by symmetry all x_i^* are the same for $i = 1, 2, \ldots, L$), their fairness components $f_\beta(\mathbf{x}^*)$, and their efficiency components $\sum_i x_i^*$, all against $1/\lambda$. As $1/\lambda$, the weighting of the efficiency component, grows, the efficiency component's value $\sum_i x_i^*$ increases and skews the optimal rate allocation away from the equal allocation: the long session x_0 in the linear network gets penalized, while short sessions x_i are favored. At the same time, the fairness component of the objective function decreases.

20.3.2 Taxation

An interesting application of fairness is in evaluating the fairness of taxation schemes. Let us denote the pre-tax personal income in the population by vector

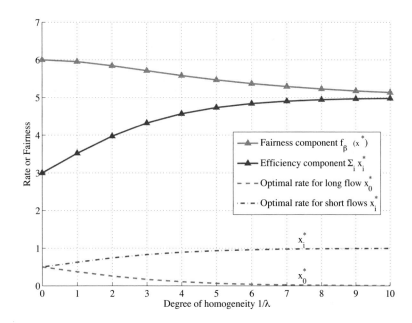

Figure 20.3 An example of a generalized objective for capacity allocation. As the degree of homogeneity $1/\lambda$ increases, λ and thus the emphasis on fairness component drop. Consequently, the capacity allocated to the long session becomes smaller.

$\mathbf{x}$, and the tax amount by vector $\mathbf{c}$. We can evaluate taxation fairness in two ways. An obvious one is by comparing the fairness of the after-tax income distribution with that of the pre-tax income distribution:

$$\frac{f_\beta(\mathbf{x} - \mathbf{c}(\mathbf{x}))}{f_\beta(\mathbf{x})}. \tag{20.8}$$

But this is not the *only* question of fairness involved. If a few people shoulder most of the tax burden, and most people pay very little tax, that can be unfair too. This second view looks at the fairness of the tax distribution itself:

$$\frac{f_\beta(\mathbf{c}(\mathbf{x}))}{f_\beta(\mathbf{x})}. \tag{20.9}$$

If you read the editorials of the *The New York Times*, it is often the first metric that is mentioned: more redistribution of wealth is needed by taxing the high-income population more, for otherwise it is too far from an equal distribution. If you read the editorials in *The Wall Street Journal*, it is often the second one mentioned: almost half of the US population does not even pay any federal income tax at all (and some pay a negative amount in receiving benefits checks), while the top earners pay a disproportionately large share of the overall income tax.

Tax rate (%)	Income bracket ($)	Number of people
10	0 – 8700	264664
15	8700 – 35350	468988
25	35350 – 85650	207803
28	85650 – 178650	35735
33	178650 – 388350	13665
35	388350+	9170

Table 20.3 US federal income tax rates in 2012 for individuals and distribution of personal income (based on a sample of 1 million taxpayers) in 2009. We uniformly distribute the people over each income bracket in our illustrative example.

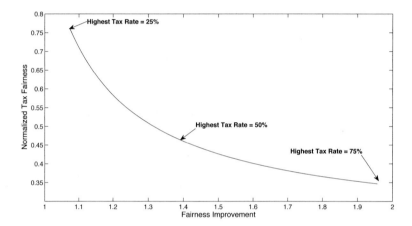

Figure 20.4 Fairness of tax vs. fairness of post-tax income for $\beta = 0.4$. The x-axis is $\frac{f_\beta(\mathbf{x} - \mathbf{c}(\mathbf{x}))}{f_\beta(\mathbf{x})}$. The y-axis is $\frac{f(\mathbf{c}(\mathbf{x}))}{f_\beta(\mathbf{x})}$. Each point on the curve corresponds to a different tax rate on the highest bracket. The slope between the point of 25% tax rate and that of 50% tax rate is about 1, highlighting one of the challenges in this debate: for an improvement by 1% of the income distribution fairness, there needs to be about 1% reduction in the taxation fairness. The scales on both axes further quantify the difficulty of the debate.

To visualize the tradeoff between these two metrics, (20.8) and (20.9), we can put them on two axes and look at the impact of different personal income tax rates (e.g., on the highest income bracket) on both merics. We use the US income and tax distributions detailed in Table 20.3. Of course, the tradeoff curves are different for different β. We show a typical curve for $\beta \in [0, 1]$ in Figure 20.4, since most of the common fairness metrics concentrate around that range of β.

Over the range of 25%–50% for the highest tax bracket in Figure 20.4, the slope of the tradeoff curve, *i.e.*, the "exchange rate" between these two axes, is around 1. This illustrates one reason why this debate is hard: for every 1% increase in income distribution fairness, there is about 1% decrease in tax burden distribution fairness.

Another reason is the scale of the two axes: the y-axis value drops quickly: more than 25% of the fairness in tax revenue contribution is lost, yet the x-axis value is very small. It is difficult to improve fairness in income distribution through taxation.

Of course, this discussion is inadequate for the purpose of assessing taxation fairness, for many reasons.

- There are other taxes whose effects are intertwined with federal income tax's impact, e.g., state income tax, local income tax, payroll taxes, corporate tax, dividend tax, alternative minimum tax, sales tax, value added tax, property tax, estate tax, healthcare mandate tax, tourism tax, tobacco tax, real estate transaction tax... Some of these are also taxed multiple times.
- Fairness of taxation should be positioned in a feedback loop with people's incentives and reactions considered. Raising the tax rate beyond a certain point *reduces* tax revenue.
- How the collected tax is spent is as important as how much can be collected, including questions on the effectiveness, efficiency, accountability, and flexibility of individual choices in the spending decisions made by government bureaucrats.
- Fairness is often tied to the difference between income derived from merits and that derived from corruption. It is tied to how much upward mobility there is in a society, where one's own effort can realize each person's different potential to the fullest.
- An even deeper debate is on the roles of the government versus individuals (and private institutions) in providing solutions to society's problems. Tax dollars increase the power of government's decisions and reduce the self-reliance of individuals.

20.4 Advanced Material

Across many disciplines, fairness has been extensively studied by raising different kinds of questions and answering them with different methodologies.

- Different indices, from the Atkinson index to the Gini index, have been studied at the intersection of economics, statistics, and sociology.
- Bargaining has been studied at the intersection of economics, sociology, and political philosophy.
- The ultimatum game, dictator game, divide-a-dollar game, and their extension of fair cake-cutting have been studied at the intersection of economics, computer science, and psychology. This subject, together with opinion aggregation and voting theory, forms the field of social choice theory.
- Fairness is not just about the outcome, but also the process. Procedural fairness has been studied extensively in law and psychology. Dictatorship

have long been practiced in the Orwellian name of fairness, and individual choices curtailed by governments under the disguise of social welfare maximization.

We will see some examples of the above list in homework problems. Below we focus on a brief discussion of Rawls' theory of justice as distributive fairness.

20.4.1   Rawls' theory of justice and distributive fairness

In the twentieth-century political philosophy, Rawls' work on fairness has been one of the most influential since its original publication in 1971. It starts with the "original position" behind the "veil of ignorance," where each person does not know where in the society she will land. This is similar to the approach of "one cuts the cake, the other selects a slice first" in the problem of fair cake-cutting that we will see in a homework problem. You can probably also sense the connection of this assumption to the max-min fairness already.

The arguments posed by Rawls are based on two fundamental principles (axioms stated in English rather than mathematics), as described in his 2001 restatement.

1. "Each person is to have an equal right to the most extensive scheme of equal basic liberties compatible with a similar scheme of liberties for others."
2. "Social and economic inequalities should be arranged so that they are both (a) to the greatest benefit of the least advantaged persons, and (b) attached to offices and positions open to all under conditions of equality of opportunity."

The first principle governs the distribution of *liberties* and has priority over the second principle. But suppose we also interpret it as a principle of distributive fairness in allocating limited *resources* among users. It can now be captured as a theorem, rather than an axiom. We did not have the time for this, but we could have extended the fairness measure to incorporate different weights$\{q_i\}$ to each user. Function f_β becomes $f_\beta(\mathbf{x}, \mathbf{q})$. The theorem says that any fairness measures satisfying the axioms (detailed in the Advanced Material) will satisfy the following property: for any β, adding an equal amount of "basic" resource $c \geq 0$ to each user will only increase the fairness value:

$$f_\beta(\mathbf{x} + c\mathbf{1}, \mathbf{q}) \geq f_\beta(\mathbf{x}, \mathbf{q}), \text{ for } q_i = \frac{1}{n}, \ \forall c \geq 0, \ \forall \beta,$$

where equal weights $q_i = 1/n$ for all i can be viewed as a quantification of "equal right" in the first principle of Rawls' theory. If the vector $\mathbf{q}$ is a general one, or if the λ weight between fairness and efficiency is introduced, the above property holds only for c up to some upper limit.

The second part is the celebrated **difference principle**: "to the greatest benefit" rather than "to the greatest relative benefit." So [100, 101] is more fair than [1, 1]. It is an approach different both from strict **egalitarianism** (since it concerns the absolute value of the least advantaged user rather than the relative

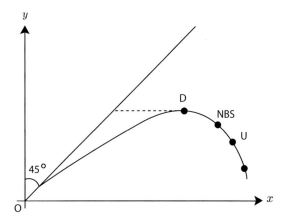

Figure 20.5 An annotated version of a quantitative graph in Rawls 2001 book. Since the efficiency frontier between a "more advantaged group" user x and a "less advantaged group" user y does not pass through the 45-degree line, the strictly egalitarian point is the origin. The difference principle generates a point D on the efficient frontier, whereas a sum utility maximization generates point U. The Nash Bargaining Solution (as discussed in Chapter 6) point N may lie between D and U. The fairness function $F_{\beta,\lambda}(\mathbf{x}, \mathbf{q})$ can generate any point on the efficient frontier when $\lambda \leq \bar{\lambda}$.

value) and from **utilitarianism** (when narrowly interpreted where the utility function does not capture fairness).

The difference principle can also be axiomatically constructed as a special case of a continuum of generalized tradeoff between fairness with efficiency. This is best illustrated by annotating Rawls' own graph in Figure 20.5. The point representing the difference principle is the consequence of concatenating two steps of pushing to the *extremum* on both β and λ in $f_{\beta,\lambda}$ (20.1): let $\beta \to \infty$, and make λ as large as possible while retaining Pareto efficiency, *i.e.*, $\lambda \to \bar{\lambda}$. If either β is finite (e.g., $\beta = 1$) or λ is smaller than $\bar{\lambda}$ (more emphasis on efficiency), we will have a fairness notion that is not as restricted as the difference principle.

Back to the introductory discussion at the beginning of this chapter. What would be your threshold x above which $(x, x+1)$ is more fair than $(1, 1)$? What constitutes "basic" resource needs changes depending on the answer. Questions like this also help reverse engineer your β and λ.

20.4.2   Axioms

We have mentioned axioms behind the construction of fairness measures. Now let us quickly summarize these axioms.

Let $\mathbf{x}$ be a resource allocation vector with n non-negative elements. A fairness measure is a sequence of mapping $\{f^n(\mathbf{x}), \forall n \in \mathcal{Z}_n\}$ from n-dimensional vectors $\mathbf{x}$ to real numbers, called fairness values, *i.e.*, $\{f^n : \mathcal{R}^n_+ \to \mathcal{R}, \forall n \in \mathcal{Z}_+\}$. To simplify the notation, we suppress n in $\{f^n\}$ and denote them simply as f. We introduce the following set of axioms about f, whose explanations are provided after the statement of each axiom.

1. *Axiom of Continuity.* Fairness measure $f(\mathbf{x})$ is continuous on $\mathcal{R}^n_+$, for all $n \in \mathcal{Z}_+$.

Axiom 1 is intuitive: A slight change in resource allocation shows up as a slight change in the fairness measure.

2. *Axiom of Homogeneity.* Fairness measure $f(\mathbf{x})$ is a homogeneous function:

$$f(\mathbf{x}) = f(t \cdot \mathbf{x}), \quad \forall\, t > 0.$$

Without loss of generality, for $n = 1$, we take $|f(x_1)| = 1$ for all $x_1 > 0$, *i.e.*, fairness is a constant for a one-user system. Axiom 2 says that the fairness measure is independent of the unit of measurement or the magnitude of resource allocation. Therefore, for an optimization formulation of resource allocation, the fairness measure $f(\mathbf{x})$ alone cannot be used as the objective function if efficiency (which depends on the magnitude $\sum_i x_i$) is to be captured. As we saw in this chapter, this axiom can be removed, leading to $F_{\beta,\lambda}(\mathbf{x})$ in (20.6).

3. *Axiom of Saturation.* Equal allocation's fairness value is independent of the number of users as the number of users becomes large:

$$\lim_{n \to \infty} \frac{f(\mathbf{1}_{n+1})}{f(\mathbf{1}_n)} = 1.$$

This axiom is a technical condition used to help ensure the *uniqueness* of the fairness measure. Note that it is *not* stating that equal allocation is the most fair.

4. *Axiom of Starvation.* For $n = 2$ users, we have $f(1,0) \leq f(\frac{1}{2},\frac{1}{2})$, *i.e.*, starvation is no more fair than equal allocation.

Axiom 4 is the only axiom that involves a *value* statement on fairness: starvation is no more fair than equal distribution for two users. It specifies an increasing direction of fairness. We could have picked a different axiom to achieve similar effects, but the above axiom for just two users and involving only starvation and equal allocation is the weakest such statement, thus the "strongest axiom."

A primary motivation for quantifying fairness is to allow a comparison of fairness values. Therefore, we must ensure well-definedness of the ratio of fairness measures as the number of users in the system increases. Axiom 5 states that fairness comparison is independent of the way the resource allocation vector is reached as the system grows.

5. *Axiom of Partition.* Consider a partition of a system into two subsystems. Let $\mathbf{x} = [\mathbf{x}^1, \mathbf{x}^2]$ and $\mathbf{y} = [\mathbf{y}^1, \mathbf{y}^2]$ be two resource allocation vectors, each partitioned and satisfying $\sum_j x^i_j = \sum_j y^i_j$ for $i = 1, 2$. There exists a *mean function* h such that their fairness ratio is the mean of the fairness ratios of the subsystems' allocations, for all partitions such that the sum resources of each subsystem are the same across $\mathbf{x}$ and $\mathbf{y}$:

$$\frac{f(\mathbf{x})}{f(\mathbf{y})} = h\left(\frac{f(\mathbf{x}^1)}{f(\mathbf{y}^1)}, \frac{f(\mathbf{x}^2)}{f(\mathbf{y}^2)}\right).$$

According to the axiomatic theory of mean functions, a function h is a mean function if and only if it can be expressed as follows:

$$h = g^{-1}\left(\sum_{i=1}^{2} s_i \cdot g\left(\frac{f(\mathbf{x}^i)}{f(\mathbf{y}^i)}\right)\right),$$

where g is any continuous and strictly monotonic function, referred to as the Kolmogorov–Nagumo generator function, and $\{s_i\}$ are the positive weights such that $\sum_i s_i = 1$.

By definition, a set of axioms is true, as long as the axioms are consistent. As we saw in this chapter, there exists a fairness measure $f(\mathbf{x})$ satisfying Axioms 1–5. Furthermore, from any function $g(y)$, there is a unique $f(\mathbf{x})$ thus generated. Such an $f(\mathbf{x})$ is a well-defined fairness measure if it also satisfies Axioms 1–5. We can further show that only *logarithm* and *power* functions are possible generator functions. Consequently, we can find, in closed form, all possible fairness measures satisfying Axioms 1–5, as shown in (20.1).

Summary

Box 20 Fairness can be measured based on axioms

Fairness is not equality of resource allocation. A function constructed from a set of axioms can be used as a quantitative measure of fairness and used as an objective function in network optimization. Special cases like Rawls' difference principle and α-fairness can be recovered. Alice and Bob can disagree on the notion of fairness, but each's fairness measure can be reverse-engineered and should be self-consistent.

Further Reading

Perhaps no chapter in this book has as much intellectual diversity as this one. There are literally tens of thousands of papers on the subject of fairness from different disciplines.

1. In political philosophy, the classic and influential book by Rawls has a second edition that we briefly mentioned in this chapter:

   J. Rawls, *Justice as Fairness: A Restatement*, Harvard University Press, 2001.

2. In computer science and economics, cake-cutting has been one of the standard problems in the study of fairness. A comprehensive survey, including connections to auction and voting theories, is provided in the following book:

S. J. Brams and A. D. Taylor, *Fair Division: From Cake-cutting to Dispute Resolution*, Cambridge University Press, 1996.

3. More than just concerning the outcome, fairness is also about the process. From the fields of psychology, sociology, and legal study, a historical account and survey of procedural fairness can be found in the following book:

E. A. Lind and T. R. Tyler, *The Social Psychology of Procedural Justice*, Springer, 1988.

4. On taxation and reaction to taxes as a dynamic system, the following paper discusses the impact of social perception and presents a differential equation model of the feedback loop between taxation by the government and reaction by the people:

A. Alesina and G. M. Angeletos, "Fairness and redistribution," *American Economic Review*, vol. 95, no. 4, pp. 960–980, 2005.

5. Our axiomatic development follows the article below, which also discusses the connections with other disciplines on the study on fairness:

T. Lan and M. Chiang, "An axiomatic theory of fairness," *Princeton University Technical Report*, 2011. www.princeton.edu/~chiangm/fairness.pdf

Problems

20.1   *Fairness–efficiency unification* ★★

The definition of fairness measure was given in Section 20.2.1 as follows:

$$
f_\beta(\mathbf{x}) = \begin{cases} \operatorname{sign}(1-\beta)\left(\sum_{i=1}^n \left(\frac{x_i}{w(\mathbf{x})}\right)^{1-\beta}\right)^{1/\beta} & , \text{ if } \beta \neq 0 \\ \exp\left(-\sum_{i=1}^n \frac{x_i}{w(\mathbf{x})}\log\frac{x_i}{w(\mathbf{x})}\right) & , \text{ if } \beta = 0. \end{cases}
$$

where $w(\mathbf{x}) = \sum_{j=1}^n x_j$.

(a) Prove that indeed $\lim_{\beta \to 0} f_\beta(\mathbf{x}) = f_0(\mathbf{x})$.

(b) Consider two allocation vectors $\mathbf{x} = [0.1\ 0.2\ 0.3\ 0.6]$, $\mathbf{y} = [0.2\ 0.2\ 0.8\ 0.9]$. Plot $f_\beta(\mathbf{x})$ and $f_\beta(\mathbf{y})$ as functions of $-10 \leq \beta \leq 10$ on the same graph.

(c) Fix $\beta = 0.5$. Plot fairness–efficiency measures $F_{\beta,\lambda}(\mathbf{x})$ and $F_{\beta,\lambda}(\mathbf{y})$ as functions of $0 \leq \frac{1}{\lambda} \leq 1$, on the same graph.

20.2   Reverse-engineering β ★★

Different values of the parameter β leads to different fairness functions. A natural question, then, is how to determine β. This problem will explore one way to do that.

Consider the following scenario: there are three people who want bandwidth on a shared link, and you have 10 Mbps to allocate to them. Assume each is equally deserving of the bandwidth. There are four possible allocations: (5, 2.5, 2.5), (4.5, 3.5, 2), (6, 2, 2), and (4.5, 4.5, 1).

We will now walk through the process of determining a possible β value. Suppose that there are three possibilities: $\beta = -10$, or 0.5, or 2. For parts (a) and (b), choose the more fair allocation of the two options provided, and then find which values of β can be eliminated given your answer.

(a) (5, 2.5, 2.5) and (4.5, 3.5, 2).

(b) (6, 2, 2) and (4.5, 4.5, 1).

(c) How many values of β are compatible with both of your answers in (a) and (b)?

(d) Suppose you chose (5, 2.5, 2.5) in part (a) and (4.5, 4.5, 1) in part (b) as the most fair allocations. How many values of β are compatible with both answers? Can you intuitively explain why? What does this tell you about quantitative modeling of fairness?

20.3   Multi-resource fairness ★★★

Recall that the fairness–efficiency functions introduced in the lecture notes have the form

$$F_{\beta,\lambda}(\mathbf{x}) = \text{sign}(1 - \beta)\left(\sum_{i=1}^{n}\left(\frac{x_i}{\sum_{j=1}^{n}x_j}\right)^{1-\beta}\right)^{1/\beta}\left(\sum_{i=1}^{n}x_i\right)^{\lambda}, \qquad (20.10)$$

where $\mathbf{x}$ is an n-dimensional resource allocation vector. But, in some contexts, the allocation of one resource is not enough. For instance, consider a data center utilized by two users. Each user runs one type of job, and the two types of jobs have different resource needs. Both require memory and CPU (processing power), but user A's jobs require 1 GB of memory and 2 MIPS (a unit of computational power) of CPUs per job, while user B's jobs require 2 GB of memory and 1.5 MIPS CPUs per job. These resources are limited: there are 8 GB of available memory and 12 MIPS of CPUs. Clearly, just allocating memory or just allocating CPUs is not enough: we need to consider the fairness of *both* resource allocations.

This homework question explores two different ways of measuring the fairness of multi-resource allocations, as in the above example about data centers. First, one could just measure the fairness of the number of jobs allocated to each user. Each user is allocated enough resources to complete this number of jobs. For instance, if user A is allocated 2 jobs, she receives 2 GB of memory and 4 MIPS of CPUs. The resource allocation vector in (20.10) is just the number of jobs assigned to each user. For instance, if user A is assigned 2 jobs and user B 3 jobs, the fairness of this allocation is

$$\operatorname{sign}(1 - \beta) \left(\left(\frac{2}{2+3} \right)^{1-\beta} + \left(\frac{3}{2+3} \right)^{1-\beta} \right)^{1/\beta} (2+3)^{\lambda}.$$

But this approach misses the *heterogeneity* of the resource requests among the users. The second way of measuring multi-resource fairness involves *dominant shares*. These are defined as the maximum fraction of each resource received by the user. For instance, if user A gets 2 jobs, she receives 2 GB of memory and 4 MIPS of CPUs. Then A receives 1/4 of the available memory but 1/3 of the CPUs. User A's *dominant resource* is CPUs, and her *dominant share* is 1/3. These dominant shares are then taken as the resource allocation vector. For example, if user A's dominant share is 1/3 and user B's is 2/3, the allocation vector **x** in (20.10) is [1/3 2/3].

(a) What is user B's dominant resource? Calculate the dominant shares for users A and B in terms of x_A and x_B, the number of jobs allocated to users A and B, respectively.

(b) Formulate the maximization problem for multi-resource fairness in the datacenter example above, using both fairness on jobs and fairness on dominant shares (*i.e.*, write down two formulations, one for fairness on jobs and one for fairness on dominant shares). Use your answers to part (a) and (20.10) to write down the objective function for maximizing fairness on dominant shares.

(c) Numerically solve for the optimal resource allocation according to your formulations in part (b), with the resource requirements given above and $\beta = 0.5$ and $\lambda = 1$. (Non-integer numbers of jobs are allowed). Are the optimal allocations the same? Which do you think is more "fair"?

20.4 *Cake-cutting fairness* ★★★

Suppose each person has a valuation function that maps a given piece of a cake into a positive number. For example, if the cake has a chocolate half and a vanilla half, Alice may like the chocolate half twice as much as the vanilla half, while Bob is the other way round. "One cuts, the other selects" is a well-known procedure of dividing a cake between two people such that both value their own share to be at least half of the whole cake. Alice first cut the cake into two pieces with each piece having the same value to her, and then Bob selects among the two pieces.

Can you come up with a procedure of dividing a cake for three participants so that they all value their own share to be at least one third of the whole cake?

(Hint: First divide the cake into three pieces whose values are equal for one participant.)

(More detail can be found in the following survey: S. J. Brams, M. A. Jones, and C. Klamler, "Better ways to cut a cake," *Notice of the American Mathematics Society*, vol. 53, no. 11, pp. 1314–1321, December 2006.)

20.5 *The ultimatum game* ⋆⋆

The ultimatum game is a game where two players interact to decide how to divide a sum of money between them. The first player proposes how to divide and the second player can either accept or reject this proposal. If the second player rejects, neither player receives anything. If the second player accepts, the money is split according to the proposal. The game is played only once, so reciprocation is not an issue.

Consider an ultimatum game where Alice and Bob, are to divide a one-foot-long sandwich. Alice knows that Bob will not accept any offer less than x foot; however, she is not certain about x and only has the following estimate about the probability density function of x:

$$p(x) = \begin{cases} 4x, & \text{if } x < 0.5 \\ 4(1-x), & \text{if } x \geq 0.5. \end{cases}$$

How will Alice propose to split the sandwich, and what is the expected share she receives?

Index

Notes